Tracy L. Hurst

DIGITAL
COMMUNICATIONS

Other Prentice-Hall books by **Dr. Kamilo Feher**

DIGITAL COMMUNICATIONS: Microwave Applications,
© 1981

DIGITAL
COMMUNICATIONS
SATELLITE/EARTH STATION
ENGINEERING

Dr. KAMILO FEHER, Ph.D., M.A.Sc., P.Eng.

Professor, Electrical Engineering
University of Ottawa, Ottawa, Canada

Advisor, Spar Aerospace Limited

Guest Editor, IEEE Transactions on Communications
Special Issue on Digital Satellite Communications

PRENTICE-HALL Inc., Englewood Cliffs, N.J. 07632 U.S.A.

Library of Congress Cataloging in Publication Data

Feher, Kamilo.
    Digital communications.

    Bibliography: p.
    Includes index.
    1. Artificial satellites in telecommunication.
2. Digital communications. I. Title.
TK5104.F47   1983   621.38′04222   82–13166
ISBN 0–13–212068–2

Editorial/production supervision
    and interior design by *Mary Carnis*
Cover design by *Christine Wolf*
Manufacturing buyer: *Anthony Caruso*

Printed in the United States of America

10   9   8   7   6   5   4   3   2

ISBN   0-13-212068-2

Prentice-Hall International, Inc., *London*
Prentice-Hall of Australia Pty. Limited, *Sydney*
Editora Prentice-Hall do Brasil LTDA, *Rio de Janiero*
Prentice-Hall Canada, Inc., *Toronto*
Prentice-Hall of India Private Limited, *New Delhi*
Prentice-Hall of Japan, Inc., *Tokyo*
Prentice-Hall of Southeast Asia Pte. Ltd., *Singapore*
Whitehall Books Limited, Wellington, *New Zealand*

CONTENTS

FOREWORD

Communicating via satellite is no longer the novelty it once was. Yet, the impact of the extraordinary growth and development of satellite technology and services during the 1960s and 1970s has rendered satellite communications an indispensible part of our daily lives as we move into the 1980s.

Soon to become another indispensible part of our daily world of communicating is the rapidly expanding technology of digital satellite communications.

In the 1960s, digital communications was largely regarded as theoretical and experimental; in the 1970s, the premise began to be realized; and now, the 1980s is clearly the decade of digital satellite communications. The analog satellite communications techniques that were the mainstay of services during the 1960s and 1970s are no longer adequate to meet the exploding communications demand for the future, and thus the implementation of digital communications systems, including a growing number of communications satellite systems, is well under way. Already millions of users are benefiting from advanced telecommunications systems that have brought digital technology from infancy to maturity in little more than a decade. The future of digital communications has already arrived.

This, then, is an opportune time for a well-reasoned, pragmatic book on digital satellite communications—and Dr. Kamilo Feher has given us just such a publication.

One of the most significant contributions of *Digital Communications: Satellite/*

Earth Station Engineering is its broad scope. It covers all aspects of digital radio communications, both space and ground.

The author's many years of industrial and research experience are reflected throughout this thorough work that undoubtedly will be of particular interest to the engineer who is seeking extensive knowledge of a complete modern satellite system, including the earth segment. Recognizing the importance of satellite earth station design, the book contains a wealth of information related to earth station operating principles, design, and operation.

The technical advantages of digital satellite communications, which are explored in depth in Professor Feher's book, give promise of efficiencies of greater than 2 bits per second per Hertz of radio-frequency spectrum. When coupled with satellite clusters, space platforms, and narrower spacing of satellites in geosynchronous orbit, these efficiencies hold promise of communications satellite systems that can provide global capabilities in excess of 100 gigabits per second—capabilities that were virtually inconceivable just a decade ago. These breakthroughs will not come easily, however. They will require the expertise of countless scientists and engineers to bring the new digital satellite communications technologies to fruition. Dr. Feher's clearly written book, which includes three chapters by other internationally acclaimed experts, should serve well as a "training tool" to enable engineers of the 1980s and 1990s to achieve these goals.

It is, of course, interesting and gratifying to see the recognition accorded the organization I represent—the International Telecommunications Satellite Organization. INTELSAT has been deeply involved in and committed to the development of digital satellite services, most notably TDMA—which we regard as essential for satisfying the vast digital communications requirements lurking on the horizon. Thanks to Professor Feher's comprehensive presentation, not only INTELSAT but a complete spectrum of domestic satellite systems, with numerous design examples, have been given excellent coverage.

My congratulations to Dr. Feher for this important contribution to the field of digital satellite communications and its future development.

S. Astrain
Director General
INTELSAT

PREFACE

By mastering this book, you acquire the tools and skills necessary to analyze and design elements of modern satellite communications systems. As you advance through the chapters and study the numerous *"real-life"* examples, you learn the principles and better understand operational and planned domestic, international, and intercontinental digital communication satellite systems. This book is for the professional engineer and manager, for the advanced student who wants a solid understanding of this field and for the researcher who needs a consolidated comprehensive up-to-date reference text of digital communications systems.

The growth of satellite communications capacity and capability has been revolutionary, a result of the flexibility provided by multiple-access, global-coverage digital satellite systems.

The *unique ability* of telecommunication satellites to cover the globe has opened a new era for regional and *global* communications. Systems using a *single* satellite offer the *flexibility* to interconnect any pair of users separated by great distances up to 17,000 km (approximately one-third of the circumference of the earth), and systems using three satellites can provide a *global coverage* with multiple-access flexibility.

14/12-GHz transmit–receive antenna of the University of Ottawa. The performance of new modems, codecs and SCPC and TDMA systems is evaluated under the research direction of Dr. K. Feher. The Canadian ANIK satellites are used in these experiments. Professor Feher is in foreground.

This *truly unique multiple-access* flexibility includes communication links between satellites and:

Fixed points on earth
Ships at sea
Airplanes
Trains
Automobiles
Other moving space vehicles
"Man-pack" terminals carried by a person and installed in 5 minutes or less

There are no other communication systems that can approach *this flexibility.* Satellite communications costs are essentially insensitive to the distances between terminals, whereas the costs of terrestrial (nonsatellite) services are dependent on distance.

The variety of data formats and services that can be provided by satellite links include:

Telephony signals
Television (vision and audio) signals
Computer-generated signals (computer communications)
Broadcast data for computer communications
Teleprinter
Large-screen teleconferencing
Interactive education
Medical data
Emergency services
Electronic mail
Newpaper broadcast
Control data for power systems and utilities
Traffic information
Weather and land surveillance
Navigational data for ships and airplanes
Military strategic data

This list is far from complete, and as each year passes, will become even less complete as new requirements are created and accommodated. The flexible multiple-access long- and short-distance satellite systems offer more and more reliable, cost-effective solutions.

A large number of U.S., Canadian, Japanese, European, and other domestic communications satellites, as well as the INTELSAT satellite systems, are now either operational or under construction. More than half of the commercial overseas long-distance telephone calls, as well as almost all of the overseas commercial live television programs, are relayed by *"synchronous"* satellites that, with the exception of slight drift due to imperfect "stationkeeping," appear to float at synchronous attitude (36,000 km) above a single point on the earth's surface. A significant number of countries are also relying on satellites for their data transmission and military communication needs.

During the late 1970s and early 1980s most major operational satellite and terrestrial line-of-sight microwave systems use *analog* FM modulation techniques. However, the trend in *new development* is such that the overwhelming majority of new satellite systems employs *digital* methods. This trend has been reinforced by recent domestic system decisions in the United States, Canada, and France (also in many other countries), where the digital approach for new transmission facilities will be predominant. The decision of the board of directors of INTELSAT, taken during 1980, to use time-division multiple-access digital modulation techniques on INTELSAT-V and future generations of satellites has a significant impact on the development of digital satellite communications—over 100 countries use the INTELSAT satellite network. Should this present trend continue, it is expected that by 1993 almost all new system additions will be digital. Consequently, engineering *students,* telecommunications professionals, and academics must become familiar with the *principles, design, applications,* and *planning* of digital satellite communications networks and systems.

In most system applications one satellite serves many earth stations. The number, type, and size of satellite earth stations is dramatically increasing. There is even more growth and *investment* in the *satellite earth station* segment than in the satellite transponder business. For this reason particular attention is focused on *earth station analysis and design criteria.*

Elaborating on the **scope** of this book, the *engineers* and *managers* employed by the operators of satellite communications networks, systems, and equipment designers employed by manufacturers of telecommunications equipment, manufacturing engineers and managers, marketing managers, product planners, consulting engineers, engineers engaged in research on telecommunications, and also those of the managerial, administrative, and technical staff of government agencies concerned with the regulation of telecommunications will find this book to be an invaluable source of information in problem-solving skill development. This book is also intended to be suitable for use as a *text* at the first-year *graduate* or *senior undergraduate* level, as well as for reference purposes, in *universities* and other technical institutions. It is expected that you have been exposed to the fundamentals of communications systems. A prior exposure to probability theory would be an asset.

You are introduced to the satellite communications challenge of the 1980s and 1990s, to illustrative earth station and satellite communications subsystems, and to link budget calculations in *Chapter 1.*

Signal processing and multiplexing techniques used in terrestrial interface subsystems are described in *Chapter 2.* Following a brief review of conventional PCM techniques you are introduced to more advanced analog-to-digital conversion techniques (adaptive differential PCM modulation, DPCM) for audio and television signals, to transmultiplexers, echo suppression/cancellation, digital speech interpolation, and energy dispersal systems.

In *Chapter 3* baseband transmission systems principles and design techniques are described. Baseband spectra, Nyquist theorems, and filtering/equalization techniques are presented which are essential for the design of spectrally efficient digital satellite systems.

The techniques presented in this and also in other chapters may be applied also to digital cable, fiber optics, line-of-sight microwave, and to telephony data transmission systems. You are exposed to numerous original data transmission, signal-processing, and modulation techniques which are presented for the first time in a book. A careful study of these innovative techniques will enhance your creativity and understanding of new, frequently competitive, system developments.

You study the principles, performance analysis, and design tools of *power-efficient digital modulation* techniques for linear and nonlinear satellite earth station and satellite system applications in *Chapter 4.* There is *no other book* which presents this important subject in an in-depth, pragmatic, comprehensive manner. Also included are the more important design aspects of conventional QPSK and also new innovative modulation techniques and the performance analysis of these systems in both adjacent and co-channel interference environment. Integrated-circuit implementation of frequently used building blocks are also presented. *Illustrative design examples taken from our experience in INTELSAT and domestic digital satellite system and hardware designs reinforce your understanding of the practical system capabilities and constraints.*

In *Chapter 5* you are exposed to *spectrally-efficient digital modulation* techniques which have a spectral efficiency of more than 2 b/s/Hz. The congestion prevailing in many regions of the radio spectrum has created the need for improved spectrum utilization techniques. With the recent discovery of nonlinearly amplified modulation techniques, it is reasonable to expect that some future generation satellite systems will have a sufficiently high signal-to-noise ratio to employ highly spectral efficient modulation techniques. This chapter includes a concise summary of the material contained in my previously published book, *Digital Communications: Microwave Applications* (Prentice-Hall, Inc., 1981).

You learn the principles and applications of error correction and detection codes in *Chapter 6.* This chapter, written by **Dr. W. H. Tranter,** *Professor, University of Missouri,* presents a down-to-earth, clear treatment of the most important principles of coding and information theory and their applications in digital satellite systems. You are guided gradually from the simplest concepts to a solid understanding of the advantages of coded systems.

Synchronization subsystems (particularly carrier recovery and symbol timing recovery systems) used in coherent digital transmission systems are described in *Chapter 7,* written by **Dr. L. E. Franks,** *Professor,* at the *University of Massachusetts.* The in-depth theoretical treatment of complex signal envelopes and of maximum likelihood receivers is presented in the appendices of this chapter, in order to enable understanding of synchronization systems, without the need to go through complex equations. The material presented in this chapter is of particular interest to engineers who have to acquire a solid theoretical foundation of synchronization systems.

Dr. S. J. Campanella, *Executive Director,* and **Dr. D. Schaeffer,** both of *COMSAT Laboratories,* describe time-division multiple-access (TDMA) satellite systems in *Chapter 8.* These systems are becoming the backbone of INTELSAT and of numerous domestic satellite systems. Both background and advanced material are presented in this chapter.

You learn about powerful signal-processing techniques used in regenerative satellite systems in *Chapter 9.* The advantages of these techniques and the additional flexibility introduced by the regenerative *"switchboard-in-the-sky"* systems are described.

In *Chapter 10* single-channel-per-carrier (SCPC) digital satellite systems are described. These systems are operational in numerous countries and are cost-effective, particularly for small data users. A novel, more powerful modulation technique invented by the author for SCPC applications is also presented. Finally, trade-offs and trends in satellite system design are presented.

Specifications and photographs of modern digital satellite systems are presented in many chapters. These are included to illustrate typical-state-of-the-art systems. In addition to the illustrative *design examples* and solutions given in the text, carefully selected problems are given at the ends of chapters, where additional emphasis on problem solving is considered desirable. Classroom instructors may obtain a *complete solutions manual* from Prentice-Hall, Inc. I believe that you will find the design examples and the problems educative and interesting. Some of these are tailored to enhance your *intuition,* which leads to **creative original designs.**

In a number of chapters *original research* material has been included. Even though this book is intended to be an *introductory text* to digital satellite communications fundamentals, design, and applications, I felt that it is worthwhile to introduce modern research concepts and ideas. To limit the size of this volume and to meet its original objective of being a *practical digital satellite communications book* that could be *understood* by readers who *do not* necessarily have the *mathematical* sophistication of research engineers, certain derivations have been omitted. Rather, particular emphasis is placed on the physical interpretation of final equations, the practical hardware, system constraints, and the digital *satellite earth station* systems applications. The numerous up-to-date references provided at the end of the text should be helpful to those who wish to study the theoretical derivations and obtain more in-depth knowledge of the material covered.

Considering the evolutionary style and philosophy of this book, I feel it is appropriate to state how this book was conceived. Throughout my background, which includes over 10 years of full-time industry research, design, and applications engineering and management, plus approximately nine years of university teaching, research, and **consulting,** I realized that the vast majority of practicing telecommunications engineers seldom use sophisticated mathematical tools. For their successful professional advancement they are required to have a solid knowledge of the principles of system and equipment operation and are expected to apply this knowledge to the design of modern cost-effective systems.

The material in my *previously published books*

---

(1) K. Feher: *Digital Modulation Techniques in an Interference Environment,* Don White Consultants, Inc., Germantown, Virginia, 1977.
(2) K. Feher: *Digital Communications: Microwave Applications,* Prentice-Hall, Englewood-Cliffs, N.J., 1981.

---

was based on material covered in short courses and seminars given in the United States, Canada, Mexico, South America, and many European countries. This material was supplemented by both graduate and undergraduate course material which I teach at the university. The positive feedback that I have been continuously receiving from telecommunications professionals, students, and professors who have been studying or teaching from my two books listed above encouraged me to have the same pragmatic and progressive approach in this book.

I would be delighted to hear from you. If you have questions, comments, or suggestions related to the content of this book, please feel free to write to me.

Dr. Kamilo Feher
Professor, Electrical Engineering
University of Ottawa
770 King Edward Avenue
Ottawa, Ontario
K1N 9B4
Tel: (613) 231-2288

ACKNOWLEDGMENTS

I wish to gratefully acknowledge my indebtedness to all individuals and all organizations who have contributed to both the content and quality of this book. In particular, I wish to thank:

Mr. S. ASTRAIN, Director General, INTELSAT, for his Foreword

Dr. S. J. CAMPANELLA, Executive Director, COMSAT Laboratories, and Dr. D. SCHAEFER, Manager, COMSAT Laboratories, for their joint chapter on TDMA systems

Dr. L. E. FRANKS, Professor, University of Massachusetts, for his chapter on synchronization systems

Dr. W. TRANTER, Professor, University of Missouri, for his chapter on coding techniques

Also, I thank the review board members of Prentice-Hall, Inc., particularly Dr. A. LENDER (GTE Lenkurt), Dr. S. L. GOLDMAN (RCA), Professor R. J. MULHOLLAND (Univ. of Oklahoma), and Mr. L. POLLACK (COMSAT) who have provided thorough and very thoughtful reviews. Mr. BERNARD GOODWIN, Editor, Prentice-Hall, Inc., has been very efficient in coordinating this project.

At the University of Ottawa I have been enjoying an excellent collaboration with my colleagues, and I have been encouraged by both them and our administrative staff to complete this project. Substantial research grants received from the Natural Sciences and Engineering Research Council of Canada (NSERC), Canadian and international industry, and government contracts have enabled us to establish at the University of Ottawa one of the most active digital communications research laboratories in Canada. A 14/12-GHz satellite transmit–receive earth station system with 5-meter antenna, a 2-GHz terrestrial line-of-sight microwave system, and a well-equipped laboratory of the Digital Communications Group (DCG) of the University of Ottawa provide an advanced research and educational facility for inventive digital communications research. A number of original research results discovered by members of our research team and our graduate students are described in this text. Particularly, the research efforts and reviews of Dr. D. H. MORAIS, Dr. J. Y. C. HUANG, Mr. T. LE-NGOC (currently Ph.D. candidate), Mr. P. VANDAMME (CNET-France—visiting research fellow), Dr. S. OSHITA (Assoc. Prof., Shinshu University, Japan—visiting professor), and Dr. S. KATO (staff engineer, NTT, Japan), who have been associated with our university, are specially acknowledged.

Being a consultant of Spar Aerospace Limited (formerly RCA Limited) for many years, I have had the opportunity to work on many challenging digital satellite communications projects. The drive and enthusiasm of Mr. M. J. MORRIS (Engineering Manager, Spar) inspired me to complete this project in a relatively short time.

As Guest Editor of the *IEEE Transactions on Communications*, Special Issue on "Digital Satellite Communications," I wish to thank my Associate Guest Editors, Dr. L. GREENSTEIN (Bell Laboratories), Mr. L. POLLACK (COMSAT), and Mr. D. LOMBARD (CNET, France). The Guest Editors, authors, and reviewers of this issue (published January 1983) have had valuable suggestions.

Participants of our numerous short courses and seminars (in the United States, Canada, Europe, Mexico, and South America) from various companies, government organizations, and universities suggested improvements. My graduate students at the University of Ottawa, Carleton University, and Concordia University "classroom-tested" the manuscript and the problems. Mr. D. H. M. Reekie of Telesat Canada improved the overall style and presentation of the manuscript.

Mrs. Suzanne Racine conscientiously typed major parts of the manuscript and its revisions.

I wish to express my deep appreciation to my wife, ELISABETH LEPAGE-FEHER, for her constant encouragement and good suggestions in the preparation of this book. Without her encouragement and consent, this book would not have been completed. Our children, Catherine, Valerie, and Antoine-Kamilo were patient during weekends and long evening hours while I was writing and editing this book. My mother fostered my ambition and love toward my engineering profession. My father, for whom I will always have fond memories, was a hard-working, honest man who always encouraged me and helped me throughout my studies and life.

Dr. Kamilo Feher
Ottawa, November 1982

1

SATELLITE SYSTEM CONFIGURATIONS AND LINK CALCULATIONS

1.1 INTRODUCTION TO THE CHALLENGE OF THE 1980S AND 1990S

In this introductory chapter, a brief description of the numerous services provided by presently operational and planned communications satellites is followed by a listing of a number of international and domestic satellite systems. From the discussion of analog versus digital satellite transmission trends, we conclude that modern satellite–earth station systems more and more frequently employ *digital modulation* techniques (systems that modulate and demodulate digital signals are known as digital communication systems).

The more important earth station and satellite communication subsystems are described. This chapter concludes with definitions of basic system parameters and an illustrative link calculation.

1.2 INTERNATIONAL AND DOMESTIC SATELLITE COMMUNICATION SYSTEMS

The commercial global satellite system owned and operated by a consortium of more than 100 nations known as the *International Telecommunications Satellite Organization,* INTELSAT, provides modern, high-quality services to its member countries.

Atlantic Ocean Region

Algeria: Lakhdaria 3
Angola: Cacuaco
Argentina: Balcarce
 1 & 2
Austria: Aflenz
Barbados: Barbados
Belgium: Lessive
Belize: Belmopan
Bolivia: Tiwanacu
Brazil: Natal, Tangua
 1 & 2
Cameroon: Zamengoe 1
Canada: Des Laurentides,
 Mill village
 1 & 2
Chile: Longovilo 1 & 2
Colombia: Choconta 1
Congo: Mougouni
Cuba: Caribe
Denmark
 Greenland: Godthaab
Dominican Republic:
 Cambita
Ecuador: Quito
Egypt: Maadi
El Savador: Izalco
Ethiopia: Sululta
France: Bercenay-en-
 Othe 1, Pleumeur-
 Bodou 1 & 2
 French Guiana: Trou-
 Biran
 Martinique: Trois Ilets
Gabon: Nkoltang
Gambia: Banjul
Germany: Raisting
 2 & 3
Greece: Thermopylae 2
Guyana: Georgetown
Haiti: J-C Duvalier
Iran: Asadabad 1
Iraq: Dujail 2
Israel: Emeq Ha'ela 1
Italy: Fucino 1, Lario
Ivory Coast: Abidjan 1
Jamaica: Prospect pen
Jordan: Baqa 2
Kuwait: Umm Al-Aish 2
Liberia: Sinkor

Libya: Tripoli 1
Mali: Sullymanbougou 1
Mexico: Tulancingo 1
Morocco: Sehouls
Mozambique: Boane
Netherlands: Burum 1
 Netherlands Antilles:
 Verdenberg
Nicaragua: Managua
Nigeria: Lanlate 2
Panama: Utibe
Paraguay: Aregua
Peru: Lurin
Portugal: Sintra
 Azores: Ponta
 Delgada
Romania: Cheia 2
Saudi Arabia: Taif
Senegal: Gandoul
Sierra Leone:
 Wilberforce
South Africa: Pretoria
 1 & 3
Spain: Buitrago 1 & 3
 Grand Canary:
 Aguimes
Sudan: Umm Haraz
Surinam: Partes,
 Santo Boma
Sweden: Tanum 1*
Switzerland: Leuk 1
Togo: Cacavelli
Trinidad & Tobago:
 Matura point
Turkey: Ankara
United Arab Emirates:
 Abu Dhabi
United Kingdom:
 Ascension Island,
 Gibraltar, Goonhilly
 1, 2 & 3
 Bermuda: Devonshire
United States: Andover 2,
 3, Etam 1 & 2
Upper Volta: Somgande
Uruguay: Manga
U.S.S.R.: Lvov, Moscow
Venezuela: Camatagua 1
Yugoslavia: Jugoslavija
Zaire: Nsele

*Tanum earth station is a joint undertaking of
 Denmark, Finland, Norway and Sweden

Indian Ocean Region

Algeria: Lakhdaria 1
Australia: Ceduna 1
Bahrain: Ras Abu Jarjur
Bangladesh: Betbunia
Brunei: Telisai
Burma: Rangoon
China: Pekign 2
China: Taipei 2
France: Pleumeur-
 Bodou 4
Germany: Raisting 1
Greece: Thermopylae 1
India: Ahmed, Vikram
Indonesia: Djatiluhur 2
Iran: Asadabad 2
Iraq: Dujail 1
Italy: Fucino 2
Japan: Yamaguchi 1
Jordon: Baqa 1
Kenya: Longonot 1
Korea: Kum San 2
Kuwait: Umm Al-Aish 1
Lebanon: Arbaniyeh 1
Madagascar: Philibert
 Tsiranana
Malawi: Kanjedza
Malasia: Kuantan 1
Maldives: Maldives

Mali: Sullymanbougou 2
Mauritius: Cassis
Netherlands: Burum 2
Niger: Niamey
Nigeria: Lanlate 1
Oman: Al Hajar 1
Pakistan: Deh Mandro
Philippines: Pinugay 2
Qatar: Doha
Romania: Cheia 1
Saudi Arabia: Riyadh 1
Seychelles: Bon Espoir
Singapore: Sentosa 1
Somalia: Kaaraan
South Africa: Pretoria 2
Spain: Buitrago 2
Sri Lanka: Padukka
Syria: Sednaya
Thailand: Si Racha 2
United Arab Emireates:
 Dubai, Ras Al-
 Khaimah
United Kingdom: Madley
 1, Hong Kong 2
Yemen A.R.: Sanaa
Zambia: Mwembeshi
Tanzania: Mwenge
U.S.S.R.: Moscow

Pacific Ocean Region

Australia: Carnarvon 2,
 Moree
Canada: Lake Cowichan
China: Peking 1,
 Shanghai
China: Taipei 1
Fiji Islands: Suva
France
 French Polynesia:
 Papenoo
 New Caledonia: L'lle
 Nou
Indonesia: Djatiluhur 1
Japan: Ibaraki 3
Korea: Kum San 1
Nauru: Nauru
New Hebrides: Port Vila

New Zealand:
 Warkworth
Philippines: Pinugay 1
Singapore: Sentosa 2
Solomon Islands:
 Honiara
Thailand: Si Racha 1
Tonga: Nuku 'Alofa
United Kingdom:
 Christmas Island,
 Hong Kong 1
United States:
 Brewster, Jamesburg,
 Paumalu, 2
 American Samoa:
 Pago Pago
Guam: Pulantat

Earth stations listed here were operating as of
December 31, 1979. Where there is more than
one antenna at an earth station, the numerical
designation stands for the specific antenna or
antennas providing service for the region

(a)

Figure 1.1 (a) Earth stations operating in the INTELSAT network as of December 31, 1979.
(With permission from the International Telecommunications Satellite Organization INTEL-
SAT.) (b) INTELSAT earth stations; antennas and countries. (With permission from INTEL-
SAT, 1980 Annual Report.)

2

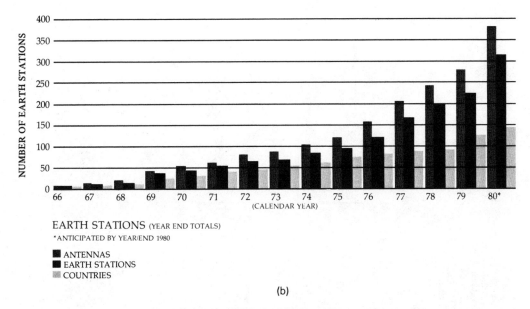

EARTH STATIONS (YEAR END TOTALS)

*ANTICIPATED BY YEAR/END 1980

■ ANTENNAS
■ EARTH STATIONS
▒ COUNTRIES

(b)

Figure 1.1 *(Continued)*

The various INTELSAT and domestic satellite systems services are listed in the preface. Earth stations in the INTELSAT system are owned and operated by the designated telecommunications entities in the countries where they are located (see Fig. 1.1).

The international and also the domestic long-distance communication traffic requirements are increasing at a fascinating speed. In Fig. 1.2 the increase in the number of telephone sets throughout the world and the increase in telecommunications traffic until the end of 1986 are shown.

The present and projected INTELSAT voice-circuit traffic and the *postulated growth* in North American domestic *satellite system demand* shown in Table 1.1 is even more fascinating. As projections go further into the future they become relatively speculative. More and more frequently, digital techniques are used for voice and also television transmission. After 1985, INTELSAT is going to use the time-division multiple-access (TDMA) digital communication method for most of its voice circuits. (Details of TDMA systems are presented in Chapter 8.) Similar trends are present in North America, Europe, and Japan for domestic satellite systems requirements.

We note that the bulk of telecommunication traffic is the transmission of conventional speech telephone signals (in analog or digitized form). New data services are more and more in demand. However, for many years to come the bulk of the *traffic* will remain the *conventional speech telephone signal.*

Although the prices of most goods and services are subject to inflation, satellite communication services become more affordable every year. These systems (in many instances) offer more flexibility than submarine cables, buried underground cables, optical fiber systems, and line-of-sight microwave systems. For these reasons it is

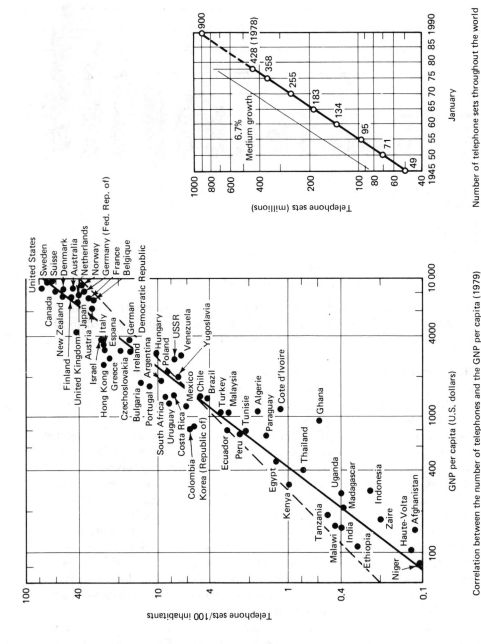

Figure 1.2 (a) Number of telephone sets throughout the world and correlation with gross national product (GNP). (With permission from the International Telecommunications Union, Ashgar, M.Y. and Senuma, R. Pinez, September 1980.) (b) International telecommuncations traffic increase, 1972–1987. (Permission from the International Telecommunications Union [Ashgar et al., September, 1980].)

4

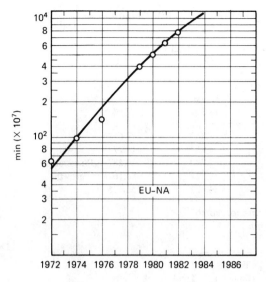

Graph 1: Traffic Europe—North America

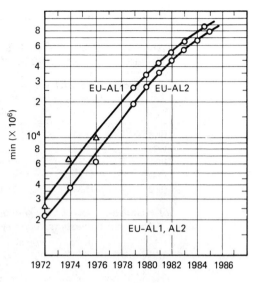

Graph 3: Traffic Europe—Latin America

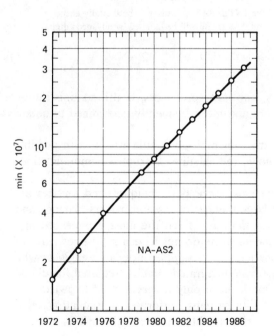

Graph 2: Traffic North America—Asian countries east of Bangladesh and Pacific countries including Australia

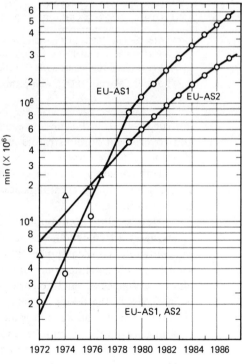

Graph 4: Traffic Near East—Middle East—Europe

(b)

Figure 1.2 (*Continued*)

TABLE 1.1 PROJECTED INTELSAT AND NORTH AMERICAN
(U.S. AND CANADIAN) VOICE-CIRCUIT GROWTH[a]

(a) Projected INTELSAT voice-circuit traffic

Year	Atlantic region	Pacific region	Indian region	Total	
1975	5,593	1,462	1,860	8,690	
1979	11,153	2,605	4,013	17,272	
1983	21,624	2,883	7,780	34,287	**Phenomenal**
1989	57,685	12,525	18,952	84,162	**growth!**
1993	95,393	23,469	34,323	153,185	

*(b) Postulated growth, voice-circuit demand, North
American domestic satellite systems*

System

Year	Westar	Americom	Comstar	Telesat (Canada)	Total	
1976	8,660	25,900	25,900	8,660	69,160	
1979	13,860	41,400	41,400	13,800	110,520	**Significant**
1983	27,720	84,800	84,800	27,720	225,040	**growth!**
1989	66,528	203,520	203,520	66,528	540,100	

[a] In the late 1980s and beyond, most of the INTELSAT circuits will be digitally encoded. The bulk of the satellite voice, television, and data traffic will be handled by advanced digital transmission techniques such as TDMA.

Source: [Marsten, 1977], with permission from the IEEE.

not surprising that business executives, university professors, health administrators, and government administrators are more and more frequently determined to obtain *their own satellite systems.*

Unfortunately, satellite systems are vulnerable to hostile attack, with possible resulting loss of the communication node, the satellite. Disrupted communications would result.

New satellite communications systems concepts, techniques, and services are emerging at a tremendous speed. The large number of research papers, patents, and innovative systems/equipment is *almost shocking.* Even the most experienced and knowledgeable satellite communications engineers do not have a solid knowledge of all of the aspects of satellite communications. For example, a research engineer who is an expert in the design of very large scale integrated circuits (required for digital satellite communications subsystems) might have only a very limited knowledge of traveling-wave-tube amplifiers and satellite systems performance simulation techniques. Alternatively, an expert in analog FM satellite systems might be only a novice in digital satellite transmission.

In this book we focus on the engineering concepts, techniques, and applications of digital satellite communication systems in general and earth stations in particular.

To enable you, the reader, to understand advanced digital satellite communication systems, we devote considerable effort to the study of data transmission techniques. After a careful study of the material presented, you will acquire a *solid foundation* on which you may base your work in digital satellite communications. Hopefully, some of our readers will literally *"fall in love"* with this exciting field. (I did, whether you believe it or not.)

It has already been proven by economic studies that digital satellite systems offer the most cost-effective methods of signal transmission for many applications. The first communication satellites were launched around 1960. Twenty years later, in the early 1980s, there are over 60 operational and planned communication satellites. Thus it is fair to state that *the growth of satellite communication is unparalleled.* *Solve Problem 1.1.*

1.2.1 Illustrative Earth Station and Satellite Systems

In Figs. 1.3 and 1.4(a)–(d), large, medium, and small (antenna) satellite earth stations equipped with digital communication subsystems are shown. These earth stations transmit and receive modulated signals to/from satellite systems such as the ones illustrated in Fig. 1.5 and listed in Table 1.2. This table is a partial list of operational and planned digital satellite communication systems. Note that most systems use quadriphase-shift keying (four-phase-shift keying), known as QPSK modulation techniques. The entries are *approximate* and should be interpreted merely as typical system parameters. In later sections detailed definitions of these parameters are presented.

Figure 1.3   Large (30-m antenna diameter) INTELSAT standard A earth station operating in the 6/4-GHz band (the first number designates the uplink frequency, the second number the downlink frequency). This earth station, owned and operated by Teleglobe Canada, carries digital single-channel-per-carrier (SCPC) FDM FM traffic. (Courtesy of Spar Aerospace Limited and of Teleglobe Canada.)

(a)

(b)

(c)

(d)

(e)

Figure 1.4 (a) Typical standard B earth station antenna (10 to 15 m diameter). (b) Emergency services provided by a small portable earth station. (c) Transportable earth station. (d) A small earth station antenna located at permanent customer services (b, c, and d, from J. N. Sivo, "Satellites Using the 30/20 GHz Band," NASA Technical Memorandum 81600, Lewis Research Center, Cleveland, Ohio 44135, by permission.) (e) Ship earth station used on the Queen Elizabeth 2. (Courtesy of Inmarsat.)

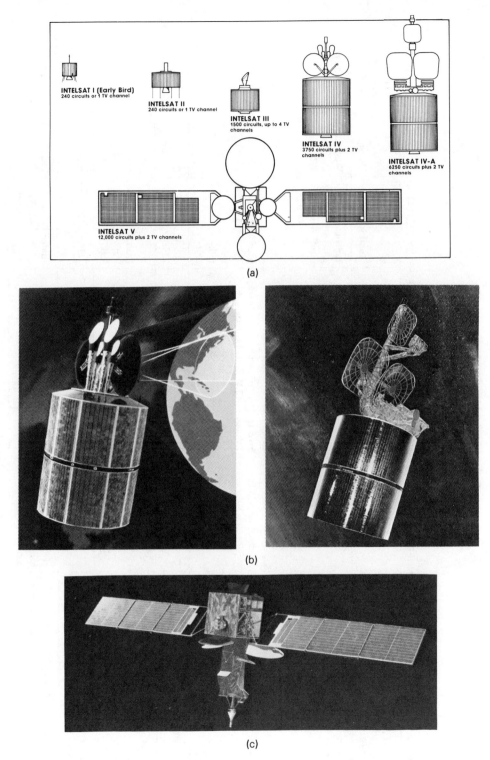

Figure 1.5 (a) Illustrative commercial INTELSAT satellites. (Courtesy of C. F. Hoeber, Ford Aerospace and Communications Corporation.) (b) INTELSAT-IV and IV-A satellites. (c) INTELSAT-V satellite.

TABLE 1.2 OPERATIONAL AND PLANNED COMMUNICATIONS SATELLITE SYSTEMS THAT EMPLOY DIGITAL MODULATION TECHNIQUES[a]

System	INTELSAT-IV	INTELSAT-IVA	INTELSAT-V	SBS	Comsat Marisat	Marecs	Insat	RCA Satcom	WU Westar	TDRS Advanced Westar
System operator	INTELSAT	INTELSAT	INTELSAT	Satellite Business Systems	Comsat General/ RCA Global Comm./ ITT World Comm./WUI	European Space Agency	Indian government	RCA American Comm.	Western Union Telegraph	Western Union Telegraph
Service[b]	fixed tel., TVD	fixed tel., TVD	fixed tel, TVD	fixed tel., TV	mobile tel., TTY	mobile tel., telegraph	fixed TV and radiometry	fixed tel., TVD	fixed tel., TV,TTY	fixed
Frequencies[c]	C	C	C / Ku	Ku	L ship C shore	L ship C shore	S / C	C / C	C	C / Ku
Mass (kg)	731	790	1020	546	326	466	1054 (into transfer)	461	297	2132
Primary power (W)	569	708	1220	1000	330	500	1250	770	260	1700
Coverage[d]	global spot	global hemisphere spot	global hemisphere zonal / spot	Conus	global	global	India	Conus, Alaska, Hawaii	Conus	Conus, Alaska, Hawaii / Conus
Number of transponders	12	20	15 / 6	10	2	1	2 / 12	24	12	4

Transponder bandwidth (MHz)	36	36	36 41 72 77	72 77 241	43	4	5.9 ship to shore 4.75 shore to ship	36	36	34	36	36	225
Number of antenna beams	6/4	7/3	5/5	2: east and west	1	1	2	1	1	2	2	2	7
Polarization	circular	circular	circular	linear	linear	circular	circular	linear		linear	linear	linear	linear and circular
G/T (dB/K) (figure of merit)	−18.6	−18 −11.6 −11.6	−18.6 −11.6 — −8.6	0.0 E 3.3 W	2 to −2	−17 −25	−12.8 ship to shore −16.5 shore to ship	—		−5 −10	−6	−7 −12.5 −12.5	−5 to 4.4
EIRP (dBW) (effective isotropic radiated power)	22.5 34.2	22 26 29	23.5, 26.5 26, 29 29	41.4 E 44.4 W	40 to 43.7	20–29.5 18.8	18.8 ship to shore 20, 26, 29.5 shore to ship	34	42	33 26	33	33 28 26	42 to 50.3
Modulation[e]	FDM/fm fm QPSK SCPC	FDM/fm fm QPSK SCPC	FDM/fm fm QPSK	FDM/fm fm QPSK	QPSK	fm BPSK	PSK fm, SCPC	fm QPSK		FDM/ fm QPSK SCPC	fm QPSK	fm QPSK	250 Mb/s TDMA
Multiple access[f]	FDMA TDMA	FDMA TDMA reuse	FDMA TDMA reuse	FDMA reuse	TDMA	TDMA FDMA	TDMA FDMA	FDMA		FDMA	FDMA TDMA	FDMA TDMA	beam switching

TABLE 1.2 (Continued)

System	ANIK B	ANIK C	ANIK D	Japan BS	Japan CS	Insat	Sirio	OTS/ECS	DSCS II	DSCS III	NATO III	Fleetsatcom
System operator	Telesat Canada	Telesat Canada	Telesat Canada	Japanese government	Japanese government	Indian government	Italian government	European Space Agency	U.S. government/Department of Defense	U.S. government/Department of Defense	NATO	U.S. government/Department of Defense
Service[b]	fixed tel.	fixed	fixed tel., TVD, telegraph	broadcast experimental	fixed experimental	fixed TV and radiometry	fixed experimental	fixed	fixed military	fixed military	fixed military	mobile military
Frequencies[c]	C/Ku	Ku	C	Ku	K, Kc, C	S | C	Ku, Kc	Ku	X	X	X	uhf, X
Mass (kg)	440	522	635	352	340	1054 (into transfer)	218	444	536	748	349	1005
Primary power (W)	840	925	1000	1000	521	1250	118	700	520	800	538	1425
Coverage[d]	Canada	Canada	Canada, northern U.S.	Japan	Japan	India	North Atlantic, Europe spot	spot Europe A Europe B	global spot	global spot	Atlantic Europe	global
Number of transponders	12/6	16	24	2	6/2	2 | 12	1	6	4	6	2	12
Transponder bandwidth (MHz)	36 72	54	36	50 80	200	36 | 36	32	120 40 5	410 total	395 total	17 85 50	0.005 0.025 0.500
Number of antenna beams	14/1	4/1	5	1	1	1 | 1	3	3	3	multiple array	2	1
Polarization	linear	linear	linear	linear	circular	linear	circular	linear, linear, circular	circular	circular	circular	circular

G/T (dB/K) (figure of merit)	−6/−1	3	−3	−8.2	noise figure = 13 dB/9 dB	—	−22.2, −17.2	4.2 −4.8 −2.2	8.5 20.2	−16 to −1	−14.1	−16.6 (uhf)
EIRP (dBW) (effective isotropic radiated power)	36 47.5	48	36	55	37 29.5	34 42	24	45.8 36.5 42.0	28 40	23–40	35 29	26–28 27
Modulation[e]	FDM/ fm QPSK, SCPC	FDM/ fm QPSK, SCPC	FDM/fm fm/TV SCPC	fm and digital	100 Mb/s digital	fm QPSK	PCM-PSK and fm video	fm PSK	fm QPSK	fm QPSK	fm QPSK	fm QPSK
Multiple access[f]	FDMA TDMA	FDMA TDMA reuse	FDMA	SCPC	TDMA	FDMA SCPC	SCPC	TDMA	FDMA TDMA CDMA	FDMA TDMA CDMA	CDMA FDMA	FDMA

[a] Approximate entries are used, merely as an illustration of typical system parameters.

[b] fixed: satellite service to specified fixed points
mobile: satellite service to ships, airplanes, and mobile ground terminals
broadcast: satellite service intended for reception by the general public
TTY: teletypewriter
tel.: telephone
TVD: television distribution

[c] P band or uhf   200 to 400 MHz   X band   7250 to 7750 MHz
L band   1530 to 2700 MHz   7900 to 8400 MHz
S band   2500 to 2700 MHz   Ku band   10.95 to 14.5 GHz
C band   3400 to 4200 MHz   Kc band   17.7 to 21.2 GHz
         4400 to 4700 MHz   K band   27.5 to 31.0 GHz
         5725 to 6425 MHz

[d] Conus: continental United States without Alaska and Hawaii

[e] SCPC: single channel per carrier   BPSK: binary phase-shift keying
FDM: frequency-division multiplexing   QPSK: quadriphase-shift keying
fm: frequency modulation   (four-phase-shift keying)
PCM: pulse-code modulation   CPSK: coherent phase-shift keying
PSK: phase-shift keying

[f] CDMA: code-division multiple access
FDMA: frequency-division multiple access
TDMA: time-division multiple access frequency reuse—same carrier frequencies assigned to different users through use of separate beams or polarizations

Source: With permission from Electronics, September 11, 1980, a McGraw-Hill publication.

13

TABLE 1.3 COMPARISON OF THE CANADIAN ANIK SERIES SPACECRAFT

	ANIK A	ANIK B		ANIK C	ANIK D
Transfer orbit weight (kg)	560	920		1080	1128
Height (on-station) (m)	3.4	3.28		6.43	6.57
Body width (m)	1.78	2.05		2.16	2.16
Maximum dimension (m) (on-station)	3.4	9.54		6.43	6.57
Stabilization	Spun	3-axis		Spun	Spun
Frequency band (GHz)	6/4	6/4	14/12	14/12	6/4
Number of channels	12	12	6	16	24
Number of TWTAs	12	12	4	20	24
Channel bandwidth (MHz)	36	36	72	54	36
Antenna coverage	All Canada	All Canada	4 spot beams	4 spot beams + switching (S. Canada)	All Canada
Saturation flux density (dBW/m²)	−80	−81	−86 (−81)	−81 (−78.5)	−81
Receive G/T (dB/K)	−7.0	−6.0	−1.0	2.0	−3.0
EIRP (dBW)	33	35.7	46.5	46.5	36
TWT rated power output (W)	5	10	20	15	11
Power Battery: nominal Capacity (AH)	7	17		17.3	17.3
Number	2	3		3	3
Array capacity (W)	235	620		800	800
Design life (yr)	7	7		10	10
Mission life (minimum) (yr)	—	—		8	8
Launch dates 1	Nov. 72	Dec. 78		1982	1982
2	Apr. 73	—		1983	TBA

The Telesat space segment consists of four satellites in geostationary orbit, three ANIK A and one ANIK B. A further three spacecraft of the ANIK C series and two of the ANIK D series are currently under construction.

All the ANIK series communications spacecraft are designed to operate in the 6/4-GHz and/or the 14/12-GHz bands designated for fixed satellite services.

Source: With permission from Telesat Canada.

In Table 1.3, specifications of the Canadian ANIK satellites, owned and operated by Telesat Canada, are highlighted. The ANIK satellites carried the first domestic time-division multiple-access commercial traffic.

1.2.2 Geostationary Orbit

Commercial communications satellites lie in the *geostationary orbit;* that is, they are stationary with respect to a particular point on the surface of the earth. Note that the alternative term *geosynchronous orbit* is also used frequently. The height of geosta-

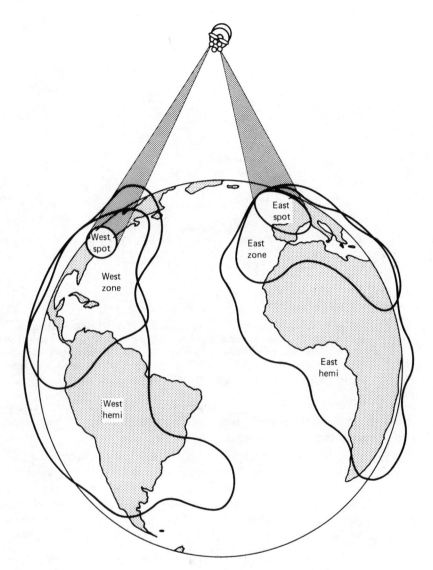

Figure 1.6   Spot beams and hemizone beams (Courtesy of [Hoeber, 1977].)

tionary satellites (distance from a particular point on the surface of the earth) is approximately 36,000 km, assuming that the elevation angle is 90°. A geosynchronous satellite with two spot beams and an earth coverage beam is illustrated in Fig. 1.6. The most important *characteristics* of these satellites include:

1. Coverage with an **earth-coverage antenna** is broad (17.34°, corresponding to approximately one-third of the earth. Earth stations (stationary or mobile) may be easily interconnected within this large area.

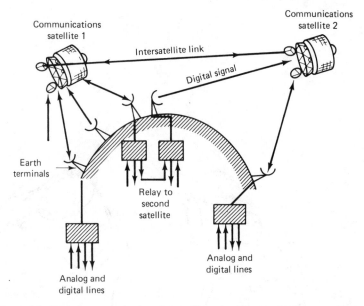

Figure 1.7 Digital communications by satellite relay. The model illustrates analog information which is relayed to a second terminal by a digital communication channel and one or more ground relays. An intersatellite link is also feasible. (J. J. Spilker, *Digital Communications by Satellite,* © 1977, p. 6, reprinted with permission of Prentice-Hall, Inc., Englewood Cliffs, N.J.)

2. The orbital period of geosynchronous satellites is 24 hours. Therefore, the geographic area covered by the satellite does not vary. Constant use of tracking equipment is not required.

3. The full-duplex round-trip delay through a synchronous satellite is approximately 600 ms. In the earlier stages of satellite system development, this delay resulted in some user dissatisfaction, particularly in the domestic service, where customers expect toll-grade-quality performance. However, echo cancelers developed in the late 1970s and early 1980s promise to reduce or almost completely eliminate the adverse effects introduced by long round-trip delays.

The coverage and interconnection flexibility of earth terminals can be increased by the use of satellite relays, such as that shown in Fig. 1.7. The transmission link may be from earth to satellite 1, back to earth and then to satellite 2, and back again to the earth terminal—or alternatively, an *intersatellite link* may be established. *Solve Problem 1.2.*

1.3 SATELLITE SYSTEMS IN THE 1980S AND 1990S

During the late 1970s and early 1980s almost all commercially operational communication satellites operate in the 6/4-GHz band (6 GHz uplink and 4 GHz downlink). These frequency bands are *shared* with many congested analog and digital *line-of-*

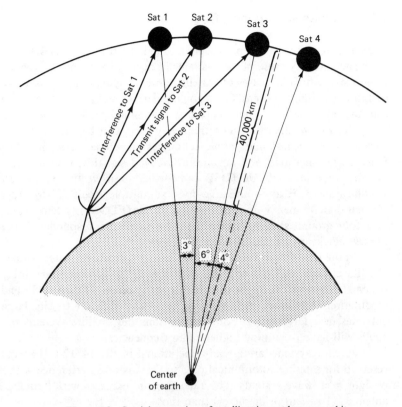

Figure 1.8 Spatial separation of satellites in synchronous orbit.

sight microwave systems, and therefore siting an earth station might pose a major problem. Finding an interference-free location close to metropolitan centers is very difficult.

Satellite systems have to share the use of limited spectrum allocations, and there is a limit to the number of satellites that can be stationed (parked) in a given arc of geostationary orbit. Satellites must have sufficient spatial separation to avoid interference; for example, the signals from one earth station to its satellite must not significantly interfere with reception at a neighboring satellite. For example, in Fig. 1.8 the earth station is transmitting a modulated signal to satellite 2 (Sat 2). However, as the earth station antenna beam width is not infinitesimally narrow, a certain portion of this signal power will also reach Sat 1 and Sat 3. This power is undesired interference power. The *required spatial separation* is between 3 and 6° depending upon:

Earth station and satellite antenna beamwidths and sidelobe level

Carrier frequency of transmission

Modulation technique used

Permissible performance degradation due to interference

1.3.1 Increased Capacity Requirements

Once the allocated frequency band is filled, further increases in satellite capacity can be achieved by *reuse* of the available frequency spectrum. By reducing the antenna beam's size, which increases the satellite antenna gain, different beams can be directed to illuminate different areas. Orthogonal polarization in a polarization diversity system can also increase the system capacity. Antennas can be made to respond to only one polarization; thus the same frequency band can be used twice within the same coverage area. *Digital modulation* techniques (which have a higher spectral efficiency), *multiplexing* methods, and advanced baseband signal processing techniques such as *digital speech interpolation* (DSI) can further improve the bandwidth utilization of satellite systems. Heavy-route time-division multiple access (TDMA) systems are more efficient than frequency-division multiple access (FDMA) systems, particularly if more than four ground stations access the same satellite transponder. The reasons for increased efficiency are described in Chapters 2 and 8.

In the time frame 1980–2000, increasing demand for international and domestic satellite circuits will outstrip the capacity of the geostationary orbit for 6/4-GHz systems, even with the application of frequency reuse, TDMA, DSI, and more efficient modulation techniques. Additionally, it is very difficult to site large earth station antennas near large cities. For these reasons the satellite systems of the 1980s and 1990s will have to employ higher radio frequencies.

Satellite systems are already operational in the 14/12-GHz band. This band is reserved for satellite communications, so there is no interference with terrestrial line-of-sight microwave systems. The antenna gain increases with frequency; thus smaller antennas (1 to 3 m in diameter) than those used in the 6/4-GHz band may be used. This makes possible the *use of 14/12-GHz band earth stations located:*

> Directly at the office
> In a factory
> In a parking lot
> In a warehouse
> In the middle of a city
> etc.

The broad categories of earth stations include:

1. *Heavy route.* High-quality, multipurpose, large stations having a capacity of thousands of voice and data channels. These earth stations cost several millions of dollars.
2. *Light (thin) route.* Special-purpose transmit/receive earth stations having a reduced capacity (5 to 300 channels) are in the range $100,000 to $500,000.
3. *Broadcast type.* Small stations used in cable television (CATV) systems in the range $10,000 to $20,000.

4. *Rooftop-receive only.* Small earth stations available in the 12-GHz band are in the $1000 to $5000 range.

The *space shuttle* introduced in the 1980s facilitates the launch of physically larger, more powerful satellites with increased capabilities. The cost of space launching of communications satellites will drop significantly in the 1980s with the availability of space shuttle vehicles. These launch vehicles return to earth for refueling and are suitable for multiple launchings rather than being lost as have been the rockets used to launch communication satellites for more than two decades. A typical rocket launching is shown in Fig. 1.9(a). In Fig. 1.9(b) the shuttle launch of a Canadian satellite is illustrated. The U.S. space shuttle and the "Canada-Arm" of this shuttle are shown in Fig. 1.9(c) and 1.9(d).

New generations of satellite systems have the following desirable characteristics:

Wider bandwidth transponders

Frequency reuse in common use

Increased sensitivity in uplink reception

Increased effective downlink power

Reduced susceptibility to interference

TDMA (time-division multiple-access digital traffic)

SS-TDMA (satellite switched TDMA)

Scanning spot beam system employing time-division switching

Regenerative satellites (on-board regeneration)

On-board switching matrix (reprogrammable by ground control)

Switchboard in the sky

Multipurpose–multiband satellites

Millimeter-wave satellite systems

Laser communication satellites

These characteristics contribute to meeting *long-term satellite requirements, which include:*

1. Increasing satellite capacity without inordinate increases in mass and cost
2. Increasing system capacity without requiring huge capital outlays for recurring earth station changes
3. Effective utilization and sharing of the orbit/spectrum resource without undue technical complexity and cost
4. Achieving an acceptable solution to rain attenuation outage problems

Satellite switching of digitized messages, on-board regeneration, and processing of multiplexed signals could increase capacity significantly over that previously possible

Figure 1.9 (a) Rocket launch of a telecommunications satellite. (With permission from INTELSAT.) (b) Artist's impression of the shuttle launch of an ANIK C/D-type Canadian satellite. (Courtesy of Hughes Aircraft Company and Spar Aerospace Limited.) (c) U.S. space shuttle. (With permission from the International Telecommunications Satellite Organization [INTELSAT], 1981 Annual Report.) (d) Photograph of the U.S. Space Shuttle with the Spar-built remote-controlled arm "Canada Arm" (Courtesy of Spar Aerospace Limited).

with conventional translation transponders. Also, advances in LSI circuits and micro-processors and their applications to message switching and digital signal processing is expected to bring *direct satellite operation within economic reach of a mass market.*

1.3.2 Advantages of Digital Over Analog Satellite Communications

The advantages of digital satellite communications over analog satellite communications include:

1. *Increased capacity in the multiple-access mode.* In satellite transponder utilization, power efficiency is of prime concern. To increase this efficiency, the high-power output amplifiers are required to operate close to saturation. Time-division multiple-access (TDMA) digital satellite systems achieve a significantly increased system capacity when compared to analog multiple-access FM systems. TDMA transponders, when used properly, relay only one signal at a time. Intermodulation in digital systems is not as critical as it is in analog systems; thus the output amplifiers in continuous digital and burst-operated TDMA systems may operate close to saturation.

2. *Immediate and long-term economical advantages.* The increased capacity, more flexible operation, and reduced production costs contribute to the economic advantages of digital satellite communication systems. The rapid advancements in the developments of ultra-high-speed (1-Gb/s range) logic circuits, high-speed digital processors, and mini- and *microcomputers* enhance the economic benefits of digital satellite communication systems.

3. *More robust to interference.* Digital systems provide excellent performance, with a carrier-to-interference ratio in the range 20 to 30 dB. Analog FDM-FM systems *frequently* require much higher carrier-to-interference ratios.

4. *Compatibility of analog/digital messages and computers.* A digital bit stream is a digital bit stream, independently of whether the source information is a color TV signal, analog voice, or digital data. Thus the signal multiplexing and processing of digital data is less costly than that of analog signals.

5. *New facilities, services.* Digital techniques, computers, and devices enable the introduction of new services which would be not practical with analog methods. Computer communication, reservation systems, and bank data transfers are among these services.

6. *Higher degree of flexibility.* "A switchboard in the sky"—a regenerative satellite with powerful signal-processing capabilities—is feasible using digital communication techniques. There is also a possibility of multibeam satellite switched operation.

7. *Direct low-cost interconnect with terrestrial microwave, cable, and optical fiber systems.* The frequently required interface of satellite earth stations with terrestrial microwave, cable, and optical fiber links is much simpler with digitally encoded signals than with analog messages.

8. *Transmission quality almost independent of distance and network topology.* Multiple switching sections, signal regeneration, and signal processing do not degrade the digital signal quality in multiple hops, whereas in analog systems noise accumulates (the uplink and downlink noise are additive in nonregenerative satellite systems).

1.4 ILLUSTRATIVE EARTH STATION AND SATELLITE COMMUNICATION SUBSYSTEMS

Illustrative block diagrams of modern digital satellite communication systems are shown in this section. In particular, simplified block diagrams of *single-channel-per-carrier* (SCPC) and *time-division multiple-access* (TDMA) earth stations are given. The block diagrams of a communication satellite equipped with 24 transponders is also presented. The main channel, that is, the signal path of the digital data, is frequently shaded in the figures to enable easier identification of the main building blocks. *These systems are discussed in detail in later chapters;* therefore, only a very brief description is presented here. This presentation is included in this introductory chapter merely to give an appreciation of the complexity of the equipment in these systems.

SCPC systems are frequency-division-multiplexed, digitally modulated carriers. In the INTELSAT standard SCPC system, each carrier is modulated in a four-phase (quadriphase) phase-shift-keyed (QPSK) modulator by a 64-kb/s analog-to-digital converted voice signal (see Fig. 1.10). The standard analog-to-digital conversion process used is *pulse-code modulation* (PCM). Each receive earth station receives the complete frequency-division-multiplexed (FDM) message, that is, the messages transmitted from all earth stations toward a specified satellite transponder. The combined received IF spectrum is also illustrated in Fig. 1.10.

In order to simplify the equipment complexity, bandsplitting filters are used (not shown in the figure) in practical earth station designs. These filters reduce the intermodulation products otherwise introduced into the receive chain.

The concept of TDMA satellite communications is illustrated in Fig. 1.11. In these systems *only one* digitally modulated carrier is transmitted through a satellite transponder at any given instant of time. Thus the generation of intermodulation products is avoided and therefore the satellite *power amplifier may be operated in a saturated* (maximal transmit power) mode. This is evidently a significant advantage of digital satellite multiple-access communication systems.

A simplified block diagram of a TDMA transmit/receive terminal and a prototype are shown in Figs. 1.12 and 1.13. The digital speech interpolation (DSI) units multiplex digital and analog messages, respectively. The system capacity advantages gained by the use of DSI subsystems will become evident in Chapter 2. The shaded units comprise the main transmission system. These units are studied in depth in a number of later chapters.

A simplified block diagram of a standard INTELSAT-V satellite is shown in

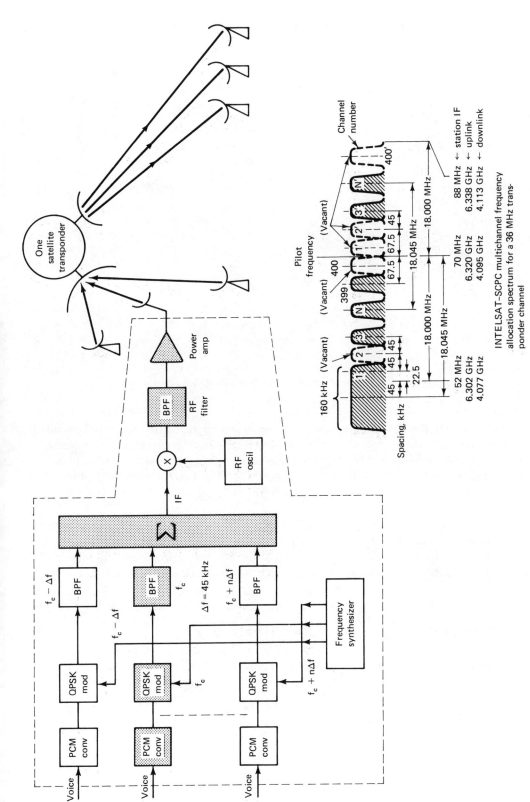

Figure 1.10 (a) Simplified block diagram of a single-channel-per-carrier (SCPC) transmit earth station. (b) IF spectrum at any one of the receive earth stations.

INTELSAT-SCPC multichannel frequency allocation spectrum for a 36 MHz transponder channel

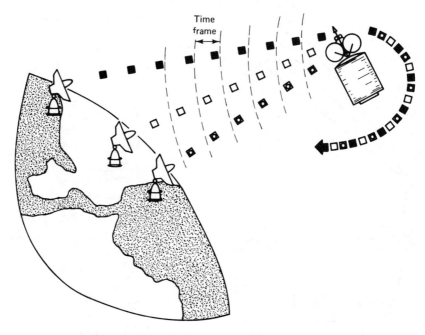

Figure 1.11   Time-division multiple-access (TDMA) system concept.

Fig. 1.14. On this satellite four frequency bands are employed: 14- and 6-GHz bands are received by the spacecraft and 11- and 4-GHz bands are transmitted to the receive earth station. Although only 1 GHz of bandwidth is allocated for satellite transmission at these frequencies, the INTELSAT-V repeater employs both polarization and spatial isolation to obtain 2137 MHz of bandwidth through frequency reuse [Hoeber, 1977]. *Solve Problem 1.3.*

The frequency spectrum shown in Fig. 1.15 is used twice at 14/11 GHz (linear polarization) and four times at 6/4 GHz (circular polarization). The wide bandwidth is used by the INTELSAT system to relay 12,000 simultaneous two-way digitized speech telephone circuits and two color television channels through each spacecraft.

1.5 SYSTEM LINK MODEL AND PARAMETERS

Typical earth stations and satellites contain a large number of radio-frequency channels. In this section the link equations of one of these channels are derived. To facilitate the comprehension of this material, a brief review of the most frequently used system definitions and parameters is presented.

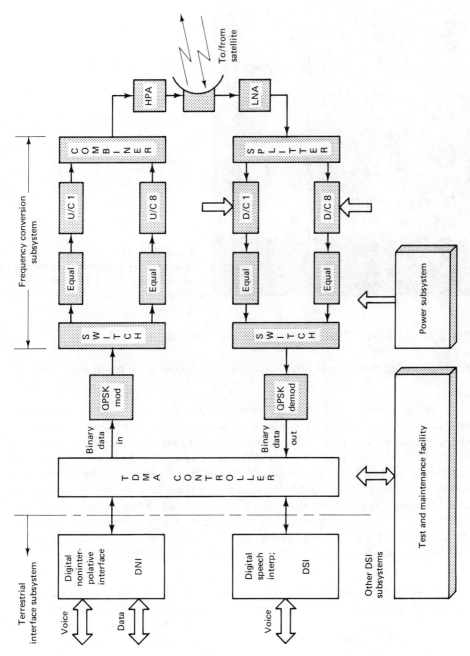

Figure 1.12 TDMA transmit–receive earth station—INTELSAT-V standard (120 Mb/s-QPSK). U/C, up-converter; D/C, down-converter; HPA, high-power amplifier; LNA, low-noise amplifier.

Figure 1.13 TDMA prototype equipment. (With permission of *Electrical Communication,* The Technical Journal of ITT.)

1.5.1 Basic Earth Station–Satellite Link Model

A somewhat simplified earth station–satellite link model is shown in Fig. 1.16.

   Uplink model. The transmit section of a typical earth station contains an intermediate frequency (IF) modulator, a bandpass filter (BPF), an up-converter, and a high-power amplifier (HPA). The modulated output power of the high-power amplifier, P_T, is expressed in watts or in dBW (decibels above 1 W). The earth station antenna gain is G_T. The transmitted signal $P_T G_T$ is attenuated by the spreading factor, atmospheric losses, and other losses. The main factors causing atmospheric absorption are: uncondensed water vapor, rain, fog and clouds, snow and hail, free electrons in the atmosphere, and molecular oxygen. The flux density at the input of the satellite receive antenna is denoted Ω_u. The uplink noise is assumed to have a flat spectral density over the receiver bandwidth of N_{ou} W/Hz.

   Satellite transponder. The signal received at the satellite P_u is amplified by the receiver antenna, filtered, and further amplified by a low-noise tunnel diode amplifier (TDA). A frequency translator is required to prevent in-band interference from the high-power satellite output to the satellite input. If the satellite transponder is not equipped with a frequency translator, then a tremendously high isolation (in the range 100 to 150 dB) between the output and input would be required (practical

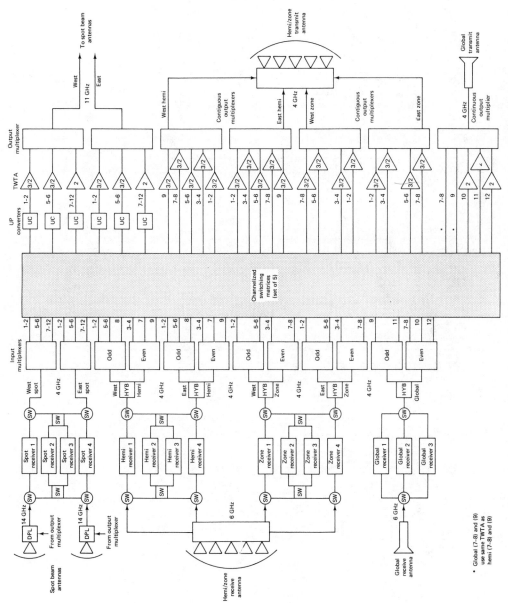

Figure 1.14 INTELSAT-V communications subsystem: simplified block diagram. (After [Hoeber, 1977], with permission.)

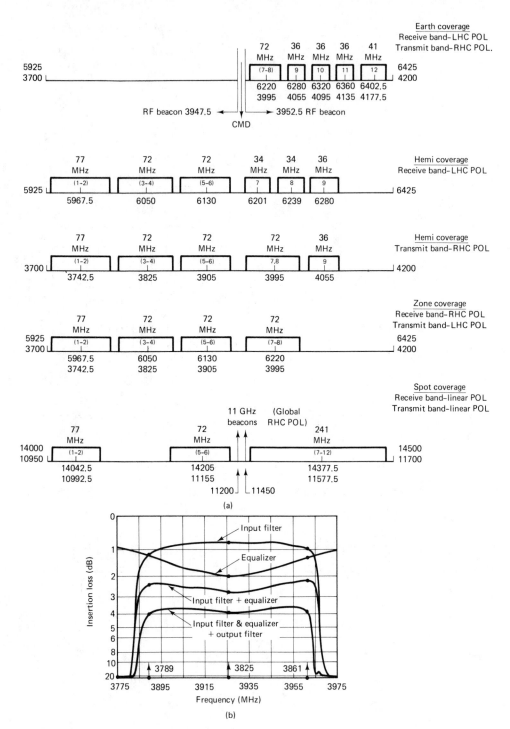

Figure 1.15 (a) INTELSAT-V frequency plan. (b) Channel response. After [Hoeber, 1977], with permission.)

radio-frequency systems do not even approach this requirement). The channel multiplexer is followed by a high-power traveling-wave-tube amplifier.

Downlink model.   The output power of a transponder is P_s. Let the satellite transmit antenna (downlink) gain equal G_{sd}. The transmitted signal $P_s G_{sd}$ attenuated by the spreading loss and other losses (L_{fs} and L_d) is received at the receiving earth station. The received signal has a flux density Ω_d, which is amplified by the receive antenna G_d and fed to the front-end low-noise amplifier (LNA). The down-converter provides the demodulator with an intermediate-frequency carrier.

1.5.2 System Parameters—Definitions

Transmit power P_T and bit energy E_b.   Typical earth station high-power amplifiers (HPA) and satellite traveling-wave-tube amplifiers are nonlinear devices. The output power/input power ratio (gain) is dependent on the input drive level. Typical input/output power and phase characteristics are shown in Fig. 1.17. To obtain efficient operation the designer should attempt to operate the power amplifiers (located in the earth station and in the transponder) as close as possible to saturation. Unfortunately, the nonlinear gain and the phase characteristics of these devices degrade the performance of modulated systems. Thus a compromise solution is required.

Let the saturated output power equal $P_{o\,\text{sat}}$. In this case the average *energy* of a transmitted bit is

$$E_b = P_{o\,\text{sat}} T_b \qquad (1.1)$$

where T_b is the bit duration. In later chapters we will see that a high bit energy E_b is a desirable feature.

From Fig. 1.17 we note that an input *backoff* of 6 dB (a 6-dB lower input power than is required to drive the amplifier into saturation) causes a 2-dB output backoff. Thus a power "compression" is evident.

Isotropic antenna.   An *isotropic antenna* is an ideal lossless antenna which radiates power equally well in all directions. An isotropic antenna can only be approximated in practice. However, for satellite system studies it is convenient to use the isotropic antenna as a reference for comparison with actual antennas. An antenna that radiates equally in all directions (i.e., an isotropic antenna) has gain = 1. For this antenna, the radiation in any direction is constant given by

$$\text{radiation intensity of isotropic antenna} = \frac{P}{4\pi} \qquad (1.2)$$

where P is the input power.

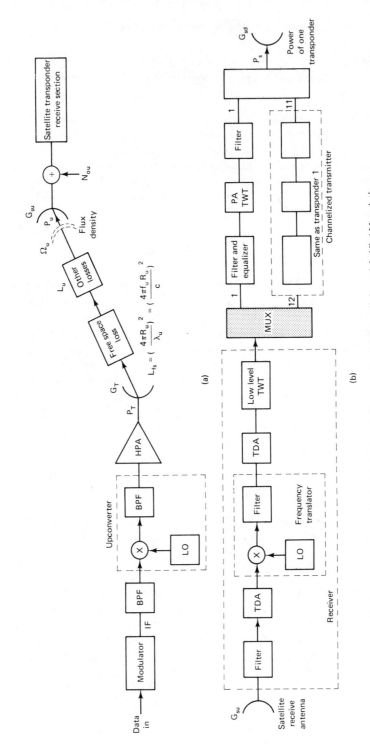

Figure 1.16 (a) Uplink model. (b) Satellite transponder. (c) Downlink model. (d) Abbreviations used in (a)–(c).

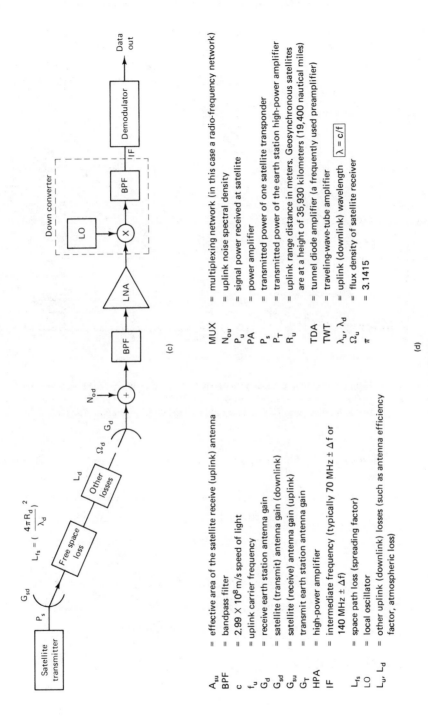

(c)

A_{su}	=	effective area of the satellite receive (uplink) antenna
BPF	=	bandpass filter
c	=	2.99×10^8 m/s speed of light
f_u	=	uplink carrier frequency
G_d	=	receive earth station antenna gain
G_{sd}	=	satellite (transmit) antenna gain (downlink)
G_{su}	=	satellite (receive) antenna gain (uplink)
G_T	=	transmit earth station antenna gain
HPA	=	high-power amplifier
IF	=	intermediate frequency (typically 70 MHz $\pm \Delta f$ or 140 MHz $\pm \Delta f$)
L_{fs}	=	space path loss (spreading factor)
LO	=	local oscillator
L_u, L_d	=	other uplink (downlink) losses (such as antenna efficiency factor, atmospheric loss)

MUX	=	multiplexing network (in this case a radio-frequency network)
N_{ou}	=	uplink noise spectral density
P_u	=	signal power received at satellite
PA	=	power amplifier
P_s	=	transmitted power of one satellite transponder
P_T	=	transmitted power of the earth station high-power amplifier
R_u	=	uplink range distance in meters. Geosynchronous satellites are at a height of 35,930 kilometers (19,400 nautical miles)
TDA	=	tunnel diode amplifier (a frequently used preamplifier)
TWT	=	traveling-wave-tube amplifier
λ_u, λ_d	=	uplink (downlink) wavelength $\boxed{\lambda = c/f}$
Ω_u	=	flux density of satellite receiver
π	=	3.1415

(d)

Figure 1.16 (*Continued*)

31

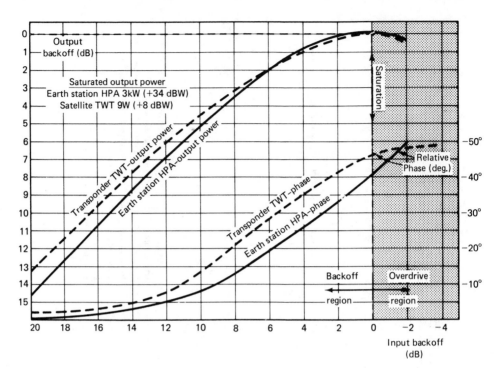

Figure 1.17 Amplitude and phase characteristics of typical earth station HPA and satellite transponder TWT—used in the INTELSAT-V TDMA 6/4-GHz system. (After International Telecommunications Satellite Organization [Intelsat], 1980.

Antenna gain.   The relative increase in power achieved by focusing the antenna is defined as the *gain of the antenna, G.* This gain may be defined as

$$G = \frac{\text{maximum radiation intensity}}{\text{radiation intensity of isotropic antenna}} \qquad (1.3a)$$

assuming the same power input. We may also define this gain as

$$G = \frac{\text{power the receiver receives from the antenna}}{\text{power the receiver would receive if the transmission were isotropic}} \qquad (1.3b)$$

Using equations (1.2) and (1.3a), we obtain

$$G = \frac{4\pi(\text{maximum radiation intensity})}{P} \qquad (1.4)$$

A useful, although approximate expression for *parabolic antenna* is

$$G = \eta \left(\frac{\pi d}{\lambda}\right)^2 = \eta \frac{4\pi f^2 A_t}{c^2} \qquad (1.5)$$

where

η = antenna efficiency (typically 0.55)

d = antenna diameter

λ = wavelength

$c = 2.99 \times 10^8$ m/s (the velocity of light)

$f = c/\lambda$ (carrier frequency)

$A_t = d^2\pi/4$ = aperture area of transmitting antenna

If the frequency is given in gigahertz and the antenna diameter in meters, then we have for the $\eta = 0.55$ antenna:

$$G = 60.7f^2d^2 \tag{1.6}$$

Antenna gains at different frequencies are shown in Fig. 1.18. From this figure we note that an advantage of selecting higher radio frequencies is that smaller antennas have the same gain. For example, a 30-m antenna used at frequencies of 4/6 GHz would give the same gain as a 15-m antenna in the 12/14-GHz band. However, in higher-frequency bands we must account for higher space path loss (spreading).

Effective isotropic radiated power EIRP.   The effective isotropic radiated power (EIRP) of an earth station or of a transponder can be expressed as

$$EIRP = PG$$

or

$$EIRP \text{ (in dB)} = P \text{ (in dBW)} + G \text{ (in dB)} \tag{1.7}$$

where

P = transmit power of the earth station, or
(transponder) high-power amplifier
G = *antenna gain*

Example 1.1

The saturated output power of an earth station amplifier is 2 kW (+33 dBW). The diameter of the earth station antenna is 15 m. The transmit frequency is 14 GHz. Calculate the saturated EIRP of the earth station. Assume that the output backoff and combining losses equal 7 dB.

Solution.   From Fig. 1.18 [or equation (1.6)] we conclude that the 15-m antenna has a 64-dB gain at 14 GHz. Thus we have:

Saturated output power	+33 dBW
Antenna gain	64 dB
Backoff and combining losses	−7 dB
EIRP =	90 dBW

[*Note:* This EIRP is typical for the European Communications Satellite (ECS) and for INTELSAT-V–TDMA earth stations.]

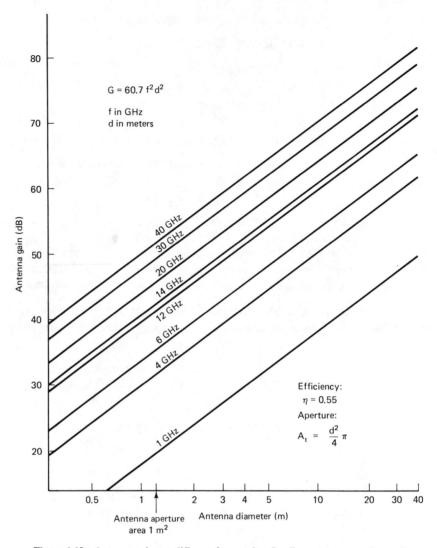

Figure 1.18 Antenna gains at different frequencies. Smaller antennas can be used when the frequency is higher. These antenna gains are for parabolic antennas having an efficiency of 0.55. For an ideal antenna, add 2.6 dB to the antenna gain.

Space path loss.   The loss of power of a radio wave in space is

$$L_{fs} = \left(\frac{4\pi R}{\lambda}\right)^2 = \left(\frac{4\pi f R}{c}\right)^2 \tag{1.8a}$$

where

$$f = \text{frequency of radio wave}$$
$$\lambda = \text{wavelength } (\lambda = c/f)$$

R = distance traveled in space; for an
elevation angle of 90° this distance is
3.593×10^7 m $\approx 3.6 \times 10^4$ km)
c = velocity of light ($c = 2.99 \times 10^8$ m/s)

The spreading or "free space loss" may also be interpreted as

$$L_{fs} = \frac{\text{power received by an isotropic antenna}}{\text{power transmitted by an isotropic antenna}} \qquad (1.8b)$$

Equation (1.8a), for an elevation angle of 90°, written in decibels, is

$$L_{fs}(\text{dB}) = 183.5 + 20 \log f \qquad (1.9)$$

where f is in gigahertz. The free space loss as a function of radio frequency is shown in Fig. 1.19. (*Note:* Water and oxygen absorption is significant near 22 GHz, and 60 GHz and should be accounted for [Spilker, 1977].)

The distance, also known as *slant range,* from an earth station to the satellite is

$$R = [(h + r_e)^2 + r_e^2 - 2r_e(h + r_e) \cos \theta]^{1/2} \qquad (1.10)$$

where h, r_e, and θ are defined in Fig. 1.20 [Yeh, 1972].

Noise temperature, T_e, and noise figure (NF).   In terrestrial microwave systems the power of noise generated in the equipment, or more generally in the transmission system, is specified in terms of the noise figure (NF). In earth station–satellite links it is frequently required to calculate and experimentally verify the accuracy of the noise budget within a fraction of a decibel. This is so because in a large satellite system having numerous earth stations a 1-dB error in the link calculations and in the system performance might be very expensive (e.g., more than \$1 million). For low-noise sources the *equivalent noise temperature, T_e,* parameter provides a more practical system parameter than the noise figure. For this reason it is a more frequently used term in satellite communications.

Noise Figure.   A well-known definition of noise figure is:

$$\boxed{\text{Noise figure (NF)} = \frac{N_{\text{practical}}}{N_{\text{ideal}}} = \frac{N_{\text{practical}}}{kT_o W' A}} \qquad (1.11)$$

where

W' = double-sided noise bandwidth of the system
being evaluated

A = gain of the system being evaluated

$k = 1.380 \times 10^{-23}$ Ws/K
    $= -198.6$ dBm/K/Hz (Boltzmann constant)

T_o = temperature of the environment in which the measurement is performed (at room temperature $T_o = 290$ K)

A noise figure measurement setup is shown in Fig. 1.21, where the measurement points of the parameters of equation (1.11) are illustrated. For a practical setup it is sufficient to have a reference amplifier which has a considerably lower noise figure than the amplifier that is being evaluated.

Noise Temperature, T_e.   If a noise source generates noise of power N, its *equivalent noise temperature, T_e,* is defined by

$$T_e = \frac{N}{kW'} \tag{1.12}$$

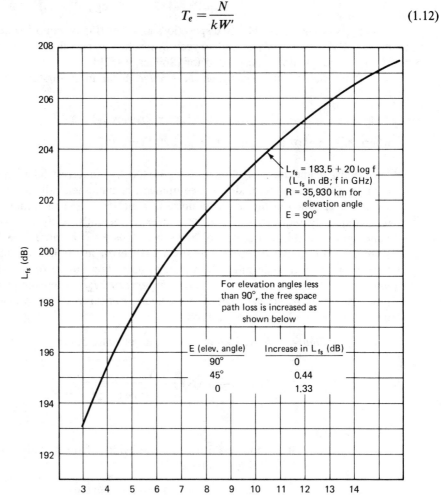

Figure 1.19   Free space path loss.

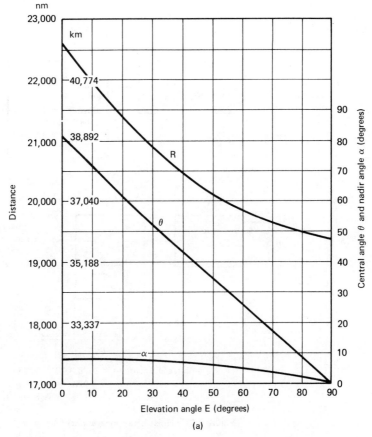

(a)

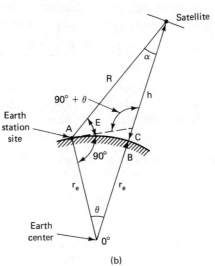

(b)

Figure 1.20 Slant range, central, nadir, and elevation angles (geostationary orbit). r_e, earth's radius: 3444 nm (6378 km); h, satellite height: 19,400 nm (35,930 km); $r_e + h$ = 22,800 nm (42,230 km); R, slant range; E, elevation angle, θ, central angle; α, nadir angle. (From L. P. Yeh, "Geostationary Satellite Orbital Geometry," © *IEEE Trans. Commun.*, April 1972, with permission from the IEEE.)

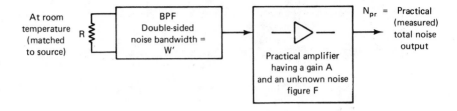

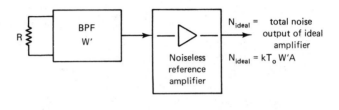

$$\text{Noise figure} = NF = \frac{N_{practical}}{N_{ideal}} = \frac{N_{practical}}{kT_o W'A}$$

$$k = \text{Boltzmann constant} = 1.38 \times 10^{-23} \text{ Ws/K}$$
$$= -228.6 \text{ dBW/K/Hz}$$
$$= -198.6 \text{ dBW/K/Hz}$$

Figure 1.21   Noise-figure measurement setup.

The noise power at the output of a receiver having a noise bandwidth W', gain A, and equivalent noise temperature T_e may be written as

$$N_{\text{practical}} = \underbrace{AkT_oW'}_{} + \underbrace{AkT_eW'}_{} = Ak(T_o + T_e)W' \qquad (1.13)$$

output noise power   output noise power due to the
assuming an ideal   noise generated in the receiver
noiseless receiver   having an equivalent noise tem-
                                perature T_e

From these equations, we obtain the noise figure, given by

$$\boxed{NF = 1 + \frac{T_e}{T_o}} \qquad (1.14)$$

The noise temperature T_e is evidently a hypothetical temperature; however, it is a very convenient and frequently used parameter. In Table 1.4 the numerical relationship between a number of noise figures and equivalent noise temperatures is presented. For low-noise temperatures (below 100 K), we have the approximate relation

$$T_e \simeq 70NF \qquad \text{(dB)} \qquad (1.15)$$

TABLE 1.4 NUMERICAL VALUES OF EQUIVALENT NOISE TEMPERATURES, T_e (K) AND CORRESPONDING NOISE FIGURES NF (dB)[a]

At room temperature (17°C) we have

$$\text{NF}_{dB} = 10 \log_{10}\left[1 + \frac{T(K)}{290}\right]$$

T_e (K) Noise temperature	7	10	35	75	300	3000	30,000
NF (dB) Noise figure	0.100	0.148	0.496	1.002	3.092	10	20

Note:

$$T(°K) = T(°C) + 273 = \tfrac{5}{9}\, T(°F) + 255$$

$$T(°C) = \tfrac{5}{9}\,(T(°F) - 32)$$

$$T(°F) = \tfrac{9}{5}\,(T(°C) + 32)$$

(°C) = Celsius   (K) = Kelvin   (°F) = Fahrenheit

[a] The noise figure (NF) is a ratio of two power levels. In the case of low-noise amplifiers (LNA), the "noisiness" is often expressed in degrees Kelvin (K). At zero degrees Kelvin all molecular motion stops and the noise figure of the amplifier is zero.

Typical equivalent noise temperatures of satellite receivers are in the 1000 K range (NF = 7 dB); those of receive earth stations are in the range 20 to 1000 K.

Noise density, N_o. The term *noise density* refers to the noise power that is present in a *normalized* 1 *Hz of bandwidth*. It is

$$N_o = \frac{N_{\text{total}}}{W'} = kT_e \tag{1.16}$$

where

$$N_{\text{total}} = \text{total noise power measured in a system}$$
$$\text{having a noise bandwidth } W'$$

$$k = \text{Boltzmann constant } (-198.6 \text{ dBm/K/Hz})$$

$$T_e = \text{equivalent noise temperature}$$

1.5.3 Frequently Used Parameters

C/N_o **Carrier-to-noise density ratio C/kT_e.** Let the average wideband carrier power equal C and the noise density N_o. The ratio C/N_o represents the average wideband carrier power-to-noise power ratio, where the noise is measured in a normalized 1-Hz bandwidth. This ratio can be written in terms of the equivalent noise temperature as

$$\frac{C}{N_o} = \frac{C}{kT_e} \tag{1.17a}$$

G/T_e **Figure of merit gain-to-equivalent noise temperature ratio.**   The efficiency of a satellite and of an earth station receive section is frequently specified in terms of the gain-to-equivalent noise temperature ratio:

$$\text{Figure of merit} = \frac{G}{T_e} \qquad (\text{dB/K or dBK}^{-1}) \tag{1.17b}$$

The link budget of the ECS satellite system studied in later sections shows that the G/T_e of the European Communication satellite transponder is specified to equal -5.3 dBK^{-1}, and that of the receive earth station to equal 37.7 dBK^{-1}.

E_b/N_o **Energy of bit-to-noise density ratio.**   The E_b/N_o ratio is one of the most frequently used parameters in digital communications systems. This ratio enables a comparison of systems having variable transmission rates (low speed, high speed) and of the performance of various modulation and coded systems in linear and nonlinear channels and in a complex interference environment.

The bit energy, E_b, is obtained by multiplying the carrier power, C, by the bit duration; that is,

$$E_b = CT_b$$

1.5.4 Link Equations

In this section we derive the link equations of a single radio-frequency (RF) carrier; that is, we assume that a single RF carrier is used per transponder [Van Trees, 1979]. If interested in more involved link calculations in which a number of RF channels are transmitted through a transponder, read [Spilker, 1977]. We follow the notation given in Fig. 1.16(a) and (b). The flux density at the input of the satellite receive antenna is

$$\Omega_u = \frac{P_T G_T}{4\pi R_u^2} L_u \qquad \text{W/m}^2 \tag{1.18}$$

The modulated carrier power received at the satellite (receive uplink power) is

$$P_u = \Omega_u A_{su} = \Omega_u \frac{G_{su}\lambda_u^2}{4\pi} \qquad \text{watts} \tag{1.19}$$

where

A_{su} = effective area of the satellite antenna

G_{su} = satellite antenna gain

λ_u = wavelength on uplink

The channel and equipment noise consists of rain-induced noise, earth background noise, and thermal noise in the receiver. For most practical systems this noise has a flat power spectral density over the receiver bandwidth of N_{ou} W/Hz. The uplink carrier power-to-noise density ratio is

$$\frac{C_u}{N_{ou}} = \frac{P_u}{N_{ou}} = \frac{P_u}{kT_e} \tag{1.20}$$

where

$$k = 1.38 \times 10^{-23} \text{ J/K} \ (10 \log k = -198.6 \text{ dBm/K/Hz}$$
$$= -228.6 \text{ dBW/K/Hz}$$

T_e = effective input noise temperature (K)

From equations (1.18) to (1.20) we obtain the *basic uplink equation:*

$$\left(\frac{C_u}{N_{ou}}\right)_{\text{dB}} = 10 \log P_T G_T - 20 \log \frac{4\pi R_u}{\lambda_u} + 10 \log \frac{G_{su}}{T_s} + 10 \log L_u - 10 \log k \tag{1.21}$$

| earth station | "free space | satellite | additional |
| EIRP | uplink loss" | G/T | uplink losses |

and by substituting C/T_e for kC/N_o [equation (1.17)], we obtain

$$\frac{C_u}{T_{eu}} = 10 \log P_T C_T - 20 \log \frac{4\pi R_u}{\lambda_u} + 10 \log \frac{G_{su}}{T_s} + 10 \log L_u \tag{1.22}$$

For the downlink model, Fig. 1.16(c), using the same procedure, we obtain the *basic downlink equations:*

$$\left(\frac{C_d}{N_{od}}\right)_{\text{dB}} = 10 \log (P_s G_{sd}) - 20 \log \frac{4\pi R_d}{\lambda_d} + 10 \log \frac{G_d}{T_d} + 10 \log L_d - 10 \log k \tag{1.23}$$

satellite	"free space	earth	additional
EIRP	downlink	station	downlink
	loss"	G/T	losses

and

$$\left(\frac{C_d}{T_{ed}}\right) = 10 \log (P_s G_{sd}) - 20 \log \frac{4\pi R_d}{\lambda_d} + 10 \log \frac{G_d}{T_d} + 10 \log L_d \tag{1.24}$$

In a *classical frequency-translating satellite* system, the overall (total) carrier power-to-noise density ratio $(C/N_o)_T$ at the receive earth station is

$$\left(\frac{C}{N_o}\right)_T = \frac{1}{(N_{ou}/C_u) + (N_{od}/C_d)} \qquad (1.25)$$

This equation indicates that the uplink and downlink noise will add up to result in an overall noise level. For digital communications systems the bit energy-to-noise density ratio E_b/N_o of the total system is obtained by inserting

$$E_{bu} = C_u T_b$$
$$E_{bd} = C_d T_b \qquad (1.26)$$

into equation (1.25). We obtain

$$\left(\frac{E_b}{N_o}\right)_T = \frac{1}{(N_{ou}/E_{bu}) + (N_{od}/E_{bd})} \qquad (1.27)$$

Note that in equations (1.25) and (1.27), the quantities are ratios and not dB-s. The digital system probability of error will be a function of the total energy per bit-to-noise density ratio $(E_b/N_o)_T$.

If a regenerative satellite transponder as described in Chapter 9 is used instead of a conventional frequency-translating transponder, there are essentially two cascaded but independent digital communications links. For these systems, equations (1.25) and (1.27) do not apply.

In the next section, an illustrative noise link budget calculation for frequency-translating satellite transponders is presented. *Solve Problem 1.4.*

1.6 LINK BUDGET CALCULATION OF AN ILLUSTRATIVE DIGITAL SATELLITE SYSTEM

The link budget of the planned European Communications Satellite (ECS) system is given in Table 1.5 [Harris, 1978]. Using the parameter definitions and link equations presented in Section 1.5, we verify the link budget of this system. The presented calculation serves as an illustrative example of link budget calculations and of conversions among dBm, dBW, and dB-s.

1.6.1 Earth Station EIRP [Expressed in dB Above 1 W (i.e., dBW)

EIRP (dBW) = *transmitter output power* (dBW) + *antenna gain* (dB)
= *transmitter output power at saturation* (dBW)
− *backoff and combining losses* (dB)
+ *antenna gain* (dB)

TABLE 1.5 LINK BUDGET OF THE 14/11-GHz EUROPEAN
COMMUNICATIONS SATELLITE (ECS) SYSTEM[a]

Uplink

1. Transmitter output power at saturation, 2 kW	33	dBW
2. Backoff and combining losses	7	dB
3. Transmit antenna gain (15 m, 14 GHz)	64	dB
4. EIRP	90	dBW
5. Free space loss (14 GHz)	207.5	dB
6. Atmospheric loss (14 GHz, clear weather)	0.6	dB
7. Satellite G/T	−5.3	dBK^{-1}
8. C/T at repeater input	−123.4	dBWK^{-1}
9. Boltzmann's constant	−228.6	dBWK^{-1} Hz^{-1}
10. Bit rate, 120 Mb/s	80.8	dBHz
11. E_b/N_o at repeater input	24.4	dB

Downlink

1. EIRP at beam edge (unmodulated carrier, saturation)	40.8	dBW
2. Modulation backoff and bandlimiting losses	0.6	dB
3. Free space loss (11.7 GHz)	205.6	dB
4. Atmospheric loss (11.7 GHz, clear weather)	0.4	dB
5. Power at receive antenna	−165.8	dBW
6. Receive antenna gain (15 m, 11.7 GHz)	62	dB
7. Receive system noise temperature (clear weather) 270 K	24.3	dBK
8. Earth station G/T	37.7	dBK^{-1}
9. C/T at receiver input	−128.1	dBWK^{-1}
10. Boltzmann's constant	−228.6	dBWK^{-1} Hz^{-1}
11. Bit rate 120 Mb/s	80.8	dBHz
12. E_b/N_o at receiver input	18.4	dB

[a] This system will provide 17,000 two-way digital telephone circuits by 1990. Individual
satellite transponders will carry 120 Mb/s of traffic.

Source: [Harris, 1978], with permission.

The antenna gain of a 15-m antenna at 14 GHz is 64 dB (see Fig. 1.18):

$$EIRP \ (dBW) = 33 \ dBW - 7 \ dB + 64 \ dB$$
$$= 90 \ dBW$$

The uplink free space loss at 14 GHz is 207.5 dB (see Fig. 1.19).

1.6.2 *C/T* of the Uplink

By insertion of the free space path loss, the EIRP, the specified satellite G/T, and
the atmospheric loss of 0.6 dB into equations (1.22) and (1.23), we obtain

$$\frac{C_u}{T_{eu}} = 90 \ dBW - 207.5 \ dB + (-5.3) \ dBK^{-1} + (-0.6) \ dB$$

$$= -123.4 \ dBWK^{-1}$$

For the E_b/N_o calculation, available at the satellite transponder input, we calculate C_u/N_{ou} first. It is

$$\left(\frac{C_u}{N_{ou}}\right)_{dB} = 90 \text{ dBW} - 207.5 \text{ dB} - 5.3 \text{ dBK}^{-1} - 0.6 \text{ dB} + 228.6 \text{ dBW}$$

$$= 105.2 \text{ dBW/Hz}$$

$$= 135.2 \text{ dBm/Hz}$$

The uplink bit energy for the 120-Mb/s system is

$$E_{bu} = C_u T_b$$

$$(E_{bu})_{dBms} = C_u \text{ (dBW)} + 10 \log T_b \text{ (s)}$$

$$= 105.2 \text{ dBW/Hz} + 10 \log \frac{1}{120.10^6} \text{ s}$$

$$= 105.2 - 80.8 \text{ dBs} = 24.4 \text{ dBs}$$

$$\frac{E_{bu}}{N_o} = \frac{C_u}{N_{ou}} T_b$$

$$= 105.2 \text{ dBW/Hz} + 10 \log \frac{1}{120.10^6} \text{ s}$$

$$= 105.2 + (-80.8) = 24.4 \text{ dB}$$

The downlink calculations follow the same pattern as the uplink calculations. *Solve Problems 1.5 to 1.7.*

PROBLEMS

1.1. Is the postulated growth in satellite communications exceeding that of telecommunications traffic in general? Draw the postulated ratio of INTELSAT satellite system to general telecommunications traffic growth for the 1975–1987 period. Try to extrapolate this ratio until the end of 1995. (*Hint:* See Figs. 1.1 and 1.2 and Table 1.1.)

1.2. What is the difference between a spot beam and an earth-coverage beam? How many satellites are required for complete earth coverage? What are the potential advantages of direct intersatellite links?

1.3. What is the purpose of intermediate-frequency band-splitting (bandpass) filters in the receive section of single-channel-per-carrier (SCPC) earth stations?

1.4. The noise figure of a receive earth station is 1.5 dB. What is the equivalent noise temperature of this earth terminal?

1.5. Verify the accuracy of the following entries in Table 1.5 (link budget of the ECS satellite): transmit and receive antenna gains, uplink and downlink atmospheric losses, and E_b/N_o at receive earth terminal input.

1.6. The transmitter output power of an earth station is 3 kW. The uplink frequency is 6 GHz and the antenna is a standard INTELSAT-A antenna (i.e., it has a diameter of 30 m). The satellite $G/T = -5.3$ dBK^{-1}. Calculate the E_b/N_o at the satellite input if the transmission rate is 60 Mb/s.

1.7. How much is the E_b/N_o at the input of a receive earth station if the transmission rate is 90 Mb/s and the C/T at the receiver input is -128.1 dBWK^{-1}?

2

SIGNAL PROCESSING AND MULTIPLEXING
IN TERRESTRIAL INTERFACE SUBSYSTEMS

2.1 DIGITAL INTERFACE SUBSYSTEMS BETWEEN SATELLITE EARTH STATIONS AND TERRESTRIAL NETWORKS

In the next few years digital transmission techniques are going to spread throughout national and international satellite networks. The introduction of digital earth station satellite systems creates *new problems* of connection with terrestrial networks; since many of these are analog systems, provision has to be made for analog-to-digital conversion.

In this chapter analog-to-digital (A/D) and digital-to-analog (D/A) conversion techniques of audio, television, and frequency-division-multiplexed (FDM) signals are reviewed. Due to the relatively long propagation delay of satellite systems, the effect of voice echo is annoying to the user. For this reason, recently developed echo suppression and cancellation subsystems are described in Section 2.4. Afterward, we study a technique known as digital speech interpolation (DSI) and show that modern DSI equipment doubles the capacity of the transmission system. The last two sections of this chapter deal with digital noninterpolated interface (DNI) systems and energy dispersal (scrambling) techniques. An important function of DNI systems is to provide a means to interface between two digital links which might be on different continents and have *plesiochronous* operation of reference clocks (they have the same nominal frequency, but are independent within a frequency tolerance). DNI systems also remove the effects of the variations of propagation time of the satellite link. The digital energy dispersal equipment reduces the variations in the transmitted power spectrum, irrespective of any patterns in the information pulse train.

Signal processing and multiplexing is a large and complex subject. Books have been written which present the theoretical fundamentals of signal processing. In this chapter we limit our presentation to a *brief review* of the principles of operation followed by applications of signal processing and multiplexing subsystems used in interfacing between terrestrial and satellite systems. The references provide detailed description of these systems.

2.2 PCM, DPCM, and DM Techniques for Analog-to-Digital Conversion of Audio and Television Signals

The most frequently employed analog-to-digital conversion methods used in communication systems are:

> **PCM:** pulse-code modulation
> **DPCM:** differential PCM
> **DM:** delta modulation

The analog-to-digital converter, located in the transmitter, is also known as the *encoder* or simply *coder.* The digital-to-analog converter, located in the receiver, is known as the *decoder.* The word **codec** is derived from coder/decoder.

In addition to the basic conversion methods listed above, more involved codecs have been developed. Frequently used acronyms include:

> **APCM:** adaptive PCM
> **ADPCM:** adaptive DPCM
> **LDM:** linear (nonadaptive) DM
> **ADM:** adaptive DM
> **CDM:** continuous DM
> **DCDM:** digitally controlled DM

The theoretical fundamentals of these systems are described in most modern textbooks on the principles of communications and signal processing. An in-depth study of these A/D conversion methods is given in [Oppenheim and Schafer, 1975, and Jayant, 1974].

2.2.1 PCM—Pulse-Code Modulation

The main functions performed by PCM encoders are illustrated in Figs. 2.1 to 2.3. These include: sampling, quantizing (linear and logarithmic-compressed), and encoding.

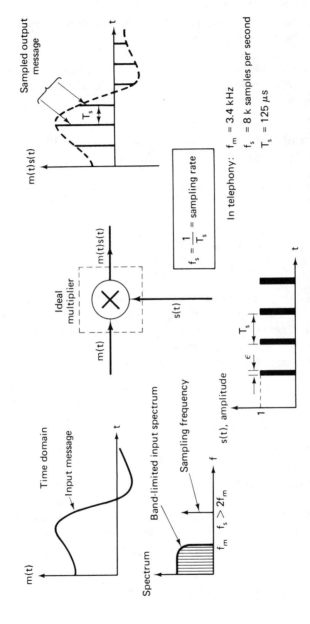

Figure 2.1 Instantaneous sampling of a bandlimited signal $m(t)$. (After [Feher, 1981], with permission from Prentice-Hall, Inc.)

48

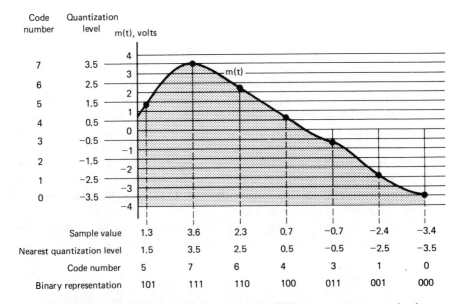

Code number	Quantization level	m(t), volts						
7	3.5							
6	2.5							
5	1.5							
4	0.5							
3	−0.5							
2	−1.5							
1	−2.5							
0	−3.5							

Sample value	1.3	3.6	2.3	0.7	−0.7	−2.4	−3.4
Nearest quantization level	1.5	3.5	2.5	0.5	−0.5	−2.5	−3.5
Code number	5	7	6	4	3	1	0
Binary representation	101	111	110	100	011	001	000

Figure 2.2 Quantization and binary encoding for PCM systems. A message signal is regularly sampled. Quantization levels are indicated. For each sample the quantized value is given and its binary representation is indicated. (After [Taub and Schilling, 1971], with permission from the McGraw-Hill Book Company.)

The *sampling theorem* in a restricted form states that: If the highest-frequency spectral component of a magnitude–time function $m(t)$ is f_m, then the instantaneous samples taken at a rate $f_s > 2f_m$ contain all the information of the original message. Figure 2.1 illustrates a typical telephony application of this theorem, where the voice signal is bandlimited to $f_m = 3.4$ kHz and is sampled at a rate of $f_s = 8$ ksamples

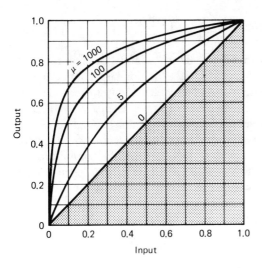

Figure 2.3 PCM signal-compressor characteristics. (After [Taub and Schilling, 1971], with permission from the McGraw-Hill Book Company.)

per second. The sampled output signal $m(t) s(t)$ has an infinite number of nonzero amplitude states. To encode this signal, the amplitude levels have to be quantized.

For simplicity, only eight *quantization levels* are shown in Fig. 2.2. The continuous signal $m(t)$ has the following sample values: 1.3, 3.6, 2.3, 0.7, . . . , -3.4 V. The quantized signal takes on the value of the nearest quantization level to the sampled value. The eight quantized levels are represented by a 3-bit *code* number. (*Note:* With 3 bits, $2^3 = 8$ distinct levels can be identified.) The amplitude difference between the sampled value and the quantized level is called the *quantization error.* This error is proportional to the step size, d, that is, the difference between consecutive quantization levels. With a higher number of quantization levels (smaller d), a lower quantization error is obtained. Experimentally, it has been found that, for an acceptable signal-to-noise ratio, 2^8 or 256 quantization levels are required. This represents 8 bits of information per quantized sample.

If the number of quantizer levels is large (>100), we may assume that the quantization error has a *uniform* probability density function given by

$$p(E) = \frac{1}{d} \qquad -\frac{d}{2} \le E < \frac{d}{2} \tag{2.1}$$

This uniform error distribution is true if the signal $m(t)$ does not overload the quantizer. For example, in a quantizer, such as that shown in Fig. 2.2, the quantizer output might saturate at level 5 for $|m\ (t)| > 5$. The quantization error during such *overload* is a linearly increasing function of $m(t)$. In the linear region of operation the mean-square value of the quantization error is

$$\int_{-d/2}^{d/2} E^2 p(E)\ dE = \int_{-d/2}^{d/2} E^2 \frac{1}{d}\ dE = \frac{d^2}{12} \tag{2.2}$$

If the root-mean-square (rms) value of the input signal $m(t)$ is M_{rms}, then the *signal-to-quantization error* ratio is

$$S/N = \frac{M_{\text{rms}}^2}{d^2/12} = 12 \frac{M_{\text{rms}}^2}{d^2} \tag{2.3}$$

From this equation we conclude that the signal-to-quantization error ratio is dependent on the rms value of the input signal M_{rms}; that is, for larger input signals a larger S/N is obtained. This is an undesirable effect in telephony systems, as some talkers have a considerably lower volume than others. For the listener it would be a nuisance if he had to listen to a very low volume signal corrupted by a relatively high quantization error (low S/N_q). To achieve the same signal-to-noise ratio for a small-amplitude signal as for a large-amplitude signal, a quantizer with a nonuniform step size is required. To achieve this nonuniform step-size quantization, given a uniform step-size quantizer such as that shown in Fig. 2.2, it is necessary to precede it with a nonlinear input–output device known as a *compandor,* or *companding device.* Note that the compandor followed by the linear quantizer amplifies the low-volume signals more than the high-volume signals. *Solve Problem 2.1.*

Frequently used compandor characteristics are shown in Fig. 2.3. The companding function is given by

$$\frac{v(x)}{V} = \pm \frac{\log(1 + \mu x/V)}{\log(1 + \mu)} \qquad -V \le x \le +V \tag{2.4}$$

where $v(x)$ is the output voltage, x is the positive or negative input voltage, V is the peak voltage of the input signal, and μ is a parameter. At the receiving terminal the inverse signal processing to that performed at the transmitter has to be done in order to recover the transmitted signal.

To summarize, we may conclude that in telephony systems the signal is bandlimited to $f_m = 3.4$ kHz. To convert this analog signal into a binary PCM data stream, a sampling rate of $f_s = 8$ ksamples per second is used. Each sample is quantized into one of the 256 quantization levels. For this number of quantization levels, 8 information bits are required ($2^8 = 256$). Thus one voice channel being sampled at a rate of 8 ksamples per second and requiring 8 bits per sample will have a transmission rate of 64 kbits/second. The signal-to-noise performance of PCM, ADPCM, and of ADM codecs is summarized in Fig. 2.4.

Broadcast-quality **color television** signals have an analog baseband bandwidth of somewhat less than 5 MHz. For conventional PCM encoding of these video signals, a sampling rate of $f_s = 10M$ samples per second and a 9-bit per sample coding scheme is used. Thus the resulting transmission rate is 90 Mb/s. Most television pictures have a large degree of correlation, which can be exploited to *reduce* the transmission rate. It is feasible to predict the color and brightness of any picture element (pel) based on values of adjacent pels that have already occurred. Digital broadcast-quality color television signals requiring only 20 to 45 Mb/s transmission rates, obtained by means of predictive techniques, have been reported [Tescher, 1980; Kamangar and Rao, 1980; Ismail and Clarke, 1980].

2.2.2 DPCM—Differential Pulse-Code Modulation

Differential pulse-code modulation is a *predictive coding scheme* which exploits the correlation between neighboring samples of the input signal to reduce statistical redundancy and thus lower the transmission rate. Instead of quantizing and coding the sample value as done in PCM, in DPCM *an estimate of the next sample value based on the previous samples is made.* This estimate is subtracted from the actual sample value. The difference of these signals is the prediction error, which is quantized, coded, and transmitted to the decoder. The decoder performs the inverse operation; that is, it reconstructs the original signal from the quantized prediction errors.

The block diagram of a DPCM system is shown in Fig. 2.5. Here the $\{s_i\}$ is the sequence of input sample values, $\{\hat{s}_i\}$ is the prediction sequence, and

$$\{e_i\} = \{s_i - \hat{s}_i\} \tag{2.5}$$

is the prediction error sequence which is quantized, encoded, and transmitted. When

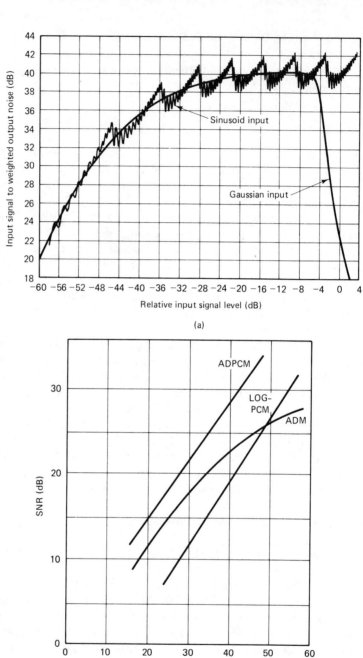

Figure 2.4 Signal-to-noise ratio of PCM, ADPCM, and ADM codecs. (a) Output SNR as a function of relative input signal level for both Gaussian and sinusoidal inputs to the 8-bit, 15-segment, μ-law PCM quantizer for C-message weighting of the noise. (After [Dammann et al., B.S.T.J., October 1972].) (b) Comparison of SNR versus bit rate functions. Bandwidth: 200 to 3200 Hz. (After [Jayant, 1974].)

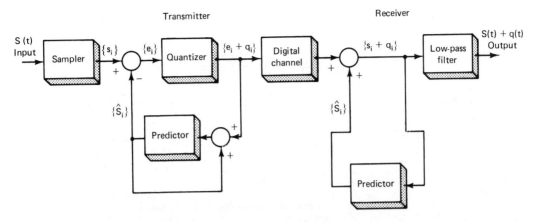

Figure 2.5 Differential PCM system, DPCM, block diagram. (After [O'Neal, 1980].)

the number of quantizing levels N is large ($N \geq 8$ is sufficiently large) and linear prediction is used, each $\{\hat{s}_i\}$ can be expressed as

$$\hat{s}_i = a_i s_{i-1} + a_2 s_{i-2} + a_3 s_{i-3} + \cdots \tag{2.6}$$

where the a_i's are predictor coefficients.

If the quantizer or both the quantizer and predictor adapt themselves to match the signal to be encoded, considerable signal-to-noise improvement can be attained. The dynamic range of the encoder can be extended by adaptive quantization if a nearly optimum step size over a wide variety of input signal conditions is generated.

The two most frequently used quantizer adaptation methods include the syllabic or slow-acting adaptation and the fast-acting or instantaneous companding with only one sample memory.

The performance of DPCM and adaptive DPCM systems is derived in [Spilker, 1977]; illustrative results are shown in Fig. 2.4(b). More advanced predictive coding techniques for digital television transmission using bandwidth compression techniques are described in the tutorial paper by [Kaneko and Ishiguro, 1980].

2.2.3 DM—Delta Modulation

The exploitation of signal correlations in DPCM suggests the further possibility of oversampling a signal to increase the adjacent sample correlations and thus to permit a simple quantizing strategy. Delta modulation (DM) is a 1-bit version of differential PCM. The DM coder approximates an input time function by a series of linear segments of constant slope. Such an analog-to-digital converter is therefore referred to as *linear delta modulator,* shown in Fig. 2.6.

At each sample time the difference between the input signal $x(t)$ and the latest staircase approximation is determined. The sign of this difference is multiplied by the step size and the staircase approximation is incremented in the direction of the

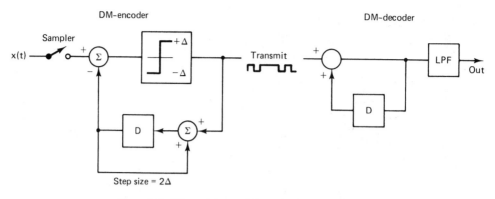

Figure 2.6   Linear delta modulator, DM. Step size, 2Δ.

input signal. Therefore, the staircase signal $y(t)$ tracks the input signal. The signs of each comparison between $x(t)$ and $y(t)$ are transmitted as pulses to the decoder, which reconstructs $y(t)$, and then low-pass filters $y(t)$ to obtain the output signal. The quantization noise is defined as

$$n(t) = x(t) - y(t) \qquad (2.7)$$

The *slope overload* distortion region (see Fig. 2.7) occurs for large and fast signal transitions. It is caused by the fact that the maximum slope that can be produced by a linear delta modulator is SS $\cdot f_r$, where SS is the step size and f_r is the sampling

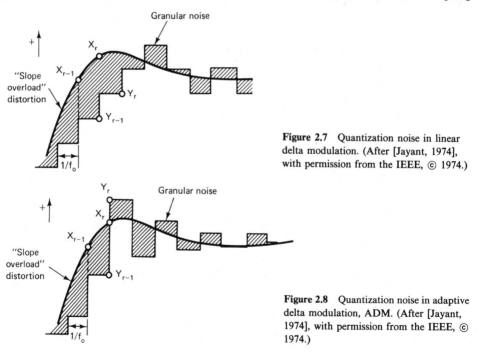

Figure 2.7   Quantization noise in linear delta modulation. (After [Jayant, 1974], with permission from the IEEE, © 1974.)

Figure 2.8   Quantization noise in adaptive delta modulation, ADM. (After [Jayant, 1974], with permission from the IEEE, © 1974.)

rate. The *granular noise* is introduced by the fact that the staircase is hunting around the input $x(t)$.

The use of adaptive techniques reduces the quantization noise and increases the dynamic range of delta modulators. The idea of adaptive step-size delta modulators is illustrated in Fig. 2.8, and an integrated circuit hardware realization is shown in Fig. 2.9. A large number of methods for suitable adaptive step-size variation exist [Jayant, 1974; O'Neal, 1980]. Monolithic integrated-circuit adaptive delta modulators using advanced digital algorithms are available from a number of manufacturers. The cost of these high-performance ADM codecs is approximately $10.

The performance of low-bit-rate adaptive delta modulators is summarized in Fig. 2.10, where the *intelligibility* of variable-slope delta modulators is presented.

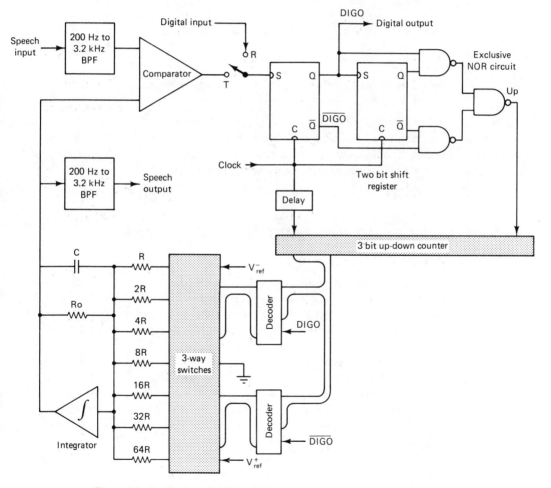

Figure 2.9 Realization of ADM with integrated-circuit hardware. (After [Jayant, 1974], with permission from the IEEE, © 1974.)

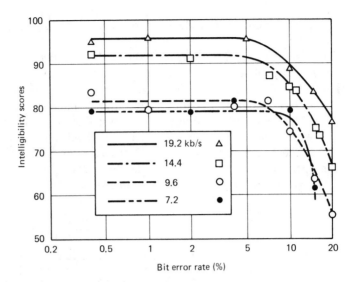

Figure 2.10 Intelligibility performance of variable-slope delta modulation. (After [Jayant, 1974], with permission from the IEEE, © 1974.)

From the intelligibility scores it is evident that adaptive delta modulators have a reasonably high intelligibility even with low bit rates (7.2 kb/s) and high channel bit error rates (10% corresponds to a channel error rate of 10^{-1}). *Solve Problem 2.2.*

2.3 TRANSMULTIPLEXERS FOR FDM-TO-TDM CONVERSION-MULTIPLEX STANDARDS

For interfacing with analog terrestrial facilities a means for conversion between analog frequency-division multiplex (FDM) and digital time-division multiplex (TDM) signals is required. Specially developed FDM/TDM converters called *transmultiplexers* integrate the conversion functions in a compact efficient form. These converters can be implemented by either analog or digital means.

In Fig. 2.11(a) the conversion between the analog FDM microwave system and the digital PCM satellite system is accomplished with the use of standard TDM and FDM multiplex equipment. Although this method requires no specially developed conversion equipment, it has serious drawbacks in terms of cost, size, and reliability because of the inherent redundancy of circuit functions. The transmultiplexer connection [Fig. 2.11(c)] performs essentially the same system functions as the voice-frequency connection. We note that the transmultiplexer performs analog-to-digital conversion, and that this conversion is equivalent to the functions performed by the **FDM-DEMUX** and **PCM-MUX** equipment.

An analog transmultiplexer employs conventional channel circuitry as in stan-

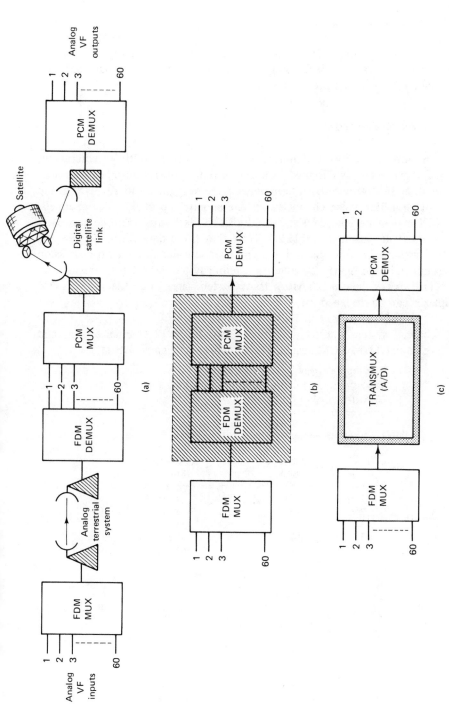

Figure 2.11 Analog and digital system connection for 60 telephony channels using the voice-frequency connection and the transmultiplexer methods. A one-way transmission model is illustrated. (a) Voice-frequency connection between an analog FDM terrestrial and TDM digital PCM satellite system. (b) Equipment required for the interconnection of analog and digital systems if the voice-frequency connection method of (a) is employed. Standard FDM and TDM multiplex set of equipment connected back to back is required. (c) Transmultiplexer connection of analog and digital transmission systems. Note that the transmux equipment is equivalent to the FDM-DEMUX-PCM-MUX equipment shown in (b) (shaded area).

dard FDM and TDM equipment: analog filters, A/D and D/A converters, and digital multiplexing circuits. A digitally implemented transmultiplexer accomplishes the FDM/TDM conversion using digital processing techniques such as digital recursive filters, nonrecursive filters, and the fast Fourier transform (FFT), and departs substantially from the analog channel processing concepts.

2.3.1 Multiplex Standards

To provide the interface between standard FDM and TDM multiplex equipment, the FDM/TDM converter is designed as a device that converts an integer number of FDM *groups*/or *supergroups*. By international agreement the standard FDM group consists of twelve 4-kHz voice channels and one supergroup of 60 voice channels. The standard North American data rates and CCITT data rates (Europe and parts of South America) are given in Table 2.1. Note that these data rates are accepted standards for terrestrial facilities. Satellite systems use these and also other transmission rates. For example, in the ANIK-C Canadian satellite system 90 Mb/s is transmitted, whereas INTELSAT-V has a 120-Mb/s transmission rate. Thus the FDM/TDM transmultiplexer can be designed as:

1. A 24-channel system interfacing between two FDM-12 channel groups and one 24-channel TDM-PCM system (1.544 Mb/s, designated DS-1)

TABLE 2.1 STANDARD DIGITAL TRANSMISSION RATES AND PCM VOICE-CHANNEL CAPACITIES

Hierarchy level no.	Standard transmission rates		
	U.S./Canada (Mb/s)	Japan (Mb/s)	Europe (CCITT) (Mb/s)
1	1.544	1.544	2.048
2	6.312	6.312	8.448
3	44.736	32.064	34.368
4	274.176	97.728	139.264
5	—	396.200	560–840

Hierarchy level no.	Number of PCM voice channels (capacity)		
	U.S./Canada	Japan	Europe (CCITT)
1	24	24	30
2	96	96	120
3	672	480	480
4	4032	1440	1920
5	—	5760	7680–11520

2. A 60-channel FDM supergroup interfacing with two 30-voice-channel TDM-PCM carriers (2.048 Mb/s each, designated CEPT-32)

3. A four-supergroup 240-channel system which either accommodates ten 1.544-Mb/s DS-1 systems or eight 2.048-Mb/s CEPT-32 systems

Alternative system capacities may also be implemented.

2.3.2 Cost Comparison of Conventional FDM/TDM and Transmultiplexer Conversion Methods

In the early 1980s the conventional FDM/TDM conversion equipment [voice-frequency connection, see Fig. 2.11(a)] is in the range $700 to $1000 per circuit end, depending on the type of signaling involved. A cost of $250 per circuit end for the digital transmultiplexer without signaling conversion is a reasonable goal for production quantities [Campanella, 1980].

In the review paper by [Freeny, 1980] you will find an excellent explanation of the principles of digital signal-processing transmultiplexer systems. A comprehensive treatment of digital filtering design methods is given in [Antoniou, 1980], and the hardware design and system evaluation results of a high-speed 120-channel transmultiplexer design are given in [Aoyama et al., 1980].

A modern 12-channel FDM/TDM transmultiplexer equipment is shown in Fig. 2.12. *Solve Problem 2.3.*

Figure 2.12   Four 12-channel FDM group transmultiplexer. This FDM/TDM converter interconnects 48 FDM channels with two 24-channel PCM digroups—Lenkurt Model 4691B. (Courtesy of Dr. A. Lender, GTE Lenkurt, Inc., San Carlos, Calif.)

2.4 ECHO SUPPRESSION AND CANCELLATION TECHNIQUES

Echo of a transmitted speech or of data signal occurs in telephone networks. The longer the echo is delayed, the more disturbing it is and the more it must be attenuated before it becomes tolerable. Synchronous satellites are placed in an orbit roughly 40,000 km above the earth's surface. Due to this large distance, the round trip of a telephone conversation relayed via satellite, including the terrestrial segment, is about 500 to 600 ms. In case of a double satellite hop, the round-trip delay can exceed 1 s.

In terrestrial U.S. telephone trunks exceeding 3500 km (approximately 35-ms echo delay), echo suppressors are required [Hatch and Ruppel, 1974]. In satellite systems *echo suppressors* or *echo cancelers*, most frequently one at each end of the long-distance connection, are required. This is known as *split* echo control.

A typical long-distance telephone circuit is shown in Fig. 2.13. At any location in this circuit, if the transmitted signal encounters an impedance mismatch, a fraction of this signal gets reflected as an echo. Telephone sets in a geographic area are connected by a two-wire line to a hybrid transformer (frequently called *hybrid*) which is located in a *central office*. For both transmission and reception to and from the central office only two wires are used, as this results in considerable savings of wire and of local switching equipment.

The circuit diagram of a hybrid located at *B* is shown in Fig. 2.14. If the two transformers are *identical* and the balancing impedance Z_n *equals* the impedance of the two-wire circuit, the signal originating on the "in" side gets transferred to the two-wire circuit of *B*, but produces no response at the "out" terminal. On the other hand, if the signal originates in the two-wire circuit (talker *B* is active), this

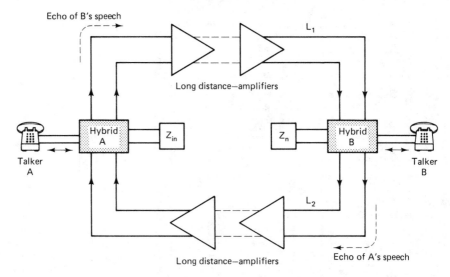

Figure 2.13 Typical long-distance telephone circuit. The hybrids are located in central offices *A* and *B*, respectively.

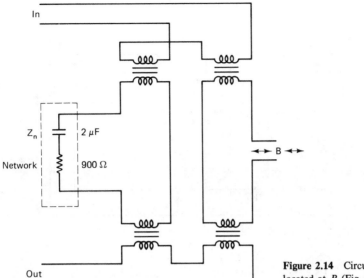

Figure 2.14 Circuit diagram of "hybrid" located at B (Fig. 2.13).

signal is transferred to both paths of the four-wire circuit. This signal has no effect on the "in" signal path, since the amplifiers shown in Fig. 2.13 amplify signals only in the opposite direction. Echoes are generated whenever the "in" side is coupled (has a leak-through) to the "out" side. Unfortunately, this occurs in almost all networks, as the Z_n network is not identical to the distributed time-variable impedance of the two-wire circuit. Also, we note that a four-wire circuit may be connected to a large number of two-wire circuits. Thus the need for echo control (suppression or cancellation) in long-distance systems is imminent.

2.4.1 Echo Suppressors

A simplified block diagram of an echo suppressor is shown in Fig. 2.15. Most of the time, talkers A and B speak alternately; thus the echo can be prevented from returning to A by disconnecting the path L_2 during the intervals when *only A* is talking. This is achieved by detecting the presence of the signal originating by talker A and activating (opening) switch S. However, interruptions are an important part of conversations, particularly of somewhat heated discussions. A statistical analysis of on–off patterns in numerous conversations demonstrates that there is a 20% probability for interruptions, that is, *simultaneous speech* [Brady, 1968]. During simultaneous speech the switch is prevented from opening. Thus during interruptions, echo is not eliminated. Additionally, the speech detector and comparator circuit design are fairly complex if good accuracy of control/comparison activities is required.

   Chopping of the initial portions of the interrupter's speech is unavoidable. When telephone conversations are conducted over satellite systems which have a round-trip delay on the order of 600 ms, the interruption rate increases and the subjective

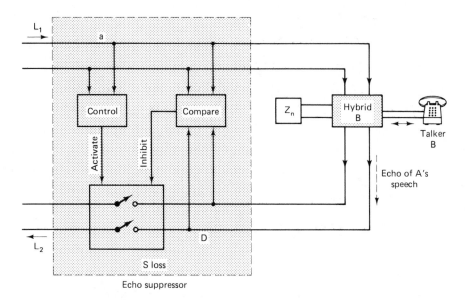

Figure 2.15 Echo suppressor—simplified diagram.

effect of echoes during interruptions worsens [Gould and Helder, 1970]. For these reasons, another generation of echo control devices, known as echo cancelers, have been developed.

2.4.2 Echo Cancelers

The conceptual diagram of an echo canceler is shown in Fig. 2.16. The basic idea of echo cancellation is to generate a *synthetic* replica of the echo and subtract it from the leaked-through echo signal returned through hybrid *B*. If an adaptive filter that perfectly matches the transfer function of the echo path were designed, complete

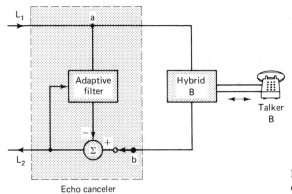

Figure 2.16 Echo canceler—conceptual diagram. (Courtesy of IEEE.)

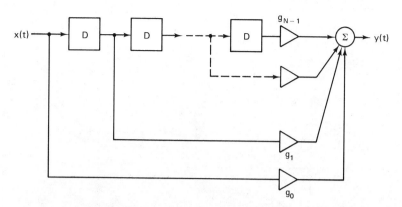

Figure 2.17 Transversal filter for the generation of synthetic echo signals. (Courtesy of IEEE.)

echo cancellation could be achieved. Adaptive filtering is required to match the time-variable distance and device-dependent impedance characteristics of the two-wire circuit. Transversal filters, such as the one shown in Fig. 2.17 and described further in Chapter 3, are frequently used for the implementation of synthetic echo signals. A simplified adaptive echo canceler block diagram is given in Fig. 2.18(a); in Fig. 2.18(b) a modern echo canceler is shown. Let the speech signal of speaker A be $x(t)$ and let $y(t)$ be the sum of A's speech echo $z(t)$ and a component $n(t)$. When A alone is talking, $n(t)$ represents low-level noise, while during periods of double talking $n(t)$ is predominantly B's speech. The transversal filter may have as many as $N = 256$ taps. It generates an estimate $\hat{z}(t)$ of the echo

$$\hat{z}(t) = \sum_{n=0}^{N-1} \hat{h}_n(t) \times (t - nD) \tag{2.8}$$

where D is typically chosen to equal 0.125 ms. For successful echo cancellation, convergence of the impulse response to the echo path is required [Sondhi and Berkeley, 1980; Sondhi, 1967]; that is, the control loop uses the error

$$e(t) = y(t) - \hat{z}(t) \tag{2.9}$$

to continuously improve the estimate $\hat{z}(t)$. The system should drive itself to the condition

$$e(t) = n(t)$$

To meet this end, convergence of the impulse response of the adaptive transversal filter toward the response of the echo path is required. If you are interested in the principles and derivation of the adaptation algorithm and in the detailed operation of modern echo cancelers, see [Sondhi and Berkeley, 1980, and Rossiter et al., 1978]. *Solve Problems 2.4 and 2.5.*

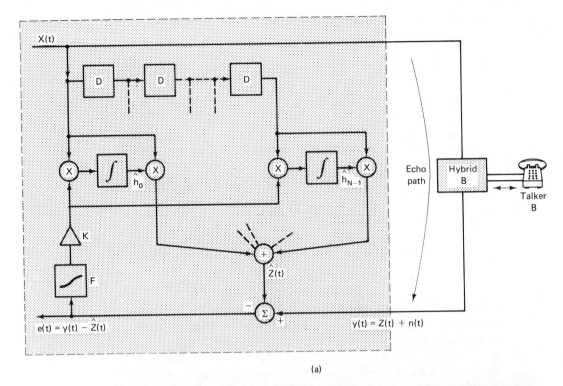

(a)

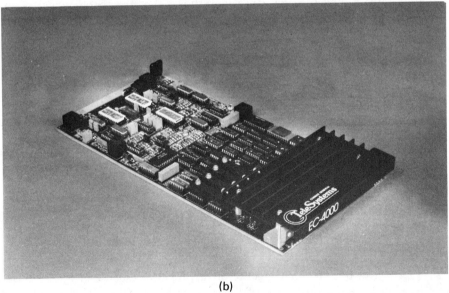

(b)

Figure 2.18 (a) Self-adaptive echo canceler—simplified block diagram. Each line represents a two-wire circuit. (After M. M. Sondhi and D. A. Berkeley, *Proc. IEEE*, August 1980, with permission from the IEEE © 1980 IEEE) (b) Modern echo canceler. (Courtesy of COMSAT General-Telesystems.)

64

2.5 DIGITAL SPEECH INTERPOLATION (DSI) SYSTEMS

The digital speech interpolation (DSI) technique exploits the fact that in telephone conversations only one speaker is usually speaking, while the other listens. There is usually a channel inactivity between calls; pauses, hesitations, and silence intervals are parts of normal conversations. On the average, speech of a particular customer is present approximately 40% of the time. Modern DSI equipment exploits these relatively low-voice-activity properties by compressing the number of *satellite* channels used to transmit a given number of telephony *terrestrial* channels. Voice detectors are used to detect speech activity on each terrestrial channel input to the DSI system. The idle time between calls and conversation pauses is used to accommodate additional terrestrial channels. Application of DSI systems in *time-division multiple-access* (TDMA) satellite systems leads to a major capacity advantage of digital TDMA satellite systems. An examination of Table 2.2 clearly indicates that the *capacity* (expressed in number of telephony channels) of a TDMA satellite system is two to *three times larger* than the capacity of a corresponding *frequency-division multiple-access* (FDMA) satellite system. This almost *three-to-one* capacity advantage warrants the additional complexity required for digital TDMA-DSI satellite systems.

TABLE 2.2 APPROXIMATE CAPACITIES FOR THE INTELSAT-V HEMISPHERIC BEAM 72-MHz TRANSPONDERS FOR INTELSAT STANDARD A (30-m antenna diameter; see Fig. 1.3) EARTH STATIONS

Transmission mode	Maximum capacity expressed in terms of number of 3.1-kHz telephony channels
FDMA (analog system)	1100
TDMA (digital system)	
Without DSI	1600
With DSI	3200

In the INTELSAT-V TDMA-DSI system, a number of DSI subsystems may be connected to the TDMA controller of an earth station (see Fig. 1.10). Each of these subsystems has a maximal capacity of N terrestrial channels utilizing up to M satellite channels.

The DSI gain G is defined by

$$G = \text{DSI}_{\text{gain}} = \frac{N}{M} = \frac{\text{number of terrestrial channels}}{\text{number of satellite channels}} \qquad (2.10)$$

In the INTELSAT-V system,

$$G = \frac{240}{127} = 1.89$$

Theoretically, the DSI gain can be adjusted over a fairly wide range. If the number of satellite channels for a given number of terrestrial channels is reduced, the DSI gain increases. However, the gain increase is limited by increased *clippings* in terrestrial channels, which cannot be transmitted for given periods of time due to nonavailability of satellite channels. Thus the terrestrial channels that do not have a satellite channel assignment are "frozen out" until one of the terrestrial channels that was previously assigned a satellite channel becomes inactive. Thus front-end clipping of speech is evident during "freeze out." The DSI gain that can be obtained with a subjectively acceptable performance is shown for the *time assigned speech interpolation* (TASI) and for the *speech predictive encoded communications* (SPEC) systems in Fig. 2.19. The development of the INTELSAT-V "freeze out" specifications was preceded by numerous subjective tests. These specifications require that the percentage of clipping be less than 2% of the speech spurts that experience clips greater than 50 ms. (*Note:* Each period of time occupied by a caller's speech is called a *speech spurt.*)

In the SPEC digital speech interpolation system the competitive clip problem experienced by the TASI-DSI system is completely avoided. The speech predictor algorithm used in the SPEC system eliminates unnecessary samples in the instantaneous speech waveform or in the short intersyllabic pauses. The SPEC predictor, similar to the adaptive DPCM system described in Section 2.2, removes more than 25% of the PCM samples during voice spurts 25% of the time. On the other hand,

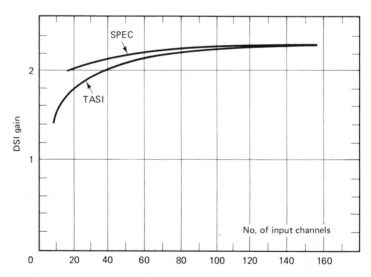

Figure 2.19 DSI gain for the TASI and SPEC systems. SPEC, speech predictive encoded communications; TASI, time assigned speech interpolation. The TASI curve is based on the assumption that less than 2% of the voice spurts will experience clipping of 50 ms or longer. (After [Campanella, 1976], with permission.)

TABLE 2.3 COMPARISON OF SPEC AND TASI TECHNIQUES

	SPEC	TASI
DSI advantage for 240 incoming channels[a]	2.2	2.2
DSI advantage for 60 incoming channels[a]	2.1	2.0
Susceptible to speech spurt clipping without special precautions	No	Yes
Requires changes in channel assignment protocol during overload	No	Yes
Uses miniprocessor for channel assignment control connectivity map	No	Yes
Susceptible to connect message channel overload	No	Yes
Relative hardware complexity	Lower	Higher
Relative cost of transmit terminal	Lower	Higher
Relative cost of receive terminal	Same	Same
Efficient application in multidestinational service	Same	Same

[a] Based on an assumed voice spurt activity of 40%.

Source: [Campanella, 1976], with permission.

the major performance degradation in the SPEC digital speech interpolation system is the production of prediction distortion (see Table 2.3).

The principle of compression of a larger number of terrestrial channels into a smaller number of satellite channels is illustrated in Fig. 2.20. The PCM encoder accepts N analog telephony inputs and converts them into N binary bit streams. The output of the DSI module provides M binary channels for satellite transmission. (*Note:* $M < N$.) One of these channels, the *assignment channel,* is used to transmit the satellite channel assignments to the receiving earth station for proper reconnection to the outgoing terrestrial channels. The satellite channel assignments are transmitted by means of assignment messages which consist of the terrestrial channel number and associated satellite channel number.

The DSI equipment is capable of *single-destination* or *multidestination* operation. A single-destination channel connectivity mapping is illustrated in Fig. 2.21. In general, single-destination DSI modules carry more traffic, containing 240 or more voice and data channels. Multidestination DSI modules combine the smaller traffic links. Note that the DSI input channels are defined as *international channels* (ICs) and are numbered in sequence. The *terrestrial channels* (TCs) are defined as the voiceband lines which connect to the DSI input channels, but may be numbered differently from the international channels. In the example, 10 terrestrial channels are compressed into five satellite channels.

For the INTELSAT-V time-division multiple-access network, digital speech interpolation using the TASI technique has been adopted. Probably the single most important reason for the choice of TASI (instead of SPEC) is the more elaborate operational experience and almost off-the-shelf availability of the TASI-DSI equipment. A brief description of this system follows; for a detailed study of the TASI and also the SPEC systems, see [Campanella, 1976]. *Solve Problem 2.6.*

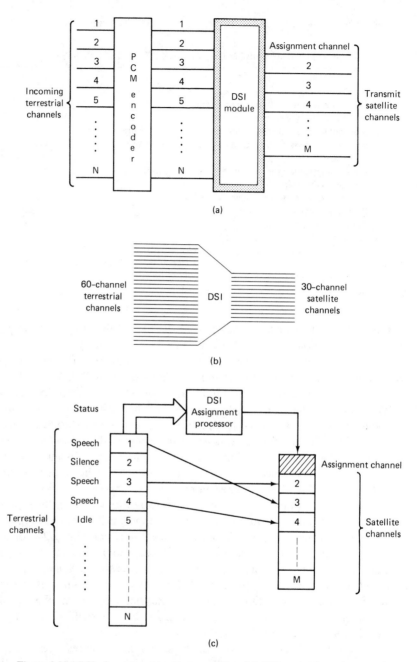

Figure 2.20 DSI channel-compression capability. (a) DSI transmit equipment: N incoming terrestrial channels are compressed into M outgoing satellite channels through DSI operation, where the ratio N/M is the DSI gain. (b) 60 incoming channels are compressed into 30 satellite channels using DSI with a gain of 2.0. (c) Mapping of terrestrial channels into satellite channels. (After Jankowski, J., 1980. Published with permission from The International Telecommunications Satellite Organization [Intelsat].)

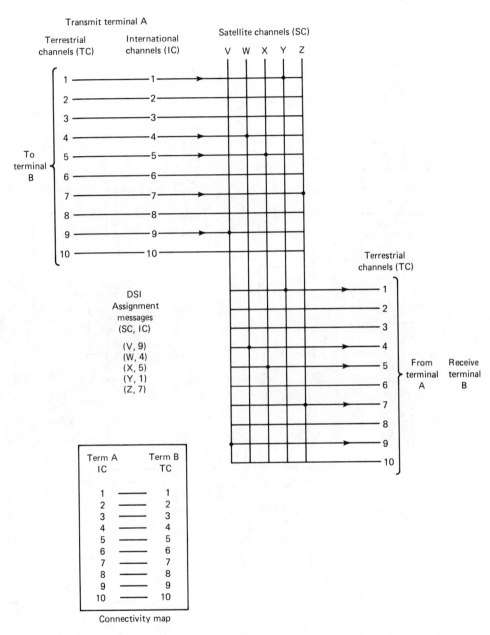

Figure 2.21 Single-destination DSI channel connectivity mapping. (After Jankowski, J., 1980. Published with permission from The International Telecommunications Satellite Organization [Intelsat].

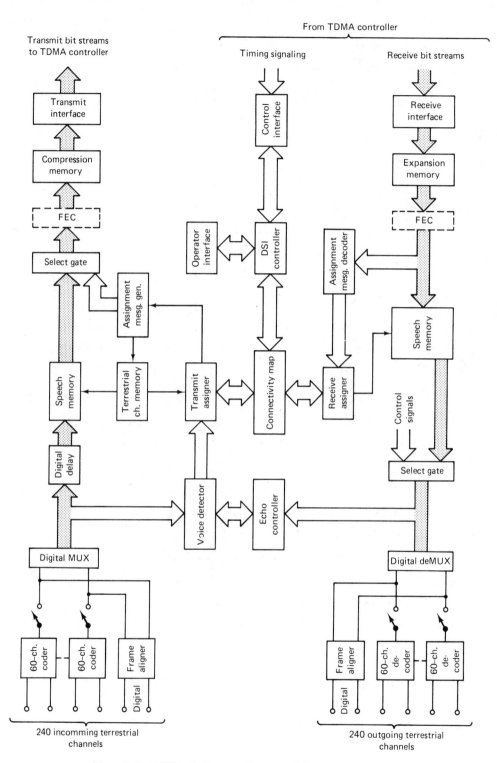

Figure 2.22 DSI block diagram. (Courtesy of Spar Aerospace Limited.)

2.5.1 The TASI-DSI System

A *time assigned speech interpolation–digital speech interpolation* TASI-DSI system block diagram is shown in Fig. 2.22. The 60 channel coders and digital multiplexer equipment analog-to-digital convert and time-division multiplex 240 incoming terrestrial voice channels. (Transmultiplexers, described in Section 2.3, are frequently used for this purpose.) The voice detector detects the speech activity of the individual digitized voice channels. The connectivity map, the terrestrial channel memory, and the assignment message generators provide the required mapping functions. The digital delay and speech memory units are required to compensate for the processing delay of the voice detection circuits and to prevent excessive front-end voice clipping. A forward error-correcting device (FEC) may be included into the DSI unit or this device may be located in the TDMA controller, shown in Fig. 1.10. The compressed bit stream is fed to a transmit interface unit which passes the bit stream to the TDMA controller for further processing and for transmission. The receive DSI system performs the inverse functions to those of the transmit section.

Detailed studies of the principles of operation, design, and applications of DSI systems are documented in numerous references, including [Rossiter et al., 1978; Campanella, 1976; Verma and Monsees, 1978; Lombard et al., 1978; and Campanella et al., 1979]. *Solve Problem* 2.7.

2.6 INTERFACING WITH DIGITAL TERRESTRIAL FACILITIES— DIGITAL NONINTERPOLATED INTERFACE (DNI) SYSTEM

Digital satellite communication systems provide links between two or more terrestrial systems. The clocks that control the terrestrial networks (frequently located on different continents) may have *time differences* which necessitate their resetting, so that information can pass from one link to another. The two main causes of the clock time differences are:

1. *Plesiochronous* (nearly synchronous) operation of reference clocks
2. *Doppler shift effect* due to the variations of link propagation time (motion of the satellite)

2.6.1 Plesiochronous Systems

The *digital noninterpolated interface* (DNI) system must be designed in such a way that digital rate differences induced by satellite motion and by the plesiochronous operation of the terrestrial networks do not cause an excessive number of slips of the *primary* (2.048 Mb/s) multiplex data frames.

International digital satellite links are required to interface between two or more *independent* synchronous terrestrial networks. The clocks of these independent networks (known as plesiochronous) are derived from physically separated and independent cesium beam oscillators of 2×10^{-11} or 1×10^{-11} accuracy. If the nominal clock rate of the primary multiplex equipment of such a plesiochronous system is 2.048 MHz and one clock is 1×10^{-11} high in frequency and the other 1×10^{-11} low, a time displacement error of 1 bit accumulates in about 7 hours. A method frequently used to reset a time displacement (reset the clock) in the case of such a plesiochronous operation is based on the generation of *frame slips*. In this method frame slips are applied to offset the discrepancy between the transmission and reception rates. Frame slips are generated by transmitting a frame twice in a given interval or by preventing a frame from being transmitted. This process, generally used in asynchronous switched networks, provides satisfactory operation in PCM telephony where a frame slip adds or eliminates one PCM sample in each voice channel affected. However, frame slips might be troublesome in high-speed data communications. The slip frequency must therefore be kept at a very low value, either by synchronizing the clocks (frequently a very difficult task) or by ensuring high clock stability. In the case of high-stability cesium beam oscillators having an accuracy of $\pm 1 \times 10^{-11}$, the primary 2.048-Mb/s rate multiplex equipment having 256 bits/frame will have a frame slip about every 72 days. This is an acceptable frame slip rate.

2.6.2 Doppler Shift Effects

The path length from earth station to satellite is not constant. For example, the operational plan for the INTELSAT-V satellites calls for stationkeeping limits of 0.1° latitude and longitude. For a portion of the satellite lifetime, this limit may be relaxed to 0.5° in latitude. The computed maximum peak-to-peak variation in station-to-station path is 1.1 ms and the maximum rate of change of path length is 40 ns/s. This change is the cause of Doppler shift, that is, the reference bit rate transmitted at the reference earth station, f_R, is received at the receive earth station at a rate $f_s = f_R(1 + d_1 + d_2)$, where d_1 and d_2 are the Doppler shifts resulting from the path length change between the reference earth station and the satellite, and the satellite and the receiving earth station, respectively.

The functions of time difference alignment (required for plesiochronous operation) and Doppler buffering are frequently combined in a common elastic buffer store and are located in the digital noninterpolated interface unit. The system specification and design of combined aligners and buffers used in TDMA satellite systems is a fairly complex task beyond the scope of this text. For the interested reader, [Campanella, 1980b, and CCIR, 1978a] provide a wealth of information.

2.7 ENERGY DISPERSAL (SCRAMBLING) IN DIGITAL COMMUNICATIONS

When the data train is random, the energy of the baseband signal and of the corresponding modulated RF carrier is sufficiently dispersed to reduce the peaks of the power flux-density (power spectral density) at the surface of the earth produced by emission from the earth station. If, however, the baseband pulse train includes a periodic pattern, discrete line components appear in the modulated RF spectrum. In this case some of the peaks of the power flux-density may exceed the permissible level. The primary purpose of *energy dispersal* (*scrambling*) techniques is to reduce the power of individual discrete spectral lines. This is achieved by producing a random-like baseband data train, irrespective of any patterns in the information baseband data train.

A digital communication system is known to be *bit sequence transparent* if it can convey any given sequence of bits. In the study of synchronization systems and particularly of symbol timing recovery systems we will note that the symbol timing recovery system might have a degraded performance or even lose lock and thereby cause a burst of data errors if a long series of 1's or a long series of 0's is transmitted. Well-designed *scramblers* preserve bit sequence transparency by *restricting* the occurrence of periodic sequences and sequences containing long strings of 1's and 0's. Scramblers are also useful **encryption** devices which assure secrecy in military and other confidential communication systems [Spilker, 1977].

Scrambling the data is a method for substantially increasing the period of the input data and of converting the original data sequence into a random-like sequence. The *randomized* binary data sequence is known as a *pseudo-random sequence, pseudo-noise sequence,* or *pseudo-random binary data sequence.* The scrambling process permits recovery of the original data at the other receiving descrambler location. With an appropriate choice of taps, an N-bit shift register is capable of generating a scrambled output sequence whose maximum periodic length is $L = 2^N - 1$ binary digits.

The detailed analysis and design of feedforward and feedback *maximal length shift registers* used in scramblers and descramblers is described in texts dealing with algebraic coding schemes and spread-spectrum systems, such as [Peterson, 1961; Hamming, 1980; Lindsey and Simon, 1973; and Dixon, 1976]. An excellent review paper related to this subject is [Forney, 1970]. Here we present the spectral properties of pseudo-random sequences and the principle of operation for scramblers and descramblers.

2.7.1 Power Spectrum of Random and of Pseudo-Random (Pseudo-Noise) Data

In Chapter 3 we show that the spectrum of equiprobable *random data,* $w_s(f)$, measured across a 1-Ω impedance, is given by [see equation (3.37)]

$$w_s(f) = 2T_b \left(\frac{\sin \pi f T_b}{\pi f T_b} \right)^2 \tag{2.11}$$

where T_b is the bit duration and ± 1 is the amplitude of the 1 and 0 state bits, respectively. From this equation we observe that the maximum spectral density occurs near dc (at the carrier frequency for the double-sideband suppressed carrier modulated signal). The dc component is zero and no discrete spectral lines are present in this spectrum.

The spectrum of a *pseudo-random* binary data (pseudo-noise or PN data) sequence, $w_{PR}(f)$, consists of discrete spectral lines separated by $1/LT_b$ Hz, where $L = 2^N - 1$ is the period of the pseudo-random sequence. It is given by [Skolnik, 1962; Lindsey and Simon, 1973]

$$w_{PR}(f) = \frac{L+1}{L^2}\left(\frac{\sin \pi f T_b}{\pi f T_b}\right)^2 \sum_{\substack{n=-\infty \\ n\neq 0}}^{\infty} \delta\left(f - \frac{n}{LT_b}\right) + \frac{1}{L^2}\delta(f) \qquad (2.12)$$

where T_b is the bit duration.

The contributions to the line spectrum occur at frequencies that are multiples of the fundamental frequency: $1/LT_b$. Since the binary waveform is a square wave having an alternating amplitude of $+1$ and -1, its average power is constant irrespec-

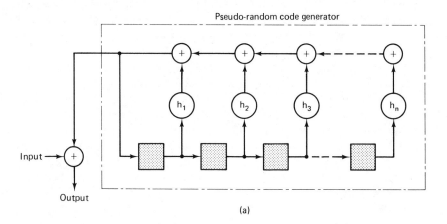

(a)

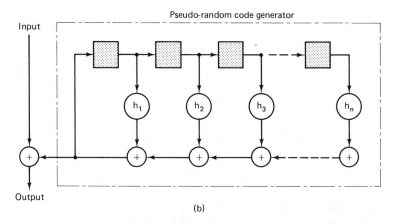

(b)

Figure 2.23   Pseudo-random energy dispersal method. (a) Scrambler. (b) Descrambler. $h_1, h_2, \ldots, h_N$ represent the shift register internal connections which determine the code sequence. (From [CCIR, 1978b], with permission from the International Telecommunications Union.)

tive of the bit duration T_b. Thus, for a given value of L, increasing the bit duration makes the spectral lines more dense and reduces their amplitudes proportionally. The envelope of the spectrum for sufficiently large L is dependent only on T_b. For a given value of T_b (constant bit rate), increasing the length of the pseudo-random sequence, L, reduces the spacing between the discrete spectral lines and their amplitudes, respectively. *Solve Problems 2.8 and 2.9.*

To ensure a desired degree of power spectral (energy) dispersal, a pseudo-random sequence of LT_b duration can be modulo-2 added to the information stream, as shown in Fig. 2.23; or a self-scrambler, as shown in Fig. 2.24, may be used.

For a reference bandwidth of 4 kHz [CCIR, 1978b], an *energy dispersal factor, D*, may be defined as

$$D = 10 \log \frac{\text{total power}}{\text{maximum power per 4 kHz}} \qquad (2.13)$$

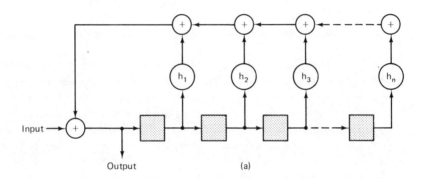

Output   (a)

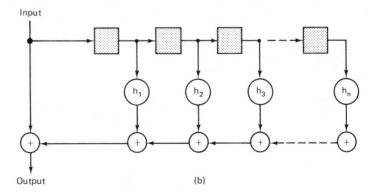

Output   (b)

Figure 2.24   Self-scrambler energy dispersal technique. (a) Scrambler. (b) Descrambler. $h_1, h_2, \ldots , h_N$ represent the shift register internal connections that determine the code sequence. (From [CCIR, 1978b], with permission from the International Telecommunications Union.)

The degree of dispersal is proportional to L as long as the sequence duration is less than the reciprocal of the reference bandwidth. There is little additional dispersal gained after the sequence duration reaches 250 μs (4-kHz reference bandwidth).

Pseudo-random energy dispersal technique.   The scrambler shown in Fig. 2.23 maintains the transmission pulse train in a state similar to that which it would have for a random pattern, irrespective of the input data. The input data are processed in an exclusive-OR gate with the data generated in the pseudo-random code generator. In the receiving unit (descrambler) a pseudo-random code generator, generating the same pattern as that in the transmitting unit, recovers the original data by processing the pseudo-random code sequence, generated by the receive code generator, with the received data stream. A merit of this technique is that energy dispersal is achieved without any degradation of the bit error rate of the information pulse train. On the other hand, a disadvantage of this scheme is that it necessitates not only bit synchronization but also synchronization of the pseudo-random code generators. In advanced time-division multiple-access (TDMA) satellite systems, synchronization between both pseudo-random code generators is achieved by means of burst synchronization signals.

Self-scrambler–energy dispersal technique.   In this energy dispersal technique, shift registers having a feedback loop and feedforward loop are utilized in the scrambler and descrambler units. In the scrambler code conversion is performed for each bit of the information pulse train. A merit of this technique is that no pseudo-random code synchronization is required between the scrambler and descrambler. After N error-free scrambled data bits have arrived, the shift register states at both the transmit and receive ends are identical. However, a single bit error in the received data (descrambler input) can produce as many errors as there are tap coefficients over an interval of N bits. The scrambler can also be used as an error rate detector for low bit error rates. For further details consult [Spilker, J.J., 1977, and Feher, K., *Digital Microwaves*, Chapter 11 on Measurement Techniques, Prentice-Hall, Inc., 1981].

PROBLEMS

2.1. A 15-kHz stereo-quality signal is PCM encoded. What is the required bit rate if a signal-to-quantization error of more than (a) 45 dB; (b) 55 dB is required?

2.2. Describe the major advantages and disadvantages of the PCM, DPCM, and DMOD techniques. If a telephony signal must be transmitted at a rate of less than 20 kb/s, which method would you use?

2.3. Describe the major advantages and applications of transmultiplexers. What are the disadvantages of these systems?

2.4. What is the purpose of the adaptive filter in an echo canceler? Why is it difficult to obtain almost perfect echo cancellation?

2.5. Why is echo cancellation or suppression more important in satellite systems than in terrestrial microwave and optical fiber systems?

2.6. Explain the functional requirements of single-destination DSI connectivity mapping. Construct a mapping diagram, assuming that eight terrestrial channels are connected to five satellite channels. Assume that the numbering systems of the terrestrial and international channels are different.

2.7. What would happen to the quality of speech signals if the digital delay line were not present in the digital-speech interpolation equipment shown in Fig. 2.22?

2.8. Plot the power spectrum density (in dB) of a pseudo-random binary data sequence of an $f_b = 1/T_b = 100$ kb/s rate source. Assume that the period of the sequence is (a) $L = 7$ bits; (b) 15 bits. Compare the spectra of (a) and (b).

2.9. Plot the power spectrum density of an $f_b = 1/T_b = 100$ kb/s random data rate. Compare this spectrum with that obtained in Problem 2.8 and the measured result shown in Fig. 3.5. How much is the difference between your calculated result and the measured result? (*Hint:* Measurement was performed in a 3-kHz bandwidth; apply a correction factor of 10 log 3 kHz/1 Hz = 34.77 dB. Why is this correction factor required?)

3

BASEBAND TRANSMISSION SYSTEMS

3.1 INTRODUCTION

Fundamental digital transmission concepts and binary baseband transmission techniques are described in this chapter. A thorough knowledge of these techniques is essential for the study of digitally modulated systems. The derivation and physical interpretation of the power spectral density (psd) function of binary signals is followed by the study of bandlimiting effects and the description of the frequently used "eye diagram" concept. The most important Nyquist transmission theorems for intersymbol interference (ISI)-free transmission with associated filter synthesis and equalization techniques are presented. Finally, the probability of error (P_e) performance in an additive white Gaussian noise environment is studied.

Particular attention is given to *binary* baseband systems, that is, systems in which only two signaling levels are used. In later chapters we will see that binary systems are more power efficient, but less spectrally efficient than multistate M-ary systems. Spectral efficiency (the alternative term "bandwidth efficiency" is also frequently used) may be expressed in terms of transmitted *bits/second/hertz* (b/s/Hz). This normalized quantity is a valuable system parameter. For instance, if a data rate of 10 Mb/s is transmitted in a 6-MHz-wide channel, the spectral efficiency is 10 Mb/s per 6 MHz, or 1.67 b/s/Hz.

The study of coherent digital modulation–demodulation (modem) systems is frequently performed in the equivalent baseband of modulated systems. The block diagram of a binary modem is shown in Fig. 3.1 and its corresponding equivalent baseband model in Fig. 3.2. A brief description of the major functional blocks of both of these systems follows, while a description of the required conditions, in regard to the validity of the equivalent baseband model, and a detailed discussion of the functional blocks are presented in Chapter 4. The baseband model is simpler and for this reason it is frequently employed in the analysis of linearly modulated systems.

The local oscillator (LO) signal $c(t)$ is multiplied by the bandlimited baseband signal $b(t)$. This multiplication is the actual modulation process. When a baseband transmit low-pass filter (LPF_T) significantly bandlimits the spectrum of the binary source, $a(t)$, we say that *premodulation* filtering has been employed. However, when the transmit bandpass filter (BPF_T) is the major bandlimiting element, we have a *postmodulation* filtered signal. The linear wideband channel simulator is provided to simulate the system noise. The receive bandpass filter eliminates out-of-band noise and interference. The carrier recovery (CR) subsystem generates a pure carrier wave from the received modulated carrier. The multiplier in the receiver, followed by the receive low-pass filter, demodulates the received signal. The demodulated, bandlimited signal is fed to a threshold comparator which is gated by the recovered clock, which is generated by the symbol timing-recovery (STR) circuitry.

The transmission characteristics of filters and of complete transmission channels may be described in terms of their respective amplitude and phase responses. The terms *filter* and *channel* may be used interchangeably. In *binary* transmissions systems the term *symbol* is synonymous with *bit* and the term *symbol rate* (f_s) with *bit*

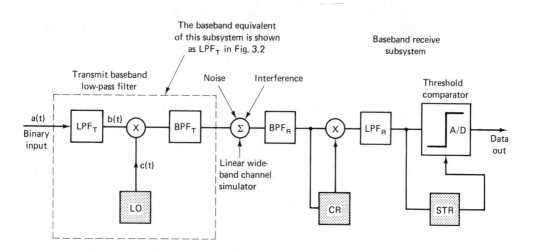

Figure 3.1   Block diagram of a coherent modulator–demodulator (modem). LO, local oscillator; LPF_T, low-pass filter (transmitter); BPF_T, bandpass filter (transmitter); BPF_R, bandpass filter (receiver); CR, carrier-recovery circuit; LPF_R, low-pass filter (receiver); STR, symbol timing-recovery circuit; A/D, one-bit analog-to-digital converter (threshold comparator).

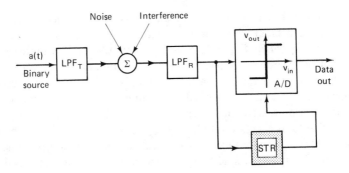

Figure 3.2   Equivalent baseband model of the coherent modem, shown in Fig. 3.1.

rate (f_b); that is, $f_s = f_b$. (Note: This is true only in binary transmissions; otherwise, a symbol may consist of several bits.)

3.2 SPECTRAL DENSITY OF RANDOM SYNCHRONOUS SIGNALS

The spectral density function of random binary signals is derived in Section 3.2.1. We assume that the data stream is *synchronous;* that is, zero crossings may occur only at integer multiples of the symbol interval T_s. Even though the derivation is tedious, the final result is of fundamental importance, since a simple extension is used in the spectral calculations of phase-shift keying (PSK), minimum shift keying (MSK), and other quadrature-modulated systems.

If your main interest is in the *applications* of baseband and of modulated signals and you do not have the time to study the following equations, then, during the first reading of this book, you may omit this derivation and continue with Section 3.2.2.

In Fig. 3.3, segments of synchronous random signal functions, $s(t, T)$, are depicted. The results of the following derivation apply to both "overlapping" and "nonoverlapping" signaling elements. A nonoverlapping signaling element, or in short, a "symbol," is confined to its own time slot of T_s seconds. The overlapping symbol occupies in addition to the unit symbol interval, T_s, the time slot of adjacent symbols. Examples of nonoverlapping and of overlapping symbols are illustrated in Fig. 3.3.

3.2.1 Derivation of the Baseband Spectrum of Binary Signals

The baseband spectral derivation methods described in both Bennett's original work [Bennett, 1958] and also in [Bennett and Davey, 1965] are followed and expanded.*

* With permission from the American Telephone and Telegraph Company and from the McGraw-Hill Book Company.

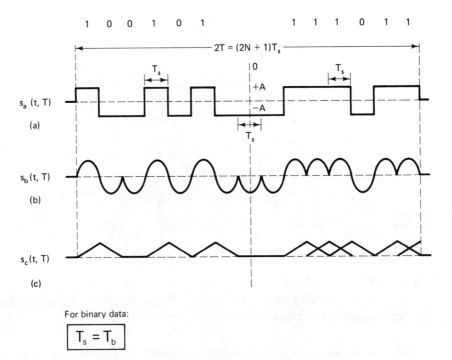

Figure 3.3 Synchronous nonoverlapping and overlapping binary baseband data streams. (a) NRZ-nonoverlapping; (b) nonoverlapping; (c) overlapping.

This expansion and the inclusion of some fundamental definitions are deemed to be necessary for the reader who does not possess an advanced knowledge of statistical communications theory.

Let $s(t, T)$, shown in Fig. 3.3, be a truncated segment of a random synchronous signal (i.e., a truncated infinite-length signal). We assume that $s(t, T)$ vanishes outside the interval $-T < t < T$, where $2T = (2N + 1)T_s$ and N is an arbitrary integer. This segment consists of a sequence of $g_1(t)$ and $g_2(t)$ pulses. Assume that

$g_1(t)$ is transmitted for an input 1 with probability p
$g_2(t)$ is transmitted for an input 0 with probability $1 - p$

Thus

$$s(t, T) = \sum_{n=-N}^{N} s_n(t) \tag{3.1}$$

where

$$s_n(t) = \begin{cases} g_1(t - nT_s) & \text{with probability } p \\ g_2(t - nT_s) & \text{with probability } 1 - p \end{cases} \tag{3.2}$$

To facilitate the derivation, let us separate that portion of $s(t, T)$ which leads to a continuous spectra from that portion which leads to a discrete power spectra.

$$s(t, T) = \underbrace{u(t, T)}_{} + \underbrace{v(t, T)}_{}$$

$$\| \qquad\qquad \|$$ (3.3)

leads to   leads to
continuous power   discrete power
spectrum   spectrum

Let the functions $u(t, T)$ and $v(t, T)$ be given by

$$u(t, T) = \sum_{n=-N}^{N} u_n(t) = \sum_{n=-N}^{N} a_n [g_1(t - nT_s) - g_2(t - nT_s)] \tag{3.4}$$

$$v(t, T) = \sum_{n=-N}^{N} [pg_1(t - nT_s) + (1 - p)g_2(t - nT_s)] \tag{3.5}$$

where

$$a_n = \begin{cases} 1 - p & \text{with prob. } p \\ -p & \text{with prob. } 1 - p \end{cases} \tag{3.6}$$

In the following section, the validity of the assumption that $s(t, T)$ is accurately represented by equations (3.3) to (3.6) is verified.

Verification.   Consider the right-hand side of equations (3.4) and (3.5), namely the series summations of $u(t, T)$ and $v(t, T)$. By inserting these series [equations (3.4) and (3.5)] into (3.3), we should obtain $s(t, T)$:

$$\sum_{n=-N}^{N} a_n [g_1(t - nT_s) - g_2(t - nT_s)] + \sum_{n=-N}^{N} [pg_1(t - nT_s) + (1 - p)g_2(t - nT_s)]$$

$$\tag{3.7}$$

$$= \sum_{n=-N}^{N} [(a_n + p)g_1(t - nT_s) + (-a_n + 1 - p)g_2(t - nT_s)]$$

From equation (3.6) we have

$$a_n = \begin{cases} 1 - p & \text{with prob. } p \\ -p & \text{with prob. } 1 - p \end{cases} \quad \text{and} \quad -a_n = \begin{cases} -(1 - p) & \text{with prob. } p \\ -(-p) & \text{with prob. } 1 - p \end{cases}$$

Therefore, the coefficients $(a_n + p)$ and $(-a_n + 1 - p)$ in (3.7) are given by

$$(a_n + p) = \begin{cases} 1 & \text{with prob. } p \\ 0 & \text{with prob. } 1 - p \end{cases} \qquad (-a_n + 1 - p) = \begin{cases} 0 & \text{with prob. } p \\ 1 & \text{with prob. } 1 - p \end{cases}$$

The right-hand side of equation (3.7) is given by

$$\sum_{n=-N}^{N} \begin{pmatrix} 1 \text{ with prob. } p \\ 0 \text{ with prob. } 1 - p \end{pmatrix} g_1(t - nT_s)$$

$$+ \sum_{n=-N}^{N} \begin{pmatrix} 0 \text{ with prob. } p \\ 1 \text{ with prob. } 1 - p \end{pmatrix} g_2(t - nT_s) \tag{3.8}$$

By inspection it is evident that this last expression is the same as equations (3.1) and (3.2). Thus

$$u(t, T) + v(t, T) = \sum_{n=-N}^{N} s_n(t) = s(t, T) \tag{3.9}$$

(End of Verification)

∎

An expression for the power spectral density of the $u(t, T)$ signal will now be derived. This continuous-power spectral density, $w_u(f)$, will be expressed as a function of the Fourier transforms of the individual signaling elements $g_1(t)$ and $g_2(t)$.

The *Fourier transform* of the signal $u(t, T)$ is given by

$$U(f, T) = \int_{-\infty}^{\infty} u(t, T) e^{-j2\pi ft} \, dt \tag{3.10}$$

and the inverse Fourier transform is given by

$$u(t, T) = \int_{-\infty}^{\infty} U(f, T) e^{j2\pi ft} \, df \tag{3.11}$$

By *Parseval's theorem* the mean-square value of the truncated signal is given by

$$\frac{1}{2T} \int_{-T}^{T} u^2(t, T) \, dt = \frac{1}{2T} \int_{-\infty}^{\infty} u^2(t, T) \, dt = \frac{1}{2T} \int_{-\infty}^{\infty} |U(f, T)|^2 \, df \tag{3.12}$$

$$= \int_{0}^{\infty} \frac{|U(f, T)|^2}{T} \, df = \int_{0}^{\infty} w_u(f, T) \, df \tag{3.13}$$

where $w_u(f, T)$ is defined by

$$w_u(f, T) = \frac{|U(f, T)|^2}{T} \tag{3.14}$$

The spectral density of the signal

$$u(t) = \lim_{T \to \infty} u(t, T) \tag{3.15}$$

could be defined as the limit of $w_u(f, T)$ as T approaches infinity. However, this limit does not exist in most cases. To circumvent this problem, the ensemble average of $|U(f, T)|^2$ will be derived and afterward its limit will be obtained. It has been established in statistical communications theory that for random **ergodic** processes the ensemble average equals the time average. Thus the ensemble average of $W_u(f, T)$ represents the continuous power spectral density of the signal $s(t) = \lim_{T \to \infty} s(t, T)$. It is given by

$$w_u(f) = \lim_{T \to \infty} \frac{E\{|U(f, T)|^2\}}{T} = \lim_{N \to \infty} \frac{E\{|U(f, T)|^2\}}{(2N + 1)T_s/2} \tag{3.16}$$

where $E\{ \cdot \}$ represents the expected value of the argument (i.e., the ensemble average).

Equipped with these background equations, we are in a position to *derive* $w_u(f)$. The Fourier transform of $u(t, T)$ is given by

$$U(f,T) = \sum_{n=-N}^{N} \int_{-\infty}^{\infty} a_n [g_1(t - nT_s) - g_2(t - nT_s)] e^{-j2\pi ft} \, dt$$

$$= \sum_{n=-N}^{N} a_n e^{-j2\pi nT_s} [G_1(f) - G_2(f)] \tag{3.17}$$

where

$$G_K(f) = \int_{-\infty}^{\infty} g_k(t) e^{-j2\pi ft} \, dt \qquad k = 1, 2 \tag{3.18}$$

The square of the absolute value of the Fourier transform $U(f, T)$ is

$$|U(f, T)|^2 = U(f, T)U^*(f, T)$$

$$= \sum_{m=-N}^{N} \sum_{n=-N}^{N} a_m a_n e^{j2\pi(n-m)T_s} [G_1(f) - G_2(f)][G_1^*(f) - G_2^*(f)] \tag{3.19}$$

$$= |G_1(f) - G_2(f)|^2 \sum_{m=-N}^{N} \sum_{n=-N}^{N} a_m a_n e^{j2\pi(n-m)T_s}$$

where "*" denotes the complex conjugate value.

To obtain the expected value of $|U(f, T)|^2$, we note that in (3.19), a_m and a_n are the only random variables. Thus it is sufficient to evaluate the expected value of $a_m a_n$. For $m = n$, from equation (3.6) we have

$$a_n^2 = \begin{cases} (1 - p)^2 & \text{with prob. } f_1(a_n) = p \\ p^2 & \text{with prob. } f_2(a_n) = 1 - p \end{cases} \tag{3.20}$$

The expected value of a_n^2 for the $m = n$ case is given by

$$E\{a_n^2\} = \sum_{k} a_n^2 f_k(a_n) = (1 - p)^2 p + p^2(1 - p) = p(1 - p) \tag{3.21}$$

where $f_k(a_n)$ is the probability density function of the random variable a_n.

For the case $m \neq n$, equation (3.6) is used for the computation of $E\{a_m a_n\}$. In this case a_m and a_n represent independent random variables. Thus

$$a_m a_n = \begin{cases} (1 - p)^2 & \text{with joint prob. } f_1(a_m, a_n) = p^2 \\ p^2 & \text{with joint prob. } f_2(a_m, a_n) = (1 - p)^2 \\ -p(1 - p) & \text{with joint prob. } f_3(a_m, a_n) = 2p(1 - p) \end{cases} \tag{3.22}$$

and

$$E\{a_m a_n\} = \sum_{k=1}^{3} a_m a_n f_k(a_m, a_n)$$

$$= (1-p)^2 p^2 + p^2(1-p)^2 + [-p(1-p)]2p(1-p) = 0$$

that is,

$$E\{a_m a_n\} = 0 \tag{3.23}$$

From equations (3.19), (3.21), and (3.23) we obtain

$$E\{|U(f, T)|^2\} = |G_1(f) - G_2(f)|^2 \sum_{n=-N}^{N} p(1-p) \tag{3.24}$$

$$= (2N+1)p(1-p)|G_1(f) - G_2(f)|^2$$

The *continuous power spectral density* of $u(t)$ is obtained from equations (3.16) and (3.24). It is given by

$$w_u(f) = \lim_{N \to \infty} \frac{E\{|U(f, T)|^2\}}{(2N+1)T_s/2} \tag{3.25}$$

$$= 2p(1-p)f_s|G_1(f) - G_2(f)|^2$$

We note that the continuous part of the spectrum is directly proportional to the difference between the Fourier transforms of the two signaling pulses $g_1(t)$ and $g_2(t)$.

We now proceed to derive the *discrete* part of the *spectrum* which corresponds to $v(t, T)$, equation (3.3). From equation (3.5) we may verify that $v(t, -T_s) = v(t)$; that is, $v(t)$ is a periodic function with a period $T_s = 1/f_s$. The Fourier series expansion of $v(t)$ can be expressed as follows:

$$v(t) = \sum_{n=-\infty}^{\infty} pg_1(t - nT_s) + (1-p)g_2(t - nT_s) \tag{3.26}$$

$$= \sum_{m=-\infty}^{\infty} c_m e^{j2\pi m f_s t} = c_0 + \sum_{m=1}^{\infty} (c_m e^{j2\pi m f_s t} + c_m^* e^{-j2\pi m f_s t})$$

where

$$c_m = \frac{1}{T_s} \int_{-T_s/2}^{T_s/2} e^{-j2\pi m f_s t} \sum_{h=-\infty}^{\infty} pg_1(t - nT_s) + (1-p)g_2(t - nT_s) \, dt$$

$$= f_s \sum_{n=-\infty}^{\infty} \int_{-nT_s-T_s/2}^{-nT_s+T_s/2} [pg_1(t) + (1-p)g_2(t)]e^{-j2\pi m f_s(t+nT_s)} \, dt \tag{3.27}$$

$$= f_s \int_{\infty}^{\infty} [pg_1(t) + (1-p)g_2(t)]e^{-j2\pi m f_s t} \, dt$$

$$= f_s[pG_1(mf_s) + (1-p)G_2(mf_s)]$$

From (3.26) we see that $v(t)$ consists of a constant term (dc component) and a series of harmonics of the signaling frequency f_s. The peak amplitude of these harmonics is $2|c_m|$, and the normalized average power, as measured across a 1-Ω impedance, is $\{2|c_m|/\sqrt{2}\}^2 = 2|c_m|^2$.

Finally, the *complete spectral density* $w(f)$ of the random data signal $s(t) = \lim_{T \to \infty} s(t, T) = \lim_{T \to \infty} [u(t, T) + v(t, T)]$ is obtained from equations (3.25) to (3.27) and is given by

$$
\begin{array}{cc}
\text{continuous} & \text{dc term and harmonics} \\
\downarrow & \downarrow
\end{array}
$$

$$w_s(f) = w_u(f) \quad + w_v(f)$$

$$= 2f_s p(1-p)|G_1(f) - G_2(f)|^2 \quad \leftarrow \text{continuous spectrum} \tag{3.28}$$

$$+ f_s^2 [pG_1(0) + (1-p)\, G_2(0)]^2\, \delta(f) \quad \leftarrow \text{dc component}$$

$$+ 2f_s^2 \sum_{m=1}^{\infty} |pG_1(mf_s)$$
$$\quad\quad\quad\quad\quad\quad\quad\quad\quad \leftarrow \text{harmonics}$$
$$+ (1-p)\, G_2(mf_s)|^2\, \delta(f - mf_s)$$

End of Derivation ∎

3.2.2 Summary and Applications of the Baseband Spectral Equations of Binary Signals

The power spectral density equation has very frequent applications. A summary of the terms and notations given in equation (3.28) follows:

f_s = signaling frequency = symbol rate (in binary systems the bit rate equals the symbol rate, i.e., $f_b = f_s$)

$T_s = 1/f_s$ = the unit symbol duration (for nonoverlapping pulses)

$w_s(f)$ = the complete power spectral density (including the continuous and the discrete spectral components)

$w_u(f)$ = the continuous part of the power spectral density

$w_v(f)$ = the discrete part of the spectrum, may include the dc term

$G_1(f)$ = the Fourier transform of the symbol $g_1(t)$

$G_2(f)$ = the Fourier transform of the symbol $g_2(t)$

m = an integer number ($m = 1, 2, 3, \ldots$)

p = probability of occurrence of the $g_1(t)$ symbol

$1 - p$ = probability of occurrence of the $g_2(t)$ symbol

In Fig. 3.4 a number of illustrative baseband waveforms are shown. The non-return-to-zero (NRZ) signal, with and without dc component, is frequently used in baseband transmission systems and in modulated systems. The return-to-zero (RZ) signal is employed for synchronization purposes. The half-rate NRZ signals, (d) and (e), are used in numerous quadrature-modulated systems. To obtain signal (d) we

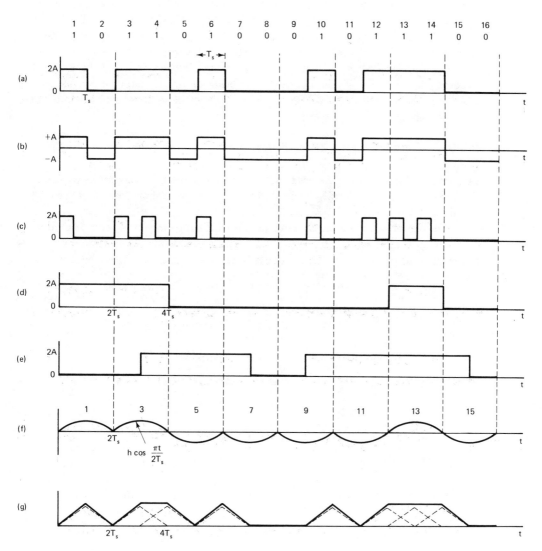

Figure 3.4 Time-domain representation of various baseband waveforms. (a) NRZ (with dc comp.); (b) NRZ (without dc); (c) RZ with dc; (d) ½ rate NRZ; (e) ½ rate NRZ; (f) ½ rate MSK element; (g) overlapping pulses. Dashed line, individual overlapping pulses; solid line, resulting wave.

select the *odd*-numbered bits in the NRZ sequence and hold the voltage values of these bits throughout two consecutive symbol intervals. Signal (e) is obtained in a similar manner; that is, the *even*-numbered bits of the (a) stream are held for two bit intervals. The $\pm h \cos(\pi t/T_s)$ signaling element is used in minimum shift keying (MSK) modulated systems, while the overlapping pulses have been included into Fig. 3.4 to remind the reader that the validity of equation (3.28) is not constrained to nonoverlapping pulses. The individual signaling elements of the overlapping pulses are shown by a dashed line, while the resultant signal is shown by a solid line. The various signaling elements, described in Fig. 3.4, are extensively used in the baseband of modulated systems. They are studied further in Chapter 4.

Discrete spectral components may be present at integer multiples of the signaling frequency $f_s = 1/T_s$. If the Fourier transforms of the signaling elements, $G_1(f)$ and $G_2(f)$ [see equation (3.28)], vanish at harmonics of f_s, there will be no discrete spectral components in the spectrum.

3.2.3 Spectral Density of NRZ Signals

Equation (3.28) applies to a large class of binary signals. As a particular example we derive the power spectral density of frequently used equiprobable balanced NRZ signals.

The equiprobable signaling states can be achieved by means of scrambler circuits, described in Chapter 2. Figure 3.4(a) and (b) provides examples of member functions of the random process. For the balanced NRZ signaling elements we have

$$g_1(t) = \begin{cases} +A \text{ volts} & -\frac{T_s}{2} < t \le \frac{T_s}{2} \\ 0 \text{ volts} & \text{elsewhere} \end{cases}$$

$$g_2(t) = \begin{cases} -A \text{ volts} & -\frac{T_s}{2} < t \le \frac{T_s}{2} \\ 0 \text{ volts} & \text{elsewhere} \end{cases} \tag{3.29}$$

and

$$p = 1 - p = 0.5 \tag{3.30}$$

From equation (3.18),

$$G_1(f) = \int_{-\infty}^{\infty} g(t) e^{-j2\pi ft} \, dt = \int_{-T_s/2}^{T_s/2} A e^{-j2\pi ft} \, dt$$

$$= \frac{A}{-j2\pi f} \int_{-T_s/2}^{T_s/2} e^{-j2\pi ft} \, d(-j2\pi ft) \tag{3.31}$$

$$= \frac{A}{\pi f} \sin(\pi f T_s)$$

Similarly,

$$G_2(f) = \int_{-T_s/2}^{T_s/2} (-A)e^{-j2\pi ft}\, dt = \frac{-A}{\pi f} \sin(\pi f T_s) \qquad (3.32)$$

As $G_1(f) = -G_2(f)$, it is evident that $G_1(f = 0) = G_1(0) = -G_2(f = 0) = -G_2(0)$. At integer multiples of the signaling frequency ($f = mf_s$), we have

$$G_1(f = mf_s) = \frac{A}{\pi f} \sin(\pi f T_s) = \frac{A}{\pi m f_s} \sin(\pi m f_s T_s)$$

$$= \frac{A}{\pi m f_s} \sin(\pi m) = 0 \qquad (3.33)$$

Similarly,

$$G_2(f = mf_s) = 0 \qquad (3.34)$$

By inserting the terms from equations (3.31) to (3.34) into (3.28), we obtain

$$w_s(f) = 2f_s p(1 - p) \left| \frac{A}{\pi f} \sin(\pi f T_s) - \frac{-A}{\pi f} \sin(\pi f T_s) \right|^2$$

$$+ zero \qquad\qquad \leftarrow \text{dc component} \qquad (3.35)$$
$$+ zero \qquad\qquad \leftarrow \text{harmonics}$$

$$= 2f_s \cdot 0.5 \cdot 0.5 \left(2\frac{A}{\pi f} \sin \pi f T_s \right)^2$$

$$= \frac{2A^2}{f_s} \left(\frac{\sin \pi f T_s}{\pi f T_s} \right)^2 = 2A^2 T_s \left(\frac{\sin \pi f T_s}{\pi f T_s} \right)^2 \qquad (3.36)$$

The complete *spectral density of equiprobable* $p(+A) = p(-A) = 0.5$ *balanced NRZ random data,* such as shown in Fig. 3.4(b), is given by

$$w_s(f) = 2A^2 T_s \left(\frac{\sin \pi f T_s}{\pi f T_s} \right)^2 \qquad (3.37)$$

Comparing this result with equation (3.28), we conclude that the dc component is zero and that there are no discrete components. The **first spectral zero** occurs for $f = 1/T_s = f_s$, that is, at the signaling frequency.

Even though the dc component is zero, the spectral density has its maximum value $2A^2 T_s$ at dc, that is, for $f = 0$. [Here, you should **stop for a minute** and read again the previous sentence and equation (3.37). It is important that you *grasp the difference* between the dc component and the spectral density at zero frequency.]

Throughout the derivation we assumed that the voltage values of the signaling elements have been specified across a 1-Ω normalized resistance.

An alternative method of power spectral density derivation is presented in [Feher, K., Prentice-Hall, Inc., Chapter 2, 1981].

Example 3.1

(a) Calculate the power spectral density of an $f_s = 100$ kb/s rate NRZ signal. Assume that logic state 1 is represented by a $+100$-mV signal and logic state 0 by a -100-mV signal. These voltages are measured across a 75-Ω termination matched to the characteristic impedance of the transmission line. The probability of a logic zero state is the same as the probability of a logic one state.

(b) Assume the same impedance and voltage levels as in part (a), but increase the signaling rate to 10 Mb/s.

Solution (a) In equation (3.37) a normalized 1-Ω resistance is assumed. For the 75-Ω system operating at a rate of 100 kb/s ($T_s = 1/f_s = 10^{-5}$ s), we have

$$w_s(f) = \frac{2A^2 T_s}{75} \left(\frac{\sin \pi f T_s}{\pi f T_s} \right)^2 = \frac{2(0.1)^2 \times 10^{-5}}{75} \left(\frac{\sin \pi f \times 10^{-5}}{\pi f \times 10^{-5}} \right)^2$$

$$= 2.666 \times 10^{-9} \left(\frac{\sin \pi f \times 10^{-5}}{\pi f\, 10^{-5}} \right)^2 \qquad \text{W/Hz}$$

For practical calculations the power spectal density is expressed in dBm/Hz. Remember: 0 dBm corresponds to 1 mW of power:

$$\boxed{0 \text{ dBm} = 1 \text{ mW} \quad \text{and} \quad 0 \text{ dBW} = 1 \text{ W}}$$

The continuous power spectral density represents the signal power in a bandwidth of 1 Hz.

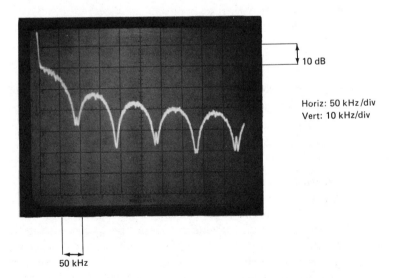

10 dB

Horiz: 50 kHz/div
Vert: 10 kHz/div

50 kHz

Figure 3.5 Measured power spectral density of an $f_s = 100$ kb/s rate equiprobable (NRZ) data source. (Courtesy of Digital Communications Research Laboratory, University of Ottawa.)

The power spectral density in a 1-Hz bandwidth centered around dc ($f = 0$ Hz) is $w_s(f = 0) = 2.66 \times 10^{-9}$ W/Hz (note that $\lim \epsilon \to 0 \sin \epsilon/\epsilon = 1$). Expressed in dBm/Hz:

$$w_s(f = 0) = 10 \log 2.66 \times 10^{-9} \text{ dBW/Hz} = 10 \log 2.66 \times 10^{-6} \text{ dBm/Hz}$$

$$= -55.76 \text{ dBm/Hz} = -85.76 \text{ dBW/Hz}$$

The measured power spectral density of the described NRZ signal is shown in Fig. 3.5. Note that the maximal density of the first sidelobe is 13.5 dB below the maximal density of the main lobe (density at dc). Because the spectrum analyser noise bandwidth was set to 3 kHz, the measured results were approximately 10 log (3 kHz/1 Hz) = 35 dB higher than the calculated spectral density results.

(b) In this case $f_s = 10$ Mb/s; that is, $T_s = 100$ ns. The total power of the signal does not change (verify this statement in the time domain!). Thus the power spectral density at zero frequency has to be 100 times lower. From equation (3.37) it follows that

$$w_s(f) = \frac{2(0.1)^2 \times 10^{-7}}{75} \left(\frac{\sin \pi f \times 10^{-7}}{\pi f \times 10^{-7}} \right)^2 \qquad \text{W/Hz}$$

∎

As additional exercises, solve Problems 3.1 and 3.2. In these problems the return-to-zero (RZ) signal and the signaling element used in the baseband of minimum-shift-keying systems are considered. Note from your solutions that the RZ signal contains discrete spectral lines in addition to the continuous signal spectrum, and that the first spectral null of the MSK signaling element is at 1.5 times the half-rate signaling frequency $1/2T_s$.

3.3 BANDLIMITED BASEBAND SYSTEMS

3.3.1 Ideal "Brick-Wall" Channels (Filters)

A bandlimited digital transmission system becomes more spectrally efficient if it obtains the capability of transmitting a greater number of bits per second (b/s) in a given bandwidth. The bandwidth is normalized to 1 Hz, and the spectral efficiency is expressed in b/s/Hz. The rectangular wave is the most frequently used signaling element. For example, the NRZ and RZ signals, appearing in Fig. 3.4, consist of rectangular, infinite bandwidth signals.

In his landmark paper on channel characteristics, Nyquist derived the minimum channel bandwidth requirements [Nyquist, 1928]. He proved that it is possible to have a deformed bandlimited wave and have a receiving device that receives and regenerates a perfect signal.

After a brief description of ideal brick-wall channels and a study of eye-diagram fundamentals in this section, we describe in Section 3.4 the most important Nyquist transmission criteria, followed by an extension of these criteria to generalized pulse shapes. In order to better visualize the practical importance of these theorems, the ensuing discussion is presented for the binary case; however, *the Nyquist transmission theorems apply to the multilevel signaling case as well.* This type of signaling is described in Chapter 5, where M-ary bandwidth efficient satellite modulation techniques are studied.

Consider the ideal brick-wall channel model of Fig. 3.6. The cutoff frequency, also known as the *Nyquist frequency,* is defined to equal $f_N = 1/2T_s = f_s/2$, where T_s is the unit symbol duration. (*Note:* In binary systems the symbol rate equals the bit rate; thus $T_s = T_b$, where T_b is the unit bit duration. In multilevel systems $T_s = T_b \log_2 M$, where M is the number of signaling levels.)

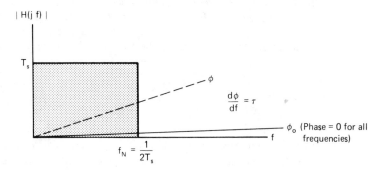

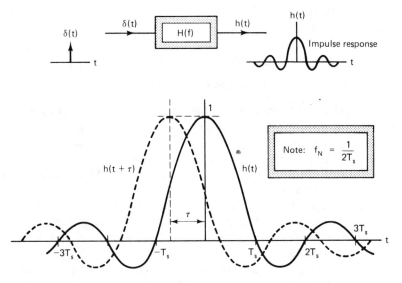

Figure 3.6 Ideal brick-wall channel and its corresponding impulse response. The anticipatory impulse response of the brick-wall channel is nonrealizable. T_s = time duration of unit signaling element.

The impulse response of the channel, $h(t)$, is given by the inverse Fourier transform of the transfer function $H(f)$:

$$H(f) = \begin{cases} T_s & |f| \leq \dfrac{1}{2T_s} \\ 0 & |f| > \dfrac{1}{2T_s} \end{cases} \tag{3.38}$$

$$h(t) = F^{-1}\{H(f)\} = \int_{-\infty}^{\infty} H(f)e^{j2\pi ft}\, df \tag{3.39}$$

$$h(t) = \frac{\sin(2\pi f_N t)}{2\pi f_N t} = \frac{\sin(\pi t/T_s)}{\pi t/T_s} \tag{3.40}$$

[*Note:* The phase of $H(f) = 0$ for all frequencies.] From this impulse response, note that

$$h(nT_s) = \begin{cases} 1 & \text{for } n = 0 \\ 0 & \text{for } n = \pm 1, \pm 2, \pm 3, \cdots \end{cases} \tag{3.41}$$

Thus the impulse response attains its full value for $t = nT_s = 0$ and has zero crossings for all other integer multiples of the symbol duration. If the ideal brick-wall channel has a nonzero but linear phase (dashed line in Fig. 3.6), the impulse response is shifted by an amount that equals the channel delay. This delay is $\tau = d\phi/d\omega$ and for linear-phase filters is constant over all frequencies. As the shape of the impulse response is the same as with $\tau = 0$, there will be no further distortion introduced.

Note that the channel input (also known as filter excitation), $\delta(t)$, has an infinitesimally short duration, whereas the output (i.e., the impulse response) has an *infinite duration*. The bandlimited channel stretches the impulse response beyond the T_s interval and deforms the input signal. (*Solve Problem 3.3.*) A desirable property of the described impulse response is that it has zero values for integer multiples of T_s. Now we should be in a position to see that in an $f_N = 1/2T_s$ wide brick-wall channel it is possible to transmit and detect synchronous, random impulses at a rate $f_s = 1/T_s = 2f_N$. Theoretically, the detection of any of these symbols can be performed without any interference from the previously sent or the subsequent impulse patterns. This situation, illustrated in Fig. 3.7, is known as *intersymbol-interference (ISI)*-free transmission.

3.3.2 Eye-Diagram Fundamentals

Channel imperfections are frequently evaluated by means of *"eye diagrams"* or *"eye patterns."* These patterns may be obtained on an oscilloscope if a signal such as $v_0(t)$ in Fig. 3.8 is fed to the vertical input y of an oscilloscope. The symbol clock $c(t - nT_s)$ is fed to the external trigger of the oscilloscope. The front trigger delay adjustment, conveniently available on most oscilloscopes, assures that the displayed eye pattern is centered on the screen. The horizontal time base is set approximately

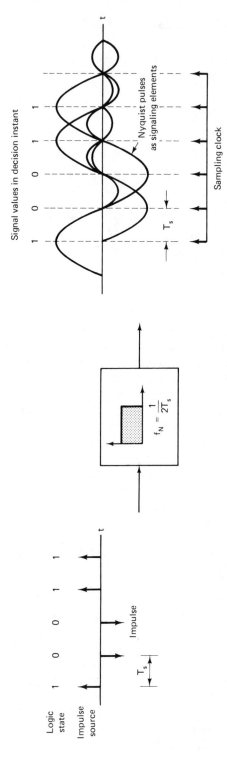

Figure 3.7 Intersymbol-interference (ISI)-free transmission of bandlimited impulses.

94

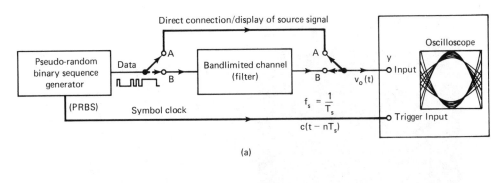

(a)

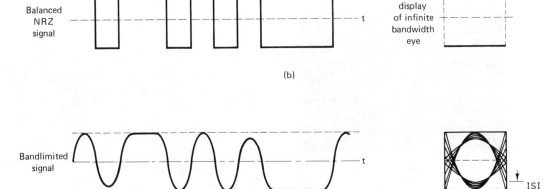

(b)

(c)

Figure 3.8   Eye-diagram measurement setup and display.

equal to the symbol duration. The inherent persistence of the cathode ray tube displays the superimposed segments of the $v_o(t)$ signal. The eye pattern of the *pseudo-random binary sequence* (PRBS) generator is displayed if the data output of this generator is directly connected to the vertical input of the oscilloscope.

The *observed* eye diagram of a *non-phase-equalized, conventional* fourth-order Butterworth filter is shown in Fig. 3.9. The applied symbol rate of the PRBS source is set at $f_s = 64$ kb/s and the 3-dB cutoff frequency for this filter is 26 kHz. Due to the ISI of the unequalized channel, the eye is only two-thirds open at the sampling instant. To maintain the same decision margin (i.e., same P_e) as in the ISI-free case, the signal level has to be increased by $1 : \frac{2}{3} = 3.52$ dB. For the systems engineer this ISI-caused degradation is a serious drawback. May I suggest that you review

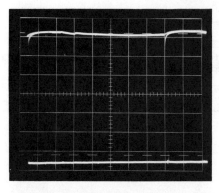

Vert: 50 mV/div
Horiz: 2 µs/div

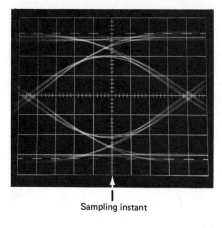

Vert: 50 mV/div
Horiz: 2 µs/div

Sampling instant

Figure 3.9 Measured eye diagram of an unequalized filter. A conventional fourth-order, unequalized-phase Butterworth filter having a 3-dB corner frequency at 26 kHz is used as channel simulator. The applied symbol rate is $f_s = 64$ kb/s. (a) Infinite-bandwidth eye diagram; (b) bandlimited eye diagram.

this situation and explain why this additional signal-level requirement (i.e., ISI degradation of 3.52 dB) is a serious drawback.

With the described technique the peak ISI can be measured. This measurement leads to an upper bound on the signal degradation and is known as *peak ISI distortion*. In addition to the peak ISI evaluation the *rms eye closure* is also an important parameter. The measurement of this parameter is more complex than the evaluation of the peak closure. For the advanced practical-minded reader, I suggest that you *solve Problem 3.4,* and read [Lucky et al., 1968].

From the observed eye diagram we also note that the overlapped signal pattern does not cross the horizontal zero line at exact integer multiples of the symbol clock. The deviation from the nominal crossing points is known as the peak-to-peak *data transition jitter, J_{pp}*. This jitter has an effect on the symbol timing recovery circuits and may significantly degrade the performance of cascaded regenerative sections [Feher, K., 1981, Prentice-Hall, Inc., Chapter 9].

3.4 NYQUIST THEOREMS

3.4.1 Nyquist's Minimum-Bandwidth Theorem

Theorem. If synchronous *impulses*, having a transmission rate of f_s symbols per second, are applied to an ideal, linear-phase brick-wall low-pass channel having a cutoff frequency of $f_N = f_s/2$ Hz, then the responses to these impulses can be observed independently, that is, without intersymbol interference.

Interpretation. From equation (3.41) and Fig. 3.7 it is evident that for the case of impulse transmission, there is no ISI in the sampling instant. Note that for ISI-free transmission the impulses need not be limited to binary values [e.g., $\pm\delta(t)$]. Intersymbol interference-free transmission is also achieved if the synchronous input stream is

$$\sum_k A_k \, \delta(t - kT_s) \tag{3.42}$$

where A_k is a multilevel discrete random variable. For example, A_k might have one of the following values: -3; -1; $+1$; $+3$. In this case we have a four-level baseband system in which every transmitted symbol is formed from two data bits.

Example 3.2

The cutoff frequency of an ideal brick-wall channel is $f_N = 2.5$ MHz. How many bits per second can we transmit through this channel if:

(a) A binary impulse train is transmitted?
(b) A four-level impulse train is transmitted?

Solution. In the binary as well as in the four-level case, the transmission rate, as specified by the Nyquist first theorem, is $2f_N = 5$ Msymbols/second. In the binary case this corresponds to 5 Mb/s, and in the four-level case to 10 Mb/s. This doubling of the information rate (not transmission rate) is achieved because two binary information bits are encoded into one four-level symbol.

∎

Note. For ISI-free transmission of f_s rate rectangular *pulses,* an $(x/\sin x)$-shaped amplitude equalizer has to be added to the ideal brick-wall channel. Thus it is possible to have an ISI-free transmission rate of 2 symbols/s/Hz.

Now, we explain why the $(x/\sin x)$-shaped amplitude equalizer is essential for the ISI-free transmission of rectangular pulses, such as used in the case of non-return-to-zero (NRZ) transmission, shown in Fig. 3.4(b). For an impulse, the amplitude of the Fourier transform is constant over all frequencies while it has a $\sin x/x$ shape for rectangular pulses (*Solve Problem 3.5.*) To keep the same system response (i.e., no ISI for both the impulse and the rectangular excitation), it is required that the Fourier transform for both cases be identical.

The output Fourier transform is obtained by multiplying the Fourier transform of the excitation by the channel transfer function. In the rectangular pulse transmission case, the brick-wall channel transfer function is modified by an $x/\sin x$ amplitude equalizer. The ideal minimum-bandwidth Nyquist channel characteristics for the conceptual synchronous impulse and for the practical rectangular (NRZ) pulse transmission case are summarized in Fig. 3.10.

Unfortunately, the described minimum-bandwidth Nyquist channels are not realizable. An infinite number of filter sections would be required to synthesize the infinite attenuation slope of the brick-wall channel. Additionally, the decay in the lobes of the time-domain response is very slow. This, in turn, would cause a prohibitively large ISI degradation (i.e., the P_e may approach 0.5) for the smallest filtering or symbol timing imperfections.

To alleviate these problems and to define more practical channel characteristics, Nyquist introduced a theorem on vestigial symmetry.

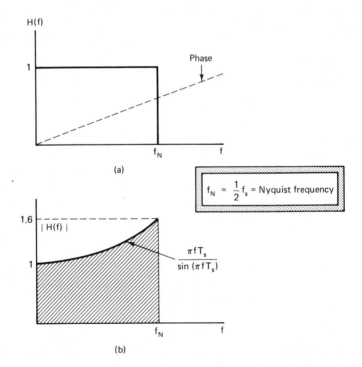

Figure 3.10 Minimum-bandwidth Nyquist channel models for impulse and pulse transmission. (a) Amplitude response of an ideal minimum-bandwidth filter that has no ISI for the conceptual case of impulse transmission. (b) Amplitude response of minimum-bandwidth channel for NRZ rectangular pulse transmission.

3.4.2 Nyquist's Vestigial Symmetry Theorem

Theorem. The addition of a skew-symmetrical, real-valued transmittance function $Y(\omega)$ to the transmittance of the ideal low-pass filter maintains the zero-axis crossings of the impulse response. These zero-axis crossings provide the necessary condition for ISI-free transmission. The symmetry of $Y(\omega)$ about the cutoff frequency ω_N (Nyquist radian frequency $\omega_N = 2\pi f_N$) of the linear-phase brick-wall filter is defined by

$$Y(\omega_N - x) = -Y(\omega_N + x) \qquad 0 < x < \omega_n \qquad (3.43)$$

where $\omega_N = 2\pi f_N$.

Heuristic proof and interpretation. Let us denote the brick-wall filter modified by the $Y(\omega)$ transmittance function as $Y_m(\omega)$. This filter is given by

$$Y_m(\omega) = \begin{cases} 1 + Y(\omega) & |\omega| < \omega_N \\ Y(\omega) & \omega_N < |\omega| < 2\omega_N \end{cases} \qquad (3.44)$$

For an illustration, see Fig. 3.11. In this derivation we assume that the phase of the filter is zero for all frequencies (i.e., the delay $\tau = 0$ in Fig. 3.6). Alternatively, we may shift the origin of time of the impulse response to remove the effect of a finite delay.

Let the impulse response of $Y_m(\omega)$ be $h_m(t)$. It is described by

$$h_m(t) = \frac{\sin(2\pi f_N t)}{\pi t} + h(t) \qquad (3.45)$$

The first term is the familiar $(\sin x / x)$* shaped impulse response of the brick-wall filter [see equation (3.40)] while the second term is given by

$$h(t) = \frac{1}{\pi} \int_0^{2\omega_N} Y(\omega) \cos \omega t \, d\omega \qquad (3.46)$$

By dividing this integral into two parts from 0 to ω_N and from ω_N to $2\omega_N$ and by substituting $\omega = \omega_N - x$ and $\omega = \omega_N + x$ in the first and second parts, respectively, it can be shown [Nyquist, 1928; Bennett and Davey, 1965, Chap. 5] that

$$h(t) = \frac{2}{\pi} \sin \omega_N t \int_0^{\omega_N} Y(\omega_N - x) \sin tx \, dx \qquad (3.47)$$

This expression has zero values guaranteed for $\sin \omega_N t = 0$, that is, for $t = m/2f_N$, where $m = 0, 1, 2, \ldots$. Thus, as both terms of the modified impulse response equation (3.45) have zero values for integer multiples of the symbol interval (note that $mT_s = m/2f_N$), we conclude that there will be no ISI in the sampling instants.

---

* This term is similar to $\sin x / x$; however, a constant f_N is not present in the denominator. The reason for this is that the amplitude of the Dirac impulse is equal to the energy of the pulse.

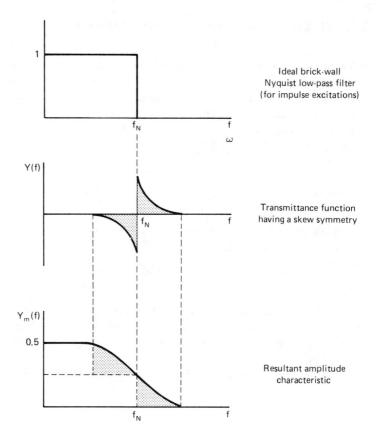

Figure 3.11 Nyquist's vestigial symmetry theorem. $\omega_N = 2\pi f_N$.

The raised cosine function satisfies Nyquist's vestigial symmetry requirement. Filter designers frequently approximate channel characteristics for ISI-free impulse transmission. The amplitude response of this channel is given by

$$|H(j\omega)| = \begin{cases} 1 & 0 \le \omega \le \dfrac{\pi}{T_s}(1-\alpha) \\[2ex] \cos^2\left\{\dfrac{T_s}{4\alpha}\left[\omega - \dfrac{\pi(1-\alpha)}{T_s}\right]\right\} & \dfrac{\pi}{T_s}(1-\alpha) \le \omega \le \dfrac{\pi}{T_s}(1+\alpha) \\[2ex] 0 & \omega > \dfrac{\pi}{T_s}(1+\alpha) \end{cases} \qquad (3.48a)$$

where $\omega = 2\pi f$ and α is the channel *roll-off* factor.

An alternative form of this raised-cosine equation, also frequently used in the literature is

$$H(f) = \begin{cases} 1 & 0 < f < f_N - f_x \\ \dfrac{1}{2}\left\{1 - \sin\dfrac{\pi}{2\alpha}\left[\dfrac{f}{f_N} - 1\right]\right\} & f_N - f_x < f < f_N + f_x \\ 0 & f > f_N + f_x \end{cases} \tag{3.48b}$$

$$\phi(f) = kf$$

where $\alpha = f_x/f_N$ is the roll-off factor and k is a constant. For practical systems that are employed for the transmission of $f_s = 1/T_s = 2f_N$ rate synchronous rectangu-

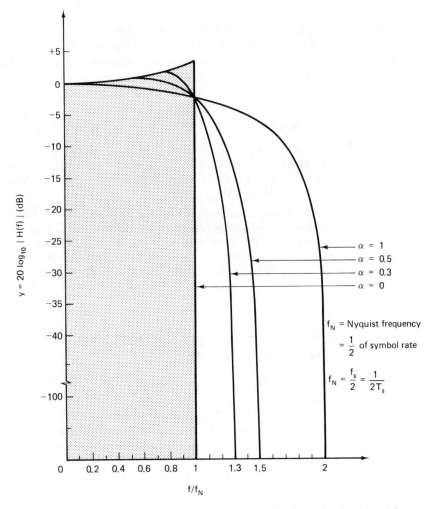

Figure 3.12 Amplitude characteristics (plotted in dB) of the Nyquist channel for rectangular pulse transmission (After [Feher, 1981], with permission from Prentice-Hall, Inc.)

lar pulses, an $(x/\sin x)$-shaped amplitude equalizer has to be added to the channel characteristics described by equation (3.48). Thus the desired channel response for ISI-free pulse transmission (such as that of NRZ signals) is given by

$$H(j\omega) = \begin{cases} \dfrac{\omega T_s/2}{\sin(\omega T_s/2)} & 0 \le \omega \le \dfrac{\pi}{T_s}(1-\alpha) \\[4mm] \dfrac{\omega T_s/2}{\sin(\omega T_s/2)} \cos^2\left\{\dfrac{T_s}{4\alpha}\left[\omega - \dfrac{\pi(1-\alpha)}{T_s}\right]\right\} & \dfrac{\pi}{T_s}(1-\alpha) \le \omega \le \dfrac{\pi}{T_s}(1+\alpha) \\[4mm] 0 & \omega > \dfrac{\pi}{T_s}(1+\alpha) \end{cases} \quad (3.49)$$

where α is the *roll-off* factor. For $\alpha = 0$ an unrealizable minimum-bandwidth filter having a bandwidth equal to $f_N = 1/2T_s$ is obtained. For $\alpha = 0.5$ a 50% excess bandwidth is used, while for $\alpha = 1$ the transmission bandwidth is double the theoretical minimum bandwidth. The amplitude characteristics for various values of the bandwidth parameter α are shown in Fig. 3.12. Theoretically, at the frequency $f = (1+\alpha)f_N$, the attenuation has an infinite value. For practical realizations an attenuation of 20 to 50 dB is specified depending on the adjacent channel interference allowance.

Example 3.3

Determine the frequency at which the theoretical raised-cosine channel has a 30-dB attenuation. Assume that the transmitted NRZ data rate is $f_s = 1$ Mb/s and that (a) $\alpha = 0.3$; (b) $\alpha = 0.5$ filters are specified.

Solution. The Nyquist frequency of the ideal brick-wall filter is at $f_N = f_s/2 = 500$ kHz. From Fig. 3.12 we see that the 30-dB attenuation point is only at a slightly lower frequency than the ∞ dB attenuation point. As a first approximation we assume that these frequencies are the same. For the $\alpha = 0.3$ case, this attenuation is at $(1 + 0.3)500$ kHz = 650 kHz, while for the $\alpha = 0.5$ case it is at $(1 + 0.5)500$ kHz = 750 kHz. Now you might want to solve equation (3.48) and obtain the exact frequencies. *Solve Problem 3.6.*

3.4.3 Nyquist's Intersymbol-Interference (ISI)- and Jitter-Free Transmission Theorem

Nyquist's minimum bandwidth and vestigial symmetry theorems stipulate the conditions for ISI-free signaling. In Fig. 3.13 the measured eye diagram of an equalized fifth-order Butterworth filter approximating a raised-cosine channel, having an $\alpha = 0.6$ roll-off factor, is illustrated. Note that in the sampling instant the ISI is virtually zero; however, the zero voltage crossings of superimposed traces that form the eye pattern do not coincide. In other words, at that instant, halfway between two adjacent pulses, the ISI is *not* negligible. The peak-to-peak deviation between the zero crossings is known as *data transition jitter,* J_{pp}. An excessive J_{pp} has harmful effects on the performance of symbol timing recovery circuits and thus, in turn, on the overall system performance. Data transition jitter tends to introduce jitter into the recovered clock. When a number of regenerative sections are cascaded, this *clock-jitter* problem

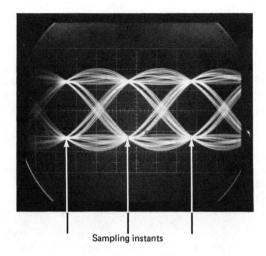

Vert: 100 mV/div
Horiz: 10 μs/div

f_s = 32 kb/s

Sampling instants

Figure 3.13 Measured eye diagram of an α = 0.6 raised-cosine channel.

cascades. In the following theorem the conditions for *simultaneous* ISI and data transition jitter-free transmission are stipulated.

Theorem. For ISI-free transmission, the null points of the ideal brick-wall low-pass filter impulse response have to be retained. To avoid data transition jitter, additional null points located halfway between adjacent null points are required. These simultaneous zero crossing conditions are satisfied by the α = 1 roll-off raised-cosine channel described by

$$H(\omega) = \frac{1}{2}\left(1 + \cos\frac{\pi\omega}{2\omega_N}\right) \qquad 0 < \omega < 2\omega_N \qquad (3.50)$$

where f_N is the Nyquist frequency, $\omega_n = 2\pi f_N$ and the signaling frequency $f_s = 1/T_s = 2 f_N$.

Note. For rectangular pulse transmission, an $x/\sin x$ equalizer has to be added to the channel.

The derivation of this theorem is fairly complex. However, the effect of data transition jitter on the probability of error performance degradation of well-designed digital satellite links is negligible in most cases.

If you are interested in the derivation of this theorem, you should consult Nyquist's original paper [Nyquist, 1928], in which a more general formulation of this theorem and a detailed derivation is presented. This paper and a number of related baseband transmission papers can be found in [Franks, 1974].

We note that for simultaneous ISI- and jitter-free impulse transmission a linear-phase channel having a bandwidth twice that of the minimum channel bandwidth is required. This you can verify by inserting $\alpha = 1$ into equation (3.48) and by simple trigonometric transformation obtaining equation (3.50). In many practical satellite channels such a wide bandwidth is not available, and therefore systems having a peak-to-peak data transition jitter on the order of only 20 to 30% of the bit duration are frequently encountered. The dependence of this jitter on the roll-off factor of several raised-cosine filters is illustrated in the computer-generated eye diagrams of Fig. 3.14. For the worst-case data pattern the peak-to-peak jitter is given by

$$J_{\text{pp}} \text{ worst case} = (x_2 - x_1) \times 100\%$$

where x_2 and x_1 are obtained by computing the roots of

$$
\begin{aligned}
s(x) - s(x - 1) &+ [s(x + 1) + s(x - 2)] - s[(x + 2) + s(x - 3)] \\
&+ [s(x + 3) + s(x - 4)] - [s(x + 4) + s(x - 5)] + \cdots = 0
\end{aligned}
\tag{3.51a}
$$

and of

$$
\begin{aligned}
s(x) - s(x - 1) &- [s(x + 1) + s(x - 2)] + [s(x + 2) + s(x - 3)] \\
&- [s(x + 3) + s(x - 4)] + [s(x + 4) + s(x - 5)] + \cdots = 0
\end{aligned}
\tag{3.51b}
$$

for $0 \le x \le 1$, where $s(x)$ is the impulse response of the raised-cosine filter given by

$$s(x) = \frac{\sin (\pi x)}{\pi x} \left[\frac{\cos (\alpha \pi x)}{1 - 4\alpha^2 x^2} \right] \tag{3.52}$$

where $x = t/T$ is the normalized time and T is the bit duration. The worst case of peak-to-peak jitter, with corresponding eye width $E_w = 1 - J_{\text{pp}}$, is given in Fig. 3.15 for different values of α. The corresponding derivation is given in [Huang et al., 1979]. *Solve Problem 3.7.*

3.4.4 Nyquist's Generalized ISI-Free Signal Transmission Theorem

The Nyquist theorems, regarding the efficient transmission of impulses, may be generalized to other pulse shapes.

Theorem. Let $Y(\omega)$ represent the transfer function of the channel which satisfies one or more Nyquist impulse transmission theorems outlined in Sections 3.4.1 to 3.4.3, and let $S(\omega)$ represent the Fourier transform of an arbitrary input (source) pulse $s(t)$. If a filter having a transfer function $M(\omega)$, which modifies $S(\omega)$ to a constant, $M(\omega) = 1/S(\omega)$, is cascaded with $Y(\omega)$, then the response to the arbitrary $s(t)$ pulse of the $M(\omega)Y(\omega)$ channel satisfies the same Nyquist theorem(s) as does the $Y(\omega)$ channel for an impulse excitation.

Note. The $x/\sin x$ amplitude equalizer mentioned earlier is a particular application of this generalized theorem.

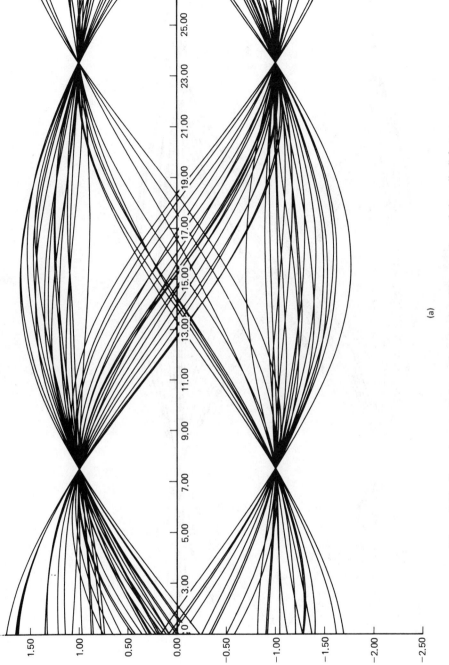

Figure 3.14 Computer-generated eye diagrams for different values of the roll-off factor α. Eye diagram is obtained at the output of the receive filter. The cascaded channel filtering for the NRZ data stream consists of an $x/\sin x$ amplitude equalizer, a transmit, and a receive filter. (a) $\alpha = 0.3$; (b) $\alpha = 0.5$; (c) $\alpha = 1$.

(a)

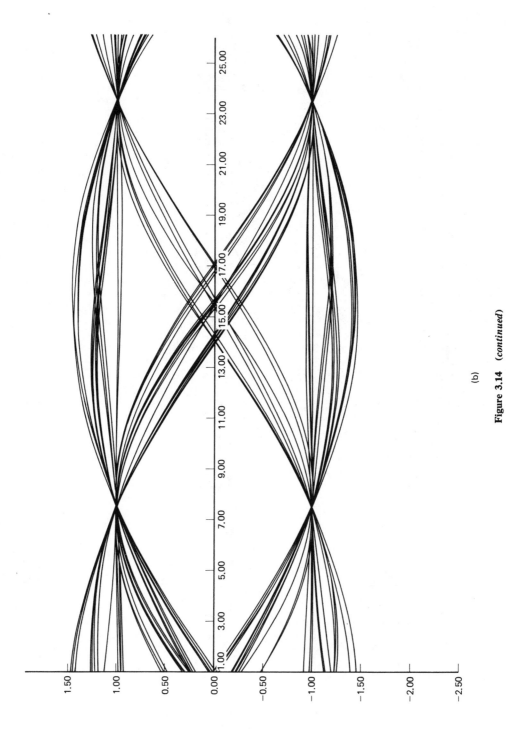

Figure 3.14 (*continued*)

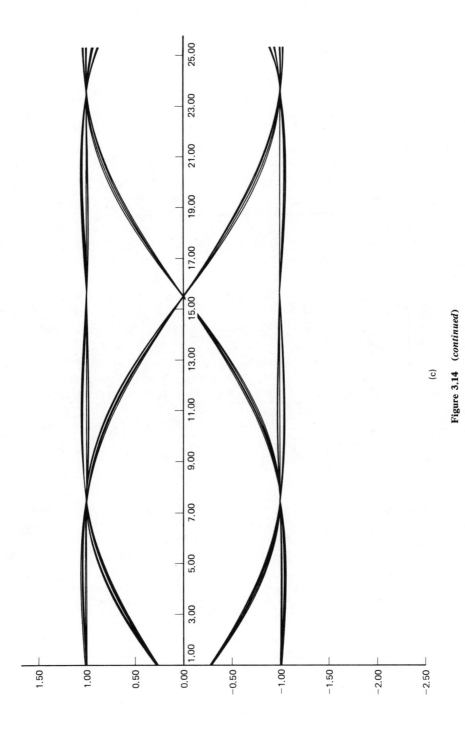

Figure 3.14 (continued)

(c)

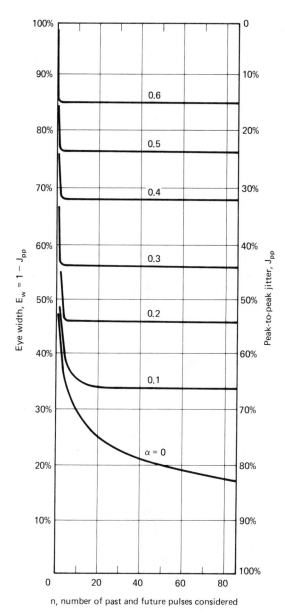

Figure 3.15   Worst-case peak-to-peak data transition jitter. (After [Huang, 1979], with permission.)

The interpretation and application of this theorem is illustrated in the following example.

Example 3.4

A synchronous $f_s = 10$ Mb/s random data source is shown in Fig. 3.16. Derive the transfer functions of the channel that produces ISI-free transmission for both the minimum bandwidth and the 50% excess bandwidth cases. Note that this signaling format is used in the baseband of **MSK** modulated systems, described in Chapter 4. In *Problem 3.8* a derivation of the power spectral density function of this baseband signal is suggested.

Solution.   The Fourier transform of the $g_1(t) = h \cos (\pi t / T_s)$ signaling element is

$$G_1(f) = \mathscr{F}[g_1(t)] = \int_{-T_s/2}^{T_s/2} h \cos \frac{\pi t}{T_s} e^{-j2\pi ft} \, dt = \frac{2hT_s}{\pi} \frac{\cos \pi f T_s}{1 - 4f^2 T_s^2} \tag{3.53a}$$

We note that $G_2(f) = -G_1(f)$. This amplitude spectrum has its first zero-amplitude point at a frequency equal to $1.5/T_s$ and its phase spectrum is always zero. Following the generalized ISI-free signal transmission theorem and also [Morais and Feher, 1979] we modify the $G_1(f)$ spectrum by $M(f)$ so as to obtain a constant spectrum. Thus the required modification (prefiltering) is

$$M(f) = \frac{1}{G_1(f)} = \left[\frac{1 - (2fT_s)^2}{\cos (\pi f T_s)} \right] \tag{3.53b}$$

To obtain both the minimum and 50% excess bandwidth cases we use the raised-cosine channel model, equation (3.48b), with $\alpha = 0$ and $\alpha = 0.5$ respectively. The resultant channel characteristics $F(f)$, consisting of the cascade of the $M(f)$ filter and the $H(f)$ raised cosine filter, is given by

$$F(f) = \begin{cases} \dfrac{1 - (f/f_N)^2}{\cos (\pi f/2f_N)} & 0 < f < f_N - f_x \\[3mm] \dfrac{1}{2} \left\{ 1 - \sin \dfrac{\pi}{2\alpha} \left(\dfrac{f}{f_N} - 1 \right) \right\} \dfrac{1 - (f/f_N)^2}{\cos (\pi f/2f_N)} & f_N - f_x < f < f_N + f_x \\[3mm] 0 & f > f_N + f_x \end{cases} \tag{3.54}$$

$$\phi(f) = kf$$

We note that when $\alpha = f_x/f_N = 0$, the spectrum shape is "brick-wall" in nature and also that its bandwidth is f_N Hz. This is the same condition as for the minimum-bandwidth NRZ transmission case; that is, the $\pm h \cos (\pi t / T_s)$ pulses can be transmitted with a spectral efficiency of 2 b/s/Hz. This pulse train is applied in MSK modulated systems, described in Chapter 4. The $F(f)$ channel amplitude characteristics for the $\alpha = 0$, $\alpha = 0.5$, $\alpha = 0.75$, and $\alpha = 1$ cases are illustrated in Fig. 3.17. For the $f_s = 10$ Mb/s transmission rate, f_N equals 5 MHz.

∎

Solve Problem 3.8: In this problem the spectral derivation of a 32-kb/s MSK signal is suggested. Your solution should confirm that the power spectral density of the sinusoidally shaped **baseband MSK** signal is

$$w_s(f) = \frac{4h^2 T_s}{\pi^2} \frac{1 + \cos 2\pi f T_s}{(1 - 4f^2 T_s^2)^2} \tag{3.55}$$

where h and T_s are as defined in Fig. 3.16(a).

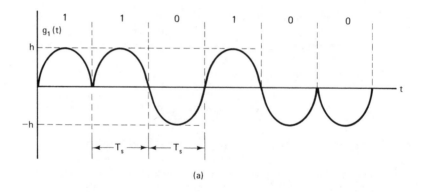

(a)

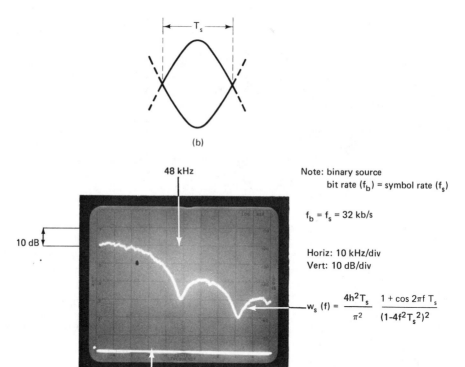

48 kHz

Note: binary source
bit rate (f_b) = symbol rate (f_s)

$f_b = f_s$ = 32 kb/s

10 dB

Horiz: 10 kHz/div
Vert: 10 dB/div

$$w_s(f) = \frac{4h^2T_s}{\pi^2} \frac{1 + \cos 2\pi f\, T_s}{(1-4f^2T_s^2)^2}$$

Symbol rate frequency
(c)

Figure 3.16 Sinusoidal baseband signaling format used in MSK modulated systems. (a) Segment of an equiprobable data pattern. (b) Eye diagram—infinite bandwidth case. $T_s = 1/f_s = 100$ ns. (c) Measured power spectral density function. (Courtesy of Digital Communiations Research Laboratory, University of Ottawa.) [*Solve Problem 3.8*]

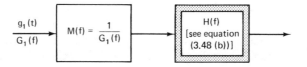

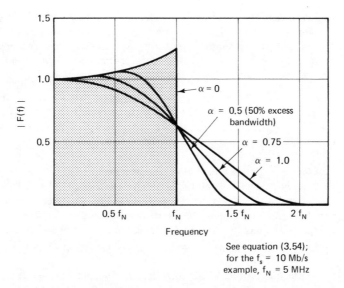

See equation (3.54);
for the f_s = 10 Mb/s
example, f_N = 5 MHz

Figure 3.17   Filtering necessary for a full-length sinusoidal pulse train to create raised-cosine ISI-free pulse at filter output. This pulse train is applied to MSK modulated systems, described in Chapter 4. (After [Morais and Feher, 1979], with permission from the IEEE, © 1979.)

3.5 FILTERING AND EQUALIZATION TECHNIQUES

Detailed study of design techniques of spectrally efficient channel filters which approach the Nyquist ISI-free transmission criteria is complex and beyond the scope of this text. In this section a *review* of the basic concepts of the classical digital transversal method and also of the nonlinearly switched filter and equalization techniques that have been used successfully in modern digital systems is presented. The numerous references in the following subsections provide detailed filter analysis and design procedures.

3.5.1 Classical (Passive and Active) Filters

The design and analysis of *classical* filters consisting of passive *LC* elements and/or operational amplifiers used in a feedback configuration with external capacitors is already well documented in books, such as [Wait et al., 1975; Irvine, 1981; Weinberg, 1957; Kuh and Pederson, 1959; Williams, 1975; and Christian and Eisenmann, 1966]. In these references the design of Butterworth, Chebyshev, Bessel, Butterworth–Thomson, elliptic, and other filters and their associated characteristics are described. However, these books are devoted to filter design and do not relate these filter characteristics to the Nyquist channel requirements. The following example illustrates a practical filter design procedure as applied to the synthesis of Nyquist channel characteristics.

Example 3.5

Design a baseband filter having a channel bandwidth within 50% of the theoretical minimum-bandwidth Nyquist channel. Assume that the transmitted NRZ data rate is $f_s = 32$ kb/s and that the designed filter is the only spectral shaping filter in the system. Your customer, based on his knowledge of the overall system interference budget (see Chapter 4) specified a 30-dB attenuation requirement for all out-of-band frequencies.

Solution. For ISI-free, or negligible ISI channel shaping we attempt to design a raised-cosine filter. To satisfy the 30-dB attenuation requirement at 50% above the minimum bandwidth Nyquist frequency (i.e. at $1.5 f_N = 1.5 \times 16$ kHz $= 24$ kHz), an $\alpha = 0.5$ roll-off filter design is required. Steps to follow:

1. Plot the $\alpha = 0.5$ theoretical filter characteristic.
2. Find in a filter handbook the amplitude characteristic of a classical filter that approximates the theoretical $\alpha = 0.5$ response. In this example the filter chosen should approximate the theoretical curve up to the 30-dB attenuation point.
3. Design the required $(x/\sin x)$-shaped amplitude and phase equalizers.

  ∎

The schematic diagram of a fourth-order Butterworth low-pass filter which approximates an $\alpha = 0.5$ raised-cosine filter is given in Fig. 4.59. A very general computer program for the synthesis of passive LC and active RC filters, detailed guidelines, and practical examples are described in [Szentirmai, 1977].

3.5.2 Transversal Filters and Equalizers

The Nyquist theorems stipulate the conditions for ISI- and jitter-free transmission of synchronous digital pulse streams. The shape of the pulse response can be specified either in the time domain or in the frequency domain by its Fourier transform. In the preceding section, dealing with classical filters, a design approach to satisfy the Nyquist ISI-free transmission theorem in the *frequency domain* (i.e., based on the Fourier transform of the input signal and channel transfer function) was described. In this and in the following section we present pulse-shaping techniques in the *time domain*.

Binary transversal filters (BTF). The principle of operation of a binary transversal filter is illustrated in Fig. 3.18. Let us assume that the desired pulse response is represented by the bold continuous line $\tilde{p}_t(t)$. Note that at the sampling

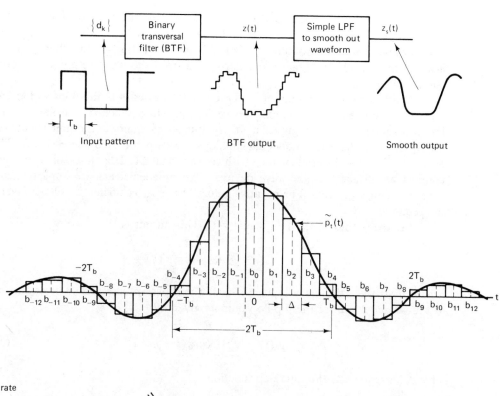

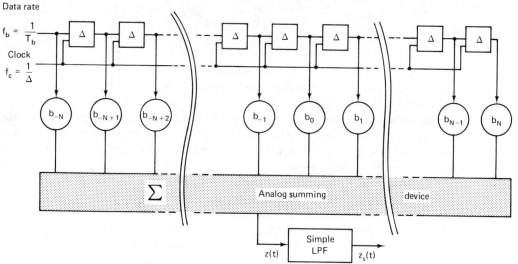

Figure 3.18 Transversal filter and corresponding pulse response. (After [Shanmugam, 1979], with permission from John Wiley & Sons, Inc., © 1979.)

time denoted zero the pulse has its *full* value and at the integer multiples of the bit duration $\pm T_b$, $\pm 2T_b$, . . . , it is *zero*. These conditions assure ISI-free transmission. The conventional low-pass filter (LPF), following the transversal filter, is merely inserted to remove the higher-order harmonics from the pulse response, smoothing out the waveform.

A simple implementation of a BTF consists of a series of flip-flops and attenuators, with a single analog summing device. The flip-flops, clocked at a rate $f_c = 1/\Delta$, are in a shift register configuration. In the illustrated example the unit bit duration T_b equals 4Δ, where the individual flip-flops are used for each successive Δ. These elements could also be implemented with conventional LC delay lines, or with charge-coupled or surface acoustic-wave devices. The b_n coefficients are implemented by resistive voltage dividers. For negative values of the b_n coefficients, voltage inverters are required.

Let the output of the BTF equal $z(t)$. This output is

$$z(t) = \sum_k d_k p_r(t - kT_b) \tag{3.56}$$

where d_k is the kth binary input bit having two permitted states, 0 or 1, and $p_r(t)$ is the weighted sum of the flip-flop outputs given by

$$p_r(t) = \sum_{m=-N}^{m=N} b_n p_u(t - m\Delta) \tag{3.57}$$

and $p_u(t)$ represents the unit step function

$$p_u(t) = \begin{cases} 1 & -\Delta/2 \leq t < \Delta/2 \\ 0 & \text{elsewhere} \end{cases} \tag{3.58}$$

From Fig. 3.18 it is evident that the designed pulse response $p_r(t)$ may provide a better approximation to the desired pulse response if the ratio of the clock frequency to data rate f_c/f_b is a larger integer.

In our example, we assume that it is sufficient to approximate the desired pulse response only in the $-n\Delta = -3T_b$ to $+n\Delta = +3T_b$ interval, where $n = 12$, and that outside this interval of 24Δ, the values of the pulse response are sufficiently small, and may thus be ignored.

Successive bits are read into the shift register at a rate of f_b seconds (i.e., once every T_b seconds). The output is obtained by overlapping the successive input bits and by superposition of the individual pulses. The multilevel staircase-like $z(t)$ signal is finally smoothed in the simple output low-pass filter. The ISI generated by the transversal filter is due both to the finite-length approximation of the pulse response and to the assumption that the values of the higher-order lobes are zero. Recall that in our example this length equals to $6T_b$.

The BTF has the property that it maintains the output pulse shape independently of the incoming bit rate. If, for example, the coefficients of a BTF are chosen such as to approximate the pulse response of an $\alpha = 0.3$ raised-cosine filter, this approxima-

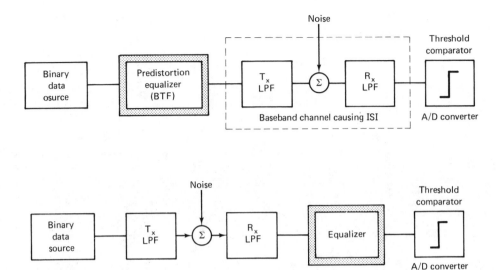

Figure 3.19 Predistortion equalizer and receive equalizer in baseband systems.

tion will hold with the same accuracy (independently of the bit rate) as long as the f_c/f_b ratio remains invariant. This important property has useful applications in variable-bit-rate modem design for satellite communications.

Transversal equalizers.   Transversal amplitude and phase equalizers have a similar configuration to binary transversal filters. The pulse response and consequently the ISI (of a practical channel which does not adhere to the Nyquist transmission criteria) may be equalized by transversal equalizers. These equalizers, if located in front of the transmission system, are known as *predistortion* equalizers; if located at the receive end, they are known simply as equalizers (see Fig. 3.19). Predistortion equalizers may use flip-flops for the implementation of the delay elements. This cost-effective hardware design is feasible in this case because the source information is in a binary form. Equalizers located at the receive end lend themselves more easily to automatic adaptive structures.

The principles and applications of many linear and nonlinear, preset, and auto-adaptive transversal equalizers are described in [Lucky et al., 1968, Chap. 6, and Clark, 1977, Chap. 4]. Design techniques particularly suited for satellite systems applications are presented in [Feher, et al., 1977, and Feher and DeCristofaro, 1976]. Here we present the principle of operation of a simple *zero-forcing* equalizer. We follow the derivation presented in [Shanmugam, 1979, Chap. 5].* This derivation is a simplified version of the one originally presented by R. W. Lucky.

The block diagram of a transversal equalizer is shown in Fig. 3.20. Let the received unequalized pulse be $p_r(t)$. In most applications this is a bandlimited pulse

---

* This material is presented by permission of John Wiley & Sons, Inc., New York; copyright 1979.

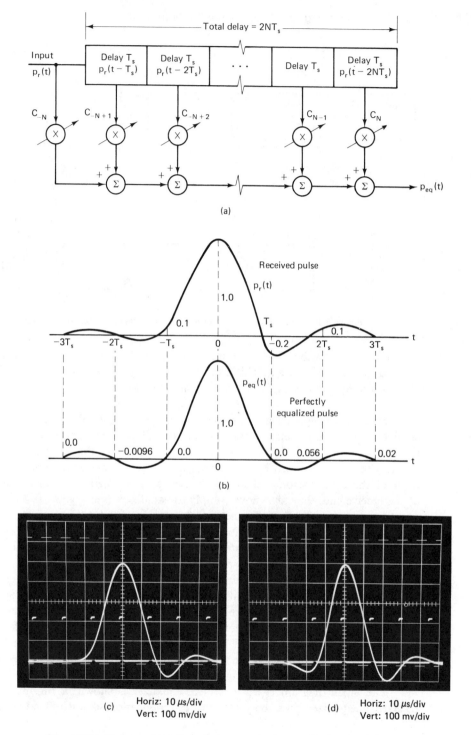

Figure 3.20 Transversal equalizer diagram, received unequalized and equalized pulse shapes. (a) Transversal equalizer. (b) Three-tap equalizer response. (c) Measured unequalized pulse. (d) Measured partly equalized pulse. (After [Shanmugam, 1979], with permission from John Wiley & Sons, Inc. © 1979.)

which has nonzero values at integer multiples of the signaling interval T_s. Thus a sequence of consecutive pulses, received in a synchronous data stream, leads to an eye diagram containing ISI. This bandlimited distorted signal is fed to the equalizer input. Note that flip-flops and conventional digital shift registers do not provide accurate delays for severely bandlimited signals. Instead, analog delay lines such as charge-coupled devices or sections of long transmission lines are used in current designs.

The equalized pulse response $p_{eq}(t)$ is obtained by first multiplying the received unequalized $p_r(t)$ pulse and its nT_s delayed replicas by the corresponding C_n coefficients and then summing these products (see Fig. 3.20).

$$p_{eq}(t) = \sum_{n=-N}^{N} C_n p_r[t - (n + N)T_s] \tag{3.59a}$$

We assume that $t = 0$ is defined for the time instant when $p_r(t)$ has its peak. In this case, sampling decisions as to whether a logic state zero or a logic state one was transmitted during a particular symbol interval are taken at $t_k = (k + N)T_s$. Thus

$$p_{eq}(t_k) = \sum_{n=-N}^{N} C_n p_r[(k - n)T_s] \tag{3.59b}$$

Using the conventional shorter notation $p_r(n)$ instead of $p_r(nT_s)$ and k instead of t_k, we have

$$p_{eq}(k) = \sum_{n=-N}^{N} C_n p_r(k - n) \tag{3.60a}$$

The value of $p_{eq}(k)$ is

$$p_{eq}(k) = \begin{cases} 1 & \text{for } k = 0 \\ 0 & k = \pm 1, \pm 2, \cdots, \pm N \end{cases} \tag{3.60b}$$

The equalized pulse will have a maximum value for $k = 0$ and is forced to equal 0 for the N preceding and N following decision instants, thus the name *zero-forcing equalizer*. This equalizer removes the ISI in $2N$ sampling instants, where N is the number of taps assuming that prior to equalization the eye is not completely closed. For an ideal equalizer N would have to be very large. Practical equalizers have taps in the range of 2 to several hundreds.

From equations (3.60a) and (3.60b), we obtain

$$\left. \begin{matrix} N \text{ zeros} \left\{ \vphantom{\begin{matrix}0\\0\\\vdots\\0\end{matrix}} \right. \\ \\ N \text{ zeros} \left\{ \vphantom{\begin{matrix}0\\\vdots\\0\end{matrix}} \right. \end{matrix} \right. \begin{bmatrix} 0 \\ 0 \\ \vdots \\ 0 \\ 1 \\ 0 \\ \vdots \\ 0 \end{bmatrix} = \begin{bmatrix} p_r(0) & p_r(-1) & \cdots & p_r(-2N) \\ p_r(1) & p_r(0) & \cdots & p_r(-2N+1) \\ p_r(2) & p_r(1) & \cdots & \\ & & & \\ p_r(2N) & \cdots & \cdots & p_r(0) \end{bmatrix} \begin{bmatrix} C_{-N} \\ C_{-N+1} \\ C_0 \\ C_{N-1} \\ C_N \end{bmatrix} \tag{3.61}$$

The C_n tap coefficients are the solutions of these $(2N + 1)$ simultaneous equations.

The sketch of the unequalized received pulse, Fig. 3.20, indicates significant signal values at the $-T_s$, $+T_s$, $+2T_s$, and $+3T_s$ sampling instants, which all contribute to the ISI. In the sketch of the equalized pulse it is assumed that all sampled values (with the exception of the desired value at the $t = 0$ instant) are forced to zero. The photographs of a laboratory-observed pulse illustrate a different example, where we note that both the unequalized and the equalized pulse responses have finite values at the sampling instants. The sampling instants are at the rising edges of the clock pulses. In this example the designed equalizer did not achieve a full zero forcing of the undesired sample values. This equalizer reduced the ISI degradation from 3 dB to 1 dB [Prendergast, 1981]. The approximate values of the pulse response are read from the photographs:

Sampling instant	$-2T_s$	$-T_s$	0	$+T_s$	$+2T_s$	$+3T_s$
Unequalized value (mV)	0	+50	+500	+140	−40	+20
Equalized value (mV)	0	−30	+500	+80	−20	+10

Solve Problem 3.9.

3.5.3 Nonlinearly Switched Filters

In a search for simpler, cost-effective hardware designs, a nonlinear switching filter has been discovered [Huang et al., 1979; Feher, 1979; Feher, 1976]. With a family of these filters known as *Feher's nonlinear processor (or filter)*, *bandlimited signals that do not contain ISI and are jitter-free* have been generated. The principle of operation of one of these filters is illustrated in Fig. 3.21.

For a −1 to +1 transition of the unfiltered NRZ input signal the rising segment of a sinusoid is switched on (connected to the transmission medium). For a +1 to −1 transition the falling segment of a sinusoid is switched on. The 0° and the 180° reference sinusoidal generators, shown in Fig. 3.21(b), provide the required sinusoidal waves. For a continuous sequence of 1's or −1's (more than one input bit without transition), a positive or negative dc segment is switched on. The decision logic provides the switch position control signals. In Fig. 3.22 the measured bandwidth-efficient, jitter- and ISI-free output eye diagram and spectrum are shown. The power spectrum of the processed signal is

$$S(x) = T\left(\frac{\sin 2\pi x}{2\pi x} \frac{1}{1 - 4x^2}\right)^2 \tag{3.62}$$

where $x = fT$ [Le-Ngoc and Feher, 1980].

(*Note.* The difference between the eye diagram and the power spectrum of this signaling format, and that of the MSK shown in Fig. 3.16.)

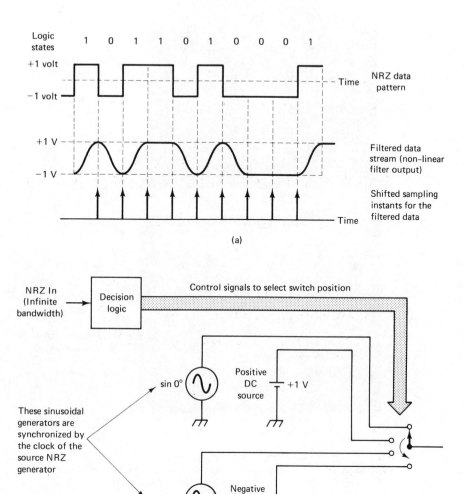

Figure 3.21 Principle of operation of Feher's bandwidth-efficient, ISI- and jitter-free nonlinear filter. (a) Generated waveshapes. (b) Block diagram. (After [Feher, 1981], with permission from Prentice-Hall, Inc.)

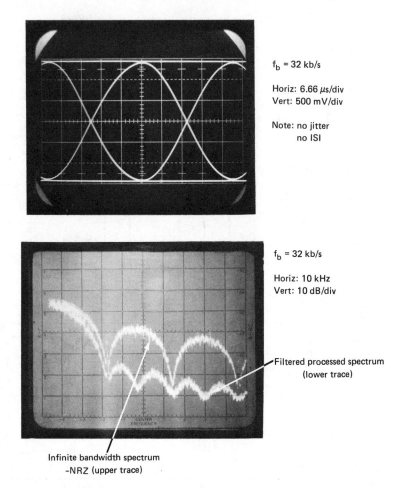

f_b = 32 kb/s

Horiz: 6.66 µs/div
Vert: 500 mV/div

Note: no jitter
       no ISI

f_b = 32 kb/s

Horiz: 10 kHz
Vert: 10 dB/div

Filtered processed spectrum
(lower trace)

Infinite bandwidth spectrum
-NRZ (upper trace)

Figure 3.22   ISI- and jitter-free eye diagrams and power spectrum of Feher's nonlinear filter. (After [Feher, 1981], with permission from Prentice-Hall, Inc.)

3.6 PROBABILITY OF ERROR, P_e, PERFORMANCE IN AN ADDITIVE WHITE GAUSSIAN NOISE (AWGN) ENVIRONMENT

In this section a simple derivation of the probability of error, P_e, performance of binary baseband systems for an additive white Gaussian noise (AWGN) channel model is presented. *Ideal raised-cosine filtering is assumed;* thus the intersymbol-interference (ISI)-caused degradation is negligible. A rigorous derivation for an optimally matched filter (correlation receiver) and its relation to the performance of the Nyquist-shaped channel is postponed until Sections 4.4 and 4.5. The performance degradations introduced by imperfect channel characteristics (i.e., channels having a significant ISI) are described in Section 4.10.

A random equiprobable NRZ data segment is illustrated in Fig. 3.23. The waveshapes of the infinite bandwidth polar signal, of the ISI-free bandlimited noise-free signal and of the bandlimited signal degraded by noise are sketched. In this example, at sampling instants 4 and 8 the noise amplitude exceeds the signal amplitude and has the opposite polarity, and thus is causing errors.

At the mth sampling instant the filtered input to the **threshold comparator** (1-bit A/D converter having a threshold level equal to zero volts) is described by

$$r(t_m) = A_m + n_o(t_m) \qquad t_m = (m + 0.5)T_b; \quad m = 0, 1, 2, \cdots \qquad (3.63)$$

where $A_m = \pm A$ represents the signal sample, for the ISI-free case, and $n_o(t_m)$ represents the noise sample. The threshold comparator provides at T_b intervals a 1 or 0 output state, depending on whether $r(t_m)$ is larger or smaller than zero, respectively. Let b_m represent the mth input bit; then the probability of error is given by

$$\begin{aligned} P_e &= P[r(t_m) > 0 \,|\, b_m = -1]P[b_m = -1] \\ &+ P[r(t_m) < 0 \,|\, b_m = 1]P[b_m = 1] \end{aligned} \qquad (3.64)$$

where $P[r(t_m) > 0 \,|\, b_m = -1]$ is the conditional probability of having a received sample larger than zero, given that at the mth sampling instant a -1 state is transmitted. For the equiprobable case of 0 and 1 logic states (-1 and $+1$ input levels), this expression becomes

$$P_e = \tfrac{1}{2}\{P[|n_o(t_m)| > A]\} \qquad (3.65)$$

The *probability density function of the Gaussian noise* is given by

$$p(v) = \frac{1}{\sigma\sqrt{2\pi}}\, e^{-(v - V_0)^2/2\sigma^2} \qquad (3.66)$$

where σ is the root-mean-square value of the noise voltage and V_o is the mean or dc value. For thermal noise this value is typically zero; hence the probability of error [substituting x for $n_o(t_m)$] is given by

$$P_e = \frac{1}{2}\int_{|x|>A}^{\infty} \frac{1}{\sqrt{2\pi N_T}}\, e^{-v^2/2N_T}\, dv \qquad (3.67)$$

$$= \int_{A}^{\infty} \frac{1}{\sqrt{2\pi N_T}}\, e^{-v^2/2N_T}\, dv$$

where N_T is the total noise power at the threshold comparator input, that is, $N_T = \sigma^2$, and is given by

$$N_T = \int_{0}^{\infty} N_o(f)\,|H_R(f)|^2\, df \qquad (3.68)$$

In this equation $N_o(f)$ is the single-sided noise density, and $H_R(f)$ is the normalized transfer function of the receive filter (see Fig. 3.24).

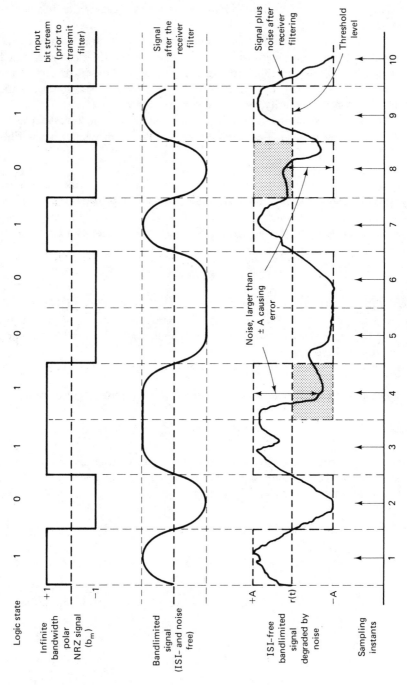

Figure 3.23 Noise-caused errors in baseband systems.

Logic state

Input bit stream (prior to transmit filter)

Infinite bandwidth polar NRZ signal (b_m)

Signal after the receiver filter

Bandlimited signal (ISI- and noise free)

Signal plus noise after receiver filtering

Threshold level

ISI-free bandlimited signal degraded by noise $r(t)$

Noise, larger than $\pm A$ causing error

Sampling instants

122

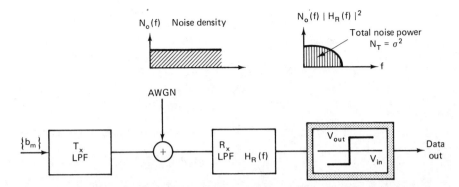

Figure 3.24   Noise in a binary baseband Nyquist channel. At the sampling instants the signal equals $+A$ or $-A$ volts. The rms noise voltage is σ volts. The noise bandwidth is limited by the receive low-pass filter. If optimal ISI-free raised-cosine filters are used, the total signal power equals A^2. (*Solve Problem 3.10.*)

A change of variable $\omega = v/\sqrt{N_T}$ in (3.67) gives

$$P_e = \int_{A/\sqrt{N_T}}^{\infty} \frac{1}{\sqrt{2\pi}} e^{-\omega^2/2}\, d\omega = Q\left(\frac{A}{\sqrt{N_T}}\right) = Q\left(\frac{A}{\sigma}\right) \tag{3.69}$$

where

$$Q(y) = \int_{y}^{\infty} \frac{1}{\sqrt{2\pi}} e^{-\omega^2/2}\, d\omega \tag{3.70}$$

The $Q(y)$ function is tabulated (see Table 3.1) and sketched in Fig. 3.25. As a frequently used reference point we should remember that for a $P_e = 10^{-4}$, a *peak* signal-to-rms noise ratio $A/\sigma = 3.7$ (i.e., 11.4 dB is required). The rms signal-to-rms noise ratio required for $P_e = 10^{-4}$ is dependent on the raised-cosine channel filter partitioning. The derivation of optimal receive filters which satisfy the Nyquist ISI-free transmission criteria is a fairly complex task beyond the scope of this text. If interested in the derivations, you should study [Bennett and Davey, 1965, Chap. 7, and Morais and Feher, 1979]. The final result: If the rms signal-to-noise ratio for the optimal receive filter ($\alpha = 1$ roll-off channel) specified *at the threshold comparator input* is

$$(S/N)_{\text{rms}_{\text{opt}}} = 10.2 \text{ dB} \quad \text{then} \quad P_e = 10^{-4} \tag{3.71}$$

We have already mentioned that noise samples exceeding the signal level A may cause errors. A time-domain illustration is given in Fig. 3.23 and an illustration in the probability domain in Fig. 3.26. The shaded area represents the P_e for a transmitted -1 state signal. For equiprobable data the P_e for a transmitted $+1$ state signal is identical to the case of a transmitted -1 signal. *The signal power at the*

TABLE 3.1 THE $Q(x)$ AND erfc (y) FUNCTIONS

Definition: $Q(y) = \int_y^\infty \frac{1}{\sqrt{2\pi}} e^{-z^2/2}\, dz$

(1) $P(X > \mu_x + y\sigma_x) = Q(y) = \int_y^\infty \frac{1}{\sqrt{2\pi}} e^{-z^2/2}\, dz$

(2) $Q(0) = \frac{1}{2};\quad Q(-y) = 1 - Q(y),\quad$ when $y \geqslant 0$

(3) $Q(y) \approx \frac{1}{y\sqrt{2\pi}} e^{-y^2/2}$ when $y \gg 1$ (approximation may be used for $y > 4$)

(4) $\text{erfc}^a\,(y) \overset{\Delta}{=} \frac{2}{\sqrt{\pi}} \int_y^\infty e^{-z^2}\, dz = 2Q(\sqrt{2}y),\ y > 0$

(5) $\text{erfc}\,(y) = 1 - \text{erf}\,(y)$

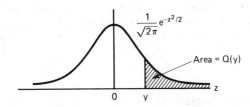

y	$Q(y)$	y	$Q(y)$	y	$Q(y)$	$Q(y)$	y
0.05	0.4801	1.05	0.1469	2.10	0.0179		
0.10	0.4602	1.10	0.1357	2.20	0.0139		
0.15	0.4405	1.15	0.1251	2.30	0.0107		
0.20	0.4207	1.20	0.1151	2.40	0.0082	10^{-3}	3.10
0.25	0.4013	1.25	0.0156	2.50	0.0062		
0.30	0.3821	1.30	0.0968	2.60	0.0047		
0.35	0.3632	1.35	0.0885	2.70	0.0035	$\dfrac{10^{-3}}{2}$	3.28
0.40	0.3446	1.40	0.0808	2.80	0.0026		
0.45	0.3264	1.45	0.0735	2.90	0.0019		
0.50	0.3085	1.50	0.0668	3.00	0.0013		
0.55	0.2912	1.55	0.0606	3.10	0.0010	10^{-4}	3.70
0.60	0.2743	1.60	0.0548	3.20	0.00069		
0.65	0.2578	1.65	0.0495	3.30	0.00048		
0.70	0.2420	1.70	0.0446	3.40	0.00034		
0.75	0.2266	1.75	0.0401	3.50	0.00023		
0.80	0.2119	1.80	0.0359	3.60	0.00016	$\dfrac{10^{-4}}{2}$	3.90
0.85	0.1977	1.85	0.0322	3.70	0.00010		
0.90	0.1841	1.90	0.0287	3.80	0.00007	10^{-5}	4.27
0.95	0.1711	1.95	0.0256	3.90	0.00005		
1.00	0.1587	2.00	0.0228	4.00	0.00003	10^{-6}	4.78

[a] In some references the error function is somewhat differently defined.

Source: [Shanmugam, 1979], with permission from John Wiley & Sons, Inc.

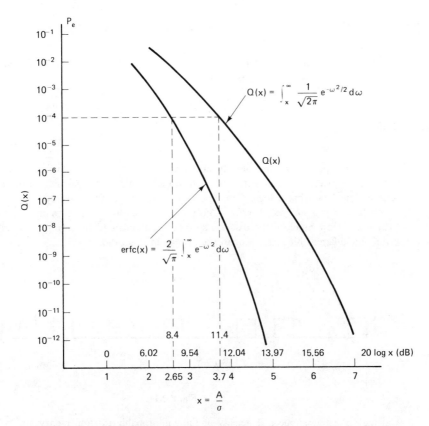

Figure 3.25 $P_e = f(S/N)$ of binary baseband signals. A, peak signal voltage; σ, rms noise voltage at threshold comparator input.

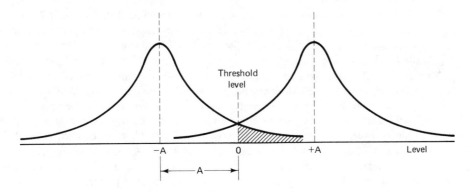

Figure 3.26 Probability of error distribution—baseband systems. A more general derivation of optimum receivers is given in Section 4.4 (see Figs. 4.9 and 4.10). Shaded area represents P_e surface for a transmitted -1 state. If there were no noise on the line, the received signal sample would equal $-A$ volts. To make an error, the noise has to be sufficiently large to change the polarity of the signal; that is, it has to be positive and larger than A.

receive filter input equals A^2. This power is independent of the roll-off factor α of the Nyquist channel. Solve Problem 3.10.

PROBLEMS

3.1. Calculate the power spectral density of a return-to-zero (RZ) random data stream as shown in Fig. P.3.1. Assume that the illustrated signal was measured across a 50-Ω termination. Note that a logic state 1 is represented by a 50% duty cycle pulse and a 0 state by a zero signal [see also Fig. 3.4(c)]. Plot the computed spectrum on a logarithmic (dB) scale. How much is the discrete spectral component to the continuous spectrum ratio? For different bit rates, explain the change in this ratio.

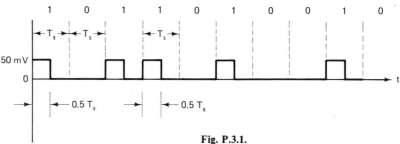

Fig. P.3.1.

3.2. Calculate the power spectral density of the equiprobable random binary bit stream of Fig. 3.4(f). This type of signaling is used in the baseband in MSK modulators. Does the spectrum of this baseband signal contain discrete spectral lines? Plot the power spectral density for an $f_s = 100$ kb/s (half-rate) signaling element. Compare this spectrum with the spectrum of a 100-kb/s NRZ signal.

3.3. Derive the output response of a brick-wall linear phase filter if the input is a T_s duration pulse having an amplitude of A volts. Are the zero crossings of the output pulse response at integer multiples of T_s? Assume that the cutoff frequency of the filter is (a) $1/2T_s$; (b) $20/T_s$.

3.4.* Assume that you have access to the received eye diagram and to an external jitter free clock of an $f_s = 10$ Mb/s system. This system operates in a high S/N environment (i.e., the noise is negligible). Present a functional diagram and describe apparatus that would measure the rms intersymbol interference and the rms data transition jitter. For additional ideas, you may consult [Feher, 1981, Chap. 11, and/or Feher, et al., 1977].

3.5. Obtain the Fourier transform of an impulse and of a T_s-second-duration rectangular pulse. Note carefully the difference in the amplitudes of these Fourier transforms. Plot the absolute value of the transfer function of an amplitude equalizer (channel) that will yield the same output pulse response as the brick-wall impulse response. Assume that the signaling rate is $f_s = 10$ Mb/s.

3.6. Calculate the exact frequency at which the raised-cosine channel has a 30-dB attenuation. Assume $\alpha = 0.3$ and $\alpha = 0.5$ roll-off filters and that the Nyquist frequency is at 500

* Advanced problem.

kHz. Compare your results with the ones obtained by the approximate method of Example 3.3.

3.7. Obtain and sketch the impulse response of raised cosine filters having $\alpha = 0$, $\alpha = 0.5$, and $\alpha = 1$ roll-off factors. Based on the obtained impulse responses, explain why a steeper filter has a larger peak-to-peak jitter. Explain why the ISI is not affected by the roll-off factor? Note that a synchronous stream of impulses (not pulses) is assumed.

3.8. Derive the power spectral density for an equiprobable random data stream, such as shown in Fig. 3.16. The $\pm h \cos (\pi t / T_s)$ signaling elements are used at baseband in MSK modulated systems, described in Chapter 4. Sketch the power spectral density in dB and compare your result to the measured power/spectral density result in Fig. 3.16. Assume that the transmitted symbol rate is 32 kb/s. *Hint:* Use equation (3.28).

3.9.* The received unequalized pulse response of a fifth-order Butterworth filter having a 3-dB cutoff frequency is shown in Fig. 3.20. The transmission rate equals $f_s = 64$ kb/s. Calculate the tap settings of a five-stage zero-forcing transversal equalizer. Sketch the equalized pulse response. Compare your sketch with the measured equalized pulse of Fig. 3.20.

3.10. The transfer function of the transmit and receive low-pass filters illustrated in Fig. 3.24 meet the Nyquist intersymbol-interference-free transmission criteria. Assume that for low frequencies, the attenuation of this baseband Nyquist channel is 0 dB. Let the transmit and receive filters be identical, with the exception of the $x/\sin x$ amplitude equalizer. This equalizer is an integral part of the transmit filter, as shown in Fig. 4.12(a). If the input data sequence $\{b_m\}$ is an equiprobable random non-return-to-zero data stream having levels of $+A$ or $-A$ volts, show that the power, measured at the receive filter input, equals A^2. Assume a normalized 1-Ω impedance. Show that this power is independent of the roll-off factor, α, of the raised-cosine channel model. Give a physical interpretation of your results. Show that the power at the receive filter output (threshold comparator input) equals $A^2[(4 - \alpha)/4]$.

4

POWER EFFICIENT MODULATION
TECHNIQUES FOR LINEAR AND NONLINEAR
SATELLITE CHANNELS

4.1 INTRODUCTION

A number of digital modulation techniques have found applications in communications satellite systems. Binary and four-state modulation techniques are more *power efficient* than the higher-state or "higher-ary" (8-state, 16-state, . . .) modulated systems, and for this reason have been more frequently employed. The two- and four-state power-efficient systems have a satisfactory probability of error, P_e, performance, even if the received carrier-to-noise, C/N, ratio is relatively low. For equal-performance smaller earth station antennas, less satellite power and systems having a higher noise figure can be employed if two- or four-state systems are used. Unfortunately, power-efficient modulation–demodulation (modem) systems are not as bandwidth efficient as the higher-ary modulated systems. Bandwidth efficiency is frequently expressed in terms of the number of transmitted bits/second/hertz (b/s/Hz).

Power efficiency may be expressed in terms of the C/N required to have an acceptable P_e performance. This efficiency is also defined in terms of the required average received bit energy-to-noise density ratio for a given P_e; that is, $P_e = f(E_b/N_o)$. (*Note:* If the noise is specified in a bandwidth which is equal to the bit rate, then $E_b/N_o = C/N$.) For the linear additive white Gaussian noise (AWGN) channel we define a modulation method as *power efficient* if to obtain a $P_e = 10^{-8}$ an $E_b/N_o < 14$ dB is sufficient. We define a modulation technique as *spectrally efficient* if it has a transmission efficiency greater than 2 b/s/Hz. In this chapter power-efficient modulation techniques are described; in Chapter 5 spectral efficient ones are treated.

A typical application of modems is illustrated in Fig. 4.1. Three independent channels are shown. However, note that the number of channels in an earth station may vary from one to several thousand. Filters F_{11} to F_{13} bandlimit the modulated signals S_1 to S_3. Usually, for common output frequency modems the first intermediate frequency (IF) may be specified to be the same for all channels. In Section 4.2 we show that this filtering can be performed with IF bandpass filters or their equivalent baseband low-pass versions.

The upconverters translate the modulated, bandlimited signal to the desired transmit radio frequency. The filters F_{21}, F_{22}, and F_{23} select the required sidebands of the upconverted signal. The center frequencies of these bandpass filters are at $f_c + \Delta f + f_{\text{IF}}$, $f_c + f_{\text{IF}}$, and $f_c - \Delta f + f_{\text{IF}}$ *or at* $f_c + \Delta f - f_{\text{IF}}$, $f_c - f_{\text{IF}}$, and $f_c - \Delta f - f_{\text{IF}}$, depending on whether the upper or the lower sidebands of the modulated and upconverted signals are used. The filters F_{31} to F_{33} have the same center frequencies as the filters F_{21} to F_{23}, respectively. The filters F_{51} to F_{53} preceding the demodulator have a common center frequency, f_{IF}. The use of the various earth station filters, and their typical specifications are described in later sections and chapters.

To obtain a high power efficiency, most transmit power amplifiers have to operate in a nonlinear mode. This mode of operation spreads the spectrum of the modulated bandlimited signal. The role of the radio-frequency filters F_{3N} is to prevent spectral spillover into the adjacent channels. The combined linearly added modulated signals are received by the satellite receive antenna and then are frequency translated and amplified in one or more transponders.

The filters F_{4N} in the receive earth station prevent excessive adjacent channel interference; that is, they prevent the overloading of the downconverters, which translate the desired radio channels to common IF frequencies before the signals are demodulated.

The heart of the satellite channel is the modulator and the demodulator (i.e., *modem*). The modulation techniques employed, the filtering strategy and demodulation method have a major impact on the performance of the system. Spectral efficiency, required power, antenna size, and overall performance are significantly influenced by the performance of the *modem* (modulator–demodulator) in both linear and nonlinear channel environments.

This chapter is devoted to a thorough study of power efficient modulation techniques and modem performance in a system such as shown in Fig. 4.1. Coherent and differentially coherent BPSK, QPSK, and MSK modems are studied in Sections 4.3 to 4.8. The modem performance in theoretical and practical (imperfect) satellite systems is described.

The performance degradations introduced by filtering and limiting (usual nonlinearity in transmit earth station power amplifiers) are described in Section 4.9. The P_e performance degradation caused by the presence of adjacent and co-channel interference in addition to that of the additive white Gaussian noise is studied in Section

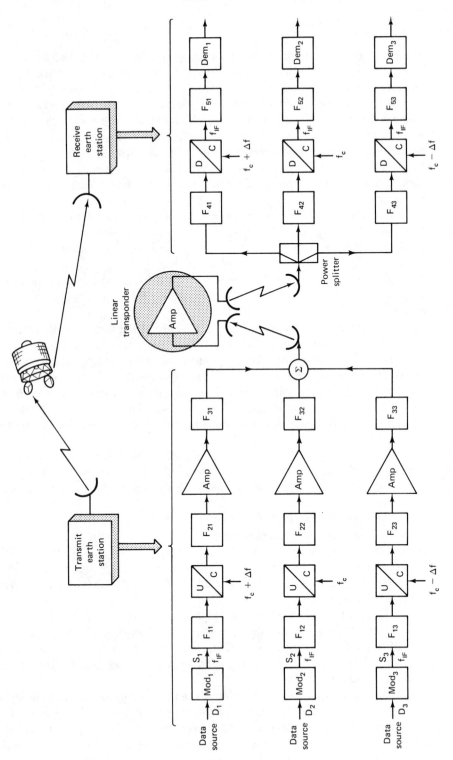

Figure 4.1 Frequently used satellite earth station configuration. Mod, modulator; Dem, demodulator; U/C, up-converter; D/C, down-converter.

4.10. Here it is assumed that the satellite transponder operates in a *linear* mode; thus the transponder filters may be lumped into the F_{3N} and F_{4N} filters. In Section 4.11 detailed circuit diagrams of typical hardware building blocks are described.

4.2 EQUIVALENCE OF LOW-PASS AND BANDPASS CHANNEL MODELS

In the brief description related to Figs. 3.1 and 3.2, we stated that an equivalent low-pass channel model can be found for the *linearly modulated signal* which is transmitted through a bandpass channel and which is also coherently demodulated. The term "linearly modulated signal" implies that the baseband spectrum is linearly translated so as to be centered around the carrier frequency. Due to the importance of (and frequent applications of) the equivalent baseband low-pass channel, a *derivation* of this model is presented in this section. A profound understanding of the appropriate equivalence conditions enables the engineer to design simple low-pass filters (LPF) instead of the more complex bandpass filters (BPF).

For *coherent* demodulation the carrier frequency and phase of the received modulated signal must be precisely established. Here it is assumed that perfect carrier and symbol timing synchronization signals are available. The description of the principles of operation of the synchronization subsystems and their possible impact on overall system performance is postponed until Chapter 7. The bandpass filter, shown in Fig. 4.2, represents the cascade of the transmit BPF, the channel filters, and the receive BPF. For conceptual simplicity assume that the noise is negligible on the channel; however, note that due to the linearity of the system, the conclusions are not restricted to the noiseless case.

If the linear modulator contains only a premodulation low-pass filter, LPF$_T$, then the modulated signal is represented by

$$s_1(t) = [a(t) * h_L(t)]c(t) \tag{4.1}$$

where * denotes the convolution, defined by

$$b(t) = a(t) * h_L(t) = \int_{-\infty}^{\infty} a(\tau)h_L(t - \tau)\, d\tau \tag{4.2}$$

Taking the Fourier transform of (4.1) and noting that convolution in the time domain corresponds to multiplication in the frequency domain, and that convolution in the frequency domain corresponds to multiplication in the time domain [Stremler, 1977; Roden, 1979] [see Fig. 4.2(b)], we obtain

$$S_1(f) = [A(f)H_L(f)] * C(f) \tag{4.3}$$

Convolution of a baseband spectrum with a sinusoidal carrier results in a double-sideband (DSB) spectrum centered around the carrier frequency. It is given by

$$S_1(f) = A(f - f_c)H_L(f - f_c) + A(f + f_c)H_L(f + f_c) \tag{4.4}$$

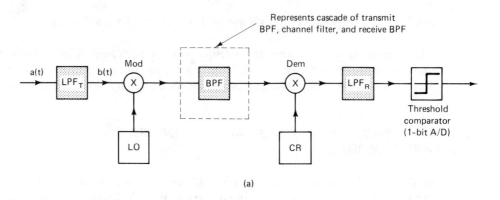

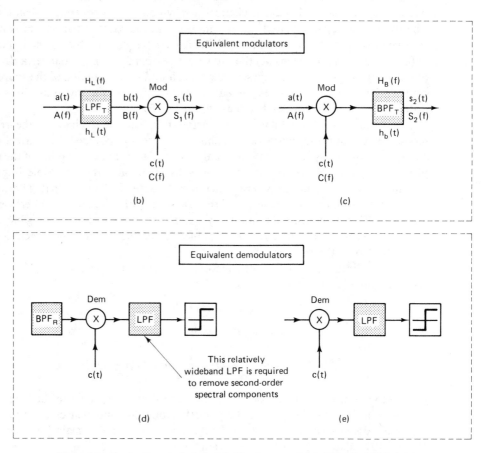

Figure 4.2 Baseband equivalent model of bandpass modulated signals. (a) Modulator, channel, demodulator. LO, local oscillator; CR, carrier recovery. (b) Linear modulator having a premodulation LPF only. (c) Modulator having a postmodulation BPF only. $h_L(t)$ and $h_b(t)$, impulse response of the low-pass and of the bandpass filter. (d) Demodulator having a predetection BPF. (e) Demodulator having a postdetection LPF only.

If the linear DSB modulator contains only a postmodulation bandpass filter [see Fig. 4.2(c)], then

$$s_2(t) = [a(t)c(t)] * h_B(t) \qquad (4.5)$$

and the corresponding Fourier transform is

$$\begin{aligned} S_2(f) &= [A(f) * C(f)]H_B(f) \\ &= [A(f - f_c) + A(f + f_c)]H_B(f) \end{aligned} \qquad (4.6)$$

The amplitude spectra of the premodulation and postmodulation filtered signals, represented by equations (4.4) and (4.6), respectively, are the *same* if $S_1(f) = S_2(f)$, that is,

$$H_L(f - f_c) + H_L(f + f_c) = H_B(f) \qquad (4.7)$$

From equation (4.7) we conclude that the equivalence condition is satisfied if the bandpass filter, $H_B(f)$, has the same transfer function as the low-pass filter, $H_L(f)$,

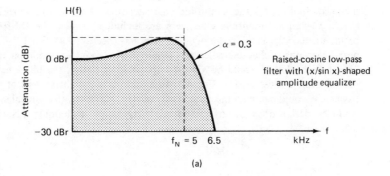

(a)

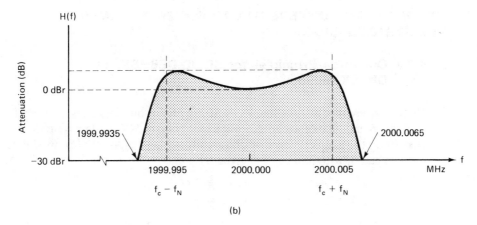

(b)

Fig 4.3   (a) Low-pass and (b) equivalent bandpass attenuation characteristics for a 30% excess bandwidth 10-kb/s double-sideband binary system. This frequency scale is in MHz.

shifted so as to be centered around the carrier frequency. *For the equivalence condition the bandpass filter must be symmetrical around the carrier frequency.*

For the receiver (i.e., coherent demodulator), the derivation of the bandpass and low-pass channel model equivalence conditions is almost the same as the derivation for the transmitter; thus it is suggested as an exercise. *Solve Problem 4.1.*

The advantage of the equivalent low-pass filtering approach for relatively low bit rate systems is illustrated next.

Example 4.1

Design a double-sideband binary modulator for a transmission rate $f_b = 10$ kb/s. Assume that the source information is in the NRZ format and that direct baseband-to-radio frequency modulation is required. The carrier frequency, f_c, is specified as 2 GHz, and a 30% excess bandwidth is permissible (30-dB attenuation point).

Solution. The DSB modulator consists of a mixer preceded by a low-pass filter or followed by a bandpass filter [see Fig. 4.2(b) and (c)]. Either a low-pass *or* a bandpass filter having a roll-off factor $\alpha = 0.3$; that is, an excess bandwidth of 30%, has to be designed. In Fig. 4.3 the amplitude characteristics of these filters are illustrated. The $(x/\sin x)$-shaped amplitude equalizers are included to meet the ISI-free transmission requirements, stipulated in the Nyquist theorems (Section 3.4). For ISI-free transmission near-linear phase characteristics are required. A careful examination of the bandpass filter reveals that it would be almost impossible to design the phase-equalized narrow bandpass filter illustrated in Fig. 4.3. In addition, the slightest drift of the filter center frequency would cause an intolerable asymmetry. However, the equivalent low-pass filter could be designed by means of the relatively simple techniques described in Section 3.5. ∎

4.3 COHERENT AND DIFFERENTIALLY COHERENT BINARY PSK SYSTEMS (BPSK)

4.3.1 Operation/Principles for BPSK, DEBPSK, and DBPSK Modems

Discrete phase modulation, known as M-ary phase-shift keying, is the most frequently used digital modulation technique. Biphase or binary PSK systems (BPSK) are considered to be the simplest form of phase-shift keying ($M = 2$). The modulated signal has two states, $m_1(t)$ and $m_2(t)$ given by

$$m_1(t) = +C \cos \omega_c t \tag{4.8}$$
$$m_2(t) = -C \cos \omega_c t \tag{4.9}$$

These signals can be generated by a system such as that shown in Fig. 4.4. The modulated signal is given by

$$m(t) = b(t)c(t) = C\, b(t) \cos \omega_c t \tag{4.10}$$

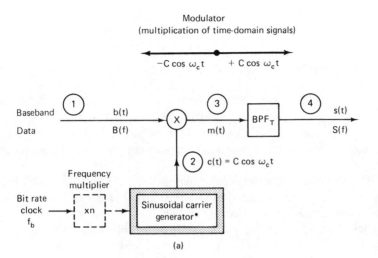

(a)

*For digital implementation a periodic
square-wave generator is used.

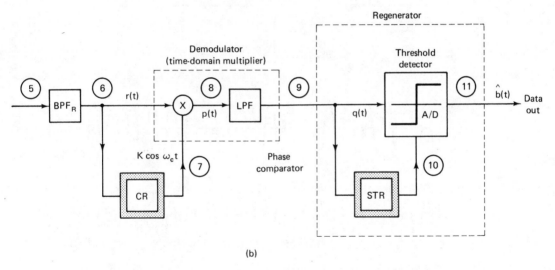

(b)

Figure 4.4 Block diagram of a coherent BPSK (a) modulator and (b) demodulator.

If $b(t)$ represents a synchronous random binary baseband signal having a bit rate, $f_b = 1/T_b$ and levels -1 and $+1$ then equation (4.10) represents the antipodal (180°) phase shifted signaling elements $m_1(t)$ and $m_2(t)$. Thus the information is contained in the *phase* of the modulated signal, that is,

$$m(t) = C \cos [\omega_c t + \theta(t)] \qquad (4.11)$$

where

$$\theta(t) = 0° \text{ or } 180°$$

For equiprobable NRZ baseband signals (+1 and −1 volt normalized input voltages) time-domain multiplication is equivalent to double-sideband suppressed-carrier amplitude modulation (DSB-SC-AM) [see equation (4.10)]. The equivalent representation in equation (4.11) implies that a phase shift-keyed (PSK) signal is obtained. We should remember that time-domain multiplication may correspond to digital-phase modulation. In other words, an **equivalence** of digital DSB-SC-AM and PSK exists.

The baseband signal $b(t)$, the unmodulated carrier wave $c(t)$, the modulated signal $m(t)$, and the modulated-bandlimited signal $s(t)$ are shown in Fig. 4.5. Note the 180° abrupt phase transition in the signals $m(t)$ and $s(t)$. The sinusoidal carrier frequency, f_c may be an integer multiple of the bit rate. This can be achieved by inserting a *frequency multiplier* in between the bit-rate clock f_b and the voltage-

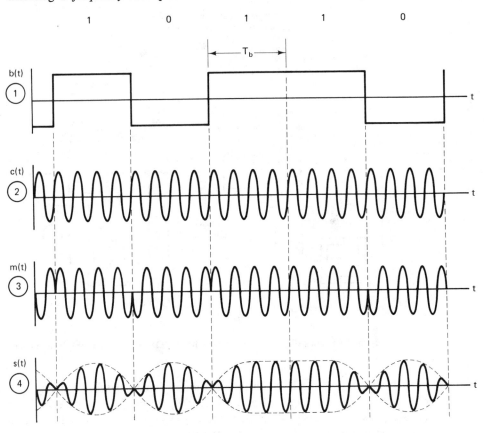

Figure 4.5 BPSK modulator signal–time domain presentation.

controlled sinusoidal carrier oscillator which operates in a phase-locked-loop mode.

The theoretical performance derivations are simpler if this *synchronism* is assumed, that is, if $f_c = nf_b$, where n is an integer. However, the source bit rate, which may be the output of a standard time-division-multiplexed equipment, has a relatively wide range of fluctuations ($f_b \pm \Delta f_b$). If the carrier frequency were to be locked to the n-multiple of the bit rate, the bit rate fluctuations would be translated into fluctuations of both the unmodulated carrier frequency ($f_c \pm n \Delta f_b$) and of the modulated spectrum. For operational systems these variations might be unduly restrictive. Fortunately, in practical earth station modem designs it is not essential to have the unmodulated carrier frequency locked to the integer multiple of the bit rate. The typical performance degradation in the *asynchronous* case ($f_c \neq nf_b$) is virtually negligible, particularly if $f_c > 10f_b$. *Solve Problem 4.2.*

For *coherent demodulation* a carrier frequency that is in synchronism with the received modulated wave is required. The carrier-recovery (CR) system provides the receive multiplier with a sinusoidal frequency that has exactly the same frequency and phase as the transmitted unmodulated carrier wave. Naturally, the propagation and equipment delay have to be taken into account.

The bandlimited received IF signal, $r(t)$, is multiplied by the recovered carrier wave $K \cos \omega_c t$ (see Fig. 4.4). The receive multiplier, followed by a low-pass filter, performs the coherent phase demodulation process (also known as coherent phase comparison). In PSK systems the information is contained in the phase of the signal. We describe $r(t)$ by

$$r(t) = C_r \cos [\omega_c t + \theta(t)] \tag{4.12}$$

where C_r is the peak amplitude of the modulated carrier. The output of the multiplier (point 8 in Fig. 4.4) is given by

$$p(t) = r(t) K \cos \omega_c t = C_r K \cos [\omega_c t + \theta(t)] \cos \omega_c t \tag{4.13}$$

Based on the trigonometric identity,

$$\cos \alpha \cos \beta \equiv \tfrac{1}{2} \cos (\alpha - \beta) + \tfrac{1}{2} \cos (\alpha + \beta)$$

we obtain

$$\begin{aligned} p(t) &= \tfrac{1}{2} C_r K \cos [\omega_c t + \theta(t) - \omega_c t] + \tfrac{1}{2} C_r K \cos [\omega_c t + \theta(t) + \omega_c t] \\ &= \tfrac{1}{2} C_r K [\cos \theta(t) + \cos [2\omega_c t + \theta(t)]] \end{aligned} \tag{4.14}$$

The receive low-pass filter removes the double-frequency spectral components. At the **threshold comparator input,** point 9, we have

$$q(t) = \tfrac{1}{2} C_r K \cos [\theta(t)] \tag{4.15}$$

In equation (4.15), $C_r K/2$ represents a gain constant, while $\cos [\theta(t)]$ is the time-variable bandlimited baseband signal. For $\theta(t) = 0°$ or $180°$ this signal equals

+1 or −1, respectively. This baseband voltage is proportional to the cosine of the phase angle differential between the received modulated carrier and the recovered carrier; thus the term *coherent phase comparator* or *phase demodulator* is justified for the multiplier–low-pass filter subsystem, shown in Fig. 4.4. Finally, the one-bit A/D converter (threshold comparator) provides the digital output $\hat{b}(t)$. If the system interference and noise are negligible, then $\hat{b}(t) = b(t - \tau)$; that is, the demodulated/regenerated output equals the source information delayed by the equipment and propagation delay, τ.

Most practical carrier-recovery circuits, as described in Chapter 7, introduce a *phase ambiguity* into the recovered carrier. In binary PSK demodulators the recovered carrier is not necessarily the required $\cos \omega_c t$ signal; it might equal $\cos \omega_c t$ or $\cos (\omega_c t + 180°)$. Thus a steady 180° phase error in the recovered carrier is possible. This error inverts (multiplies by $\cos 180° = -1$) the demodulated data stream and causes a 100% error rate. Fortunately, the insertion of a simple *differential encoder* into the transmitter and a *differential decoder* into the receiver avoids errors that could be introduced by this phase ambiguity. The basic principles of operation of such encoders are now described.

The block diagram of differentially encoded/decoded modems is shown in Fig. 4.6. In the coherent demodulator, the carrier-recovery circuit provides the required unmodulated carrier wave. This wave is multiplied by the received modulated signal and then passes through a low-pass filter. Differential decoding is performed on the regenerated data stream. The demodulator, illustrated in the lower part of the figure, performs a comparison detection (a demodulation) directly on the modulated signal, thus does not require a carrier-recovery circuit. The modulated signal is multiplied by a one-bit delayed replica and then bandlimited with a low-pass filter. The differentially encoded modem equipped with a carrier-recovery circuit is designated DEBPSK while that one without carrier recovery, DBPSK.

The differential encoding process is illustrated in Fig. 4.7. In the generation of a differentially encoded bit, d_k, of the encoded sequence $\{d_k\}$, the present bit, b_k, of the message sequence $\{b_k\}$ and the previous bit, d_{k-1}, are compared. *If there is no difference between b_k and d_{k-1}, then $d_k = 1$; otherwise, $d_k = 0$.* This can be expressed in terms of

$$d_k = \overline{b_k \oplus d_{k-1}} = b_k d_{k-1} + \bar{b}_k \bar{d}_{k-1} \tag{4.16}$$

where $\oplus$ represents the exclusive-OR operation. Note that an arbitrary reference binary digit may be assumed for the initial bit of the $\{d_k\}$ sequence.

Both differential demodulators in Fig. 4.6 perform the inverse function to that of the encoder. In the case of DEBPSK demodulation, the differential decoding is performed by means of logic circuitry. In the case of the DBPSK demodulator, the phase angles of the received signal (which may be corrupted by noise) and its one-bit delayed version are compared. Verify that in the absence of noise the message sequence $\{b_k\}$ is correctly recovered. *Solve Problem 4.3.*

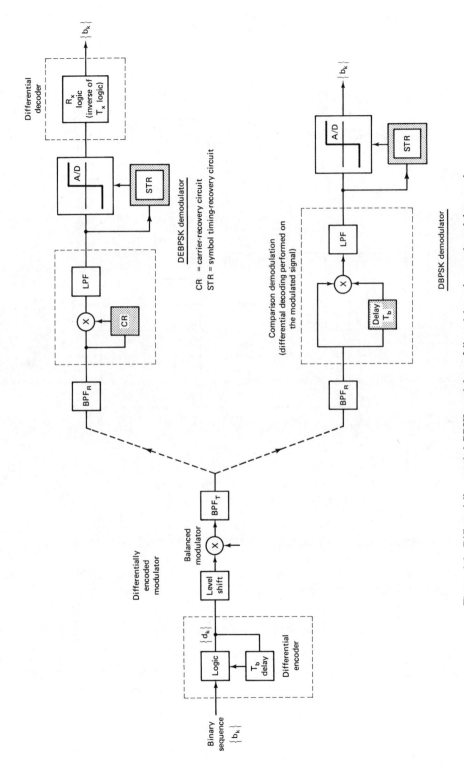

Figure 4.6 Differentially encoded BPSK modulator followed by coherent demodulator and differential decoder (DEBPSK) and differential phase demodulator (DBPSK).

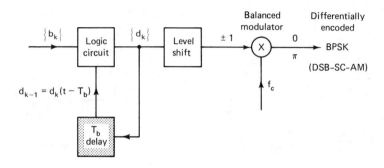

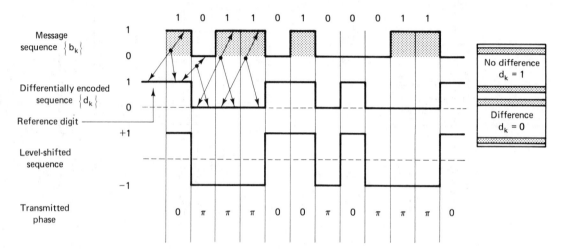

Figure 4.7 Differential encoding. If there is no difference between the signal states of b_k and d_{k-1} (same logic states), then the logic output state $d_k = 1$; otherwise, $d_k = 0$.

4.3.2 Spectrum and Spectral Efficiency of BPSK Systems

The spectral density $w_s(f)$ of equiprobable non-return-to-zero (NRZ) signals was derived in Chapter 3 [see equation (3.37)]. This density is given by

$$S_b(f) = 2A^2 T_b \left(\frac{\sin \pi f T_b}{\pi f T_b} \right)^2 \tag{4.17}$$

In binary systems the bit duration equals the symbol duration (i.e., $T_b = T_s$). In Section 4.2 we showed that for polar NRZ signals the BPSK modulator is equivalent to a DSB-SC modulator. Thus the baseband spectrum is translated so as to be centered around the carrier frequency, f_c. It is given by

$$S_m(f) = 2KA^2 T_b \left[\frac{\sin \pi (f - f_c) T_b}{\pi (f - f_c) T_b} \right]^2 \tag{4.18}$$

where K is the proportionality constant of the multiplier. The unfiltered (infinite bandwidth) and filtered (minimum bandwidth) modulated spectra are shown in Fig. 4.8.

The modulated and filtered spectrum $S_s(f)$ is given by

$$S_s(f) = S_m(f)|H_B(f)|^2 \qquad (4.19)$$

The bandlimiting bandpass filter $H_B(f)$ may include the $(x/\sin x)$-shaped amplitude equalizer, as illustrated in Fig. 4.8. For the minimum-bandwidth case, an $\alpha = 0$ roll-off filter has to be used. This filter is the bandpass equivalent of the low-pass filter described by equation (3.49).

The theoretical spectral efficiency of BPSK modems is 1 b/s/Hz. Using practical

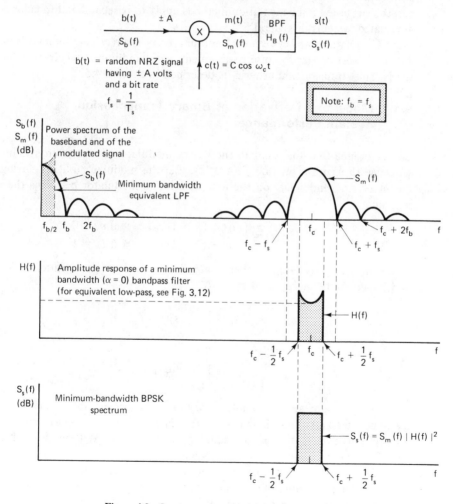

Figure 4.8 Spectrum of unfiltered and filtered BPSK.

filters with or approaching an $\alpha = 0.3$ roll-off characteristic, a spectral efficiency of 1 b/s/Hz $:$ $1.3 = 0.77$ b/s/Hz is achieved. *Solve Problem 4.4.*

4.4 OPTIMUM RECEIVER FOR BINARY DIGITAL MODULATION SYSTEMS

In this section we derive the *optimum* receiver for binary digital transmission systems. A receiver is said to be *optimum* if it yields the *minimum* probability of error, P_e. We assume that additive white Gaussian noise (AWGN) is the only system perturbance, and that the bandwidth is not constrained; that is, an *infinite bandwidth intersymbol-interference-free transmission channel is available.* The derivation that follows and the suggested implementation methods apply to baseband and to coherent binary modulated systems in general.

The relationship between the optimum receiver, a receiver that has (for a specified P_e) the lowest carrier-to-noise requirement, and bandlimited receivers, which satisfy the Nyquist transmission criteria, is described in Section 4.5.

4.4.1 Model and Derivation of Binary Transmission System Performance

Let us assume that the input to the binary modulator is a random synchronous bit sequence $\{b_k\}$ with a bit rate $f_b = 1/T_b$. Let the modulator output during the kth bit interval depend *solely* on the kth bit b_k. The modulator performs the following mapping:

$$m(t) = \begin{cases} s_1[t - (k-1)T_b] & \text{if } b_k = 0 \\ s_2[t - (k-1)T_b] & \text{if } b_k = 1 \end{cases} \tag{4.20}$$

where $s_1(t)$ and $s_2(t)$ are two elementary waveforms having a duration of T_b seconds, and having finite energy, that is,

$$E_{1b} \triangleq \int_{t_o}^{t_o + T_b} s_1^2(t) \, dt < \infty$$

$$\tag{4.21}$$

$$E_{2b} \triangleq \int_{t_o}^{t_o + T_b} s_2^2(t) \, dt < \infty$$

A modulator and a possible binary demodulator are shown in Fig. 4.9. As we assumed intersymbol-interference-free transmission, it is sufficient to study the performance of this system for an individual pulse excitation. Without loss of generality, therefore, we may assume that $t_o = 0$ and that $k = 1$.

The transfer function of the optimum receive filter, $H(f)$, its impulse response $h(t)$, and the probability of error performance of the optimum system are all obtained as a result of the following derivation.

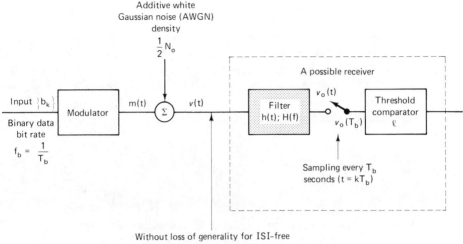

Figure 4.9   Modulator, with additive white Gaussian noise (AWGN) and demodulator.

The received carrier plus noise prior to the receive filter is

$$v(t) = \begin{cases} s_1[t - (k-1)T_b] + n(t) \\ \qquad\qquad \text{or} \\ s_2[t - (k-1)T_b] + n(t) \end{cases} \qquad (k-1)T_b \le t \le kT_b \qquad (4.22)$$

or, for the specific case of $k = 1$,

$$v(t) = \begin{cases} s_1(t) + n(t) \\ s_2(t) + n(t) \end{cases} \qquad 0 < t < T_b \qquad (4.23)$$

depending on whether the $s_1(t)$ or $s_2(t)$ signal was transmitted. (*Note:* The system propagation delay is assumed to be zero.) Now, let $s_1(t)$ and $s_2(t)$ be selected such that

$$s_{01}(T_b) < s_{02}(T_b) \qquad (4.24)$$

The *decision rule* is based on the criteria

$$s_2(t) \text{ was sent if} \qquad v_o(T_b) > \ell \qquad (4.25)$$

$$s_1(t) \text{ was sent if} \qquad v_o(T_b) < \ell \qquad (4.26)$$

where ℓ is the threshold level of $v_o(T_b)$, the sampled output, and $s_{o1}(T_b)$ and $s_{o2}(T_b)$ are the signal components of the receive filter output in the sampling instant $t = T_b$. These filtered signals are due to the $s_1(t)$ and $s_2(t)$ transmitted signals, respectively. In the sampling instant ($t = T_b$) the signal plus noise voltage at the threshold comparator input is given by

$$v_o(T_b) = s_{o1}(T_b) + n_{o1}(T_b) \qquad (4.27)$$

or

$$v_o(T_b) = s_{o2}(T_b) + n_{o2}(T_b) \qquad (4.28)$$

where $n_{o1}(t)$ and $n_{o2}(t)$ represent the filtered noise components.

Let the AWGN have a double-sided spectral density of $G_n(f) = \frac{1}{2}N_o$. Thus the filtered noise has a power spectral density $S_{no}(f)$, given by

$$\begin{aligned} S_{no}(f) &= |H(f)|^2 G_n(f) \\ &= |H(f)|^2 \tfrac{1}{2} N_o \end{aligned} \qquad (4.29)$$

The receive filter is a linear network; hence the additive Gaussian channel noise remains Gaussian after the receiver filter. (*Note:* The probability density function of Gaussian noise remains Gaussian even if it is filtered by a time-invariant *linear* filter. Here the term "linear" implies that the superposition theorem of classical system theory applies.)

The noise is stationary; thus at the sampling instant, $N = n_o(T_b)$, it is a random variable with a probability density given by

$$f_N(\eta) = \frac{e^{-\eta^2/2\sigma_0^2}}{\sqrt{2\pi\sigma_0^2}} \qquad (4.30)$$

where the total noise power N_T (i.e., variance, σ_0^2) at the receive filter output is given by

$$N_T = \sigma_0^2 = \int_{-\infty}^{\infty} |H(f)|^2 G_n(f) df \qquad (4.31)$$

The mean value of the noise is zero, by symmetry.

The sampler output has two possible states:

$$V_1 = v_1(T_b) = s_{o1}(T_b) + N \qquad \text{if } s_1(t) \text{ is transmitted} \qquad (4.32)$$

or

$$V_2 = v_2(T_b) = s_{o2}(T_b) + N \qquad \text{if } s_2(t) \text{ is transmitted} \qquad (4.33)$$

The conditional probability density functions (pdf) of the receive filter output in the sampling instants $t = kT_b$ are illustrated in Fig. 4.10. Note that the conditional

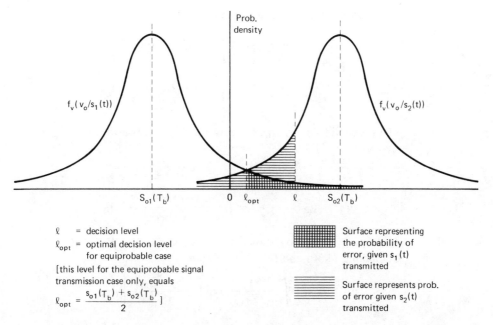

Figure 4.10   Conditional probability density functions of the receive filter output in the sampling instant $(t = T_b)$.

pdf's described in Fig. 3.26 for the specific case of binary baseband transmission have the same shape.

Probability of error derivation.   For the general binary case, we have:

Probability of transmission of an $s_1(t)$ signal is p_1.
Probability of transmission of an $s_2(t)$ signal is p_2.

The average probability of error is

$$P_e = p_1 P[e|s_1(t)] + p_2 P[e|s_2(t)] \tag{4.34}$$

In most *practical* digital transmission systems, data-stream randomizers, known as *scramblers,* assure that the transmitted signals have *equal probabilities.* The theory of operation and practical design guidelines of scramblers used for the generation of maximum length equiprobable pseudo-random sequences can be found in [Golomb, 1964, and Forney, 1970]. For equiprobable randomized data streams we have

$$p_1 = p_2 = 0.5 \tag{4.35}$$

and

$$P_e = 0.5 P[e|s_1(t)] + 0.5 P[e|s_2(t)] \tag{4.36}$$

The probability of error, assuming that $s_1(t)$ is transmitted, is given by

$$P(e|s_1(t)) = \int_{\ell}^{\infty} f_{v_0}(v_0|s_1(t)) \, dv$$

(4.37a)

$$= \int_{\ell}^{\infty} \frac{e^{-[v_0 - s_{01}(T)]^2/2\sigma_0^2}}{\sqrt{2\pi\sigma_0^2}} \, dv_0$$

The probability of error, assuming that $s_2(t)$ is transmitted, is given by

$$P(e|s_2(t)) = \int_{-\infty}^{\ell} f_{v_0}(v_0|s_2(t)) \, dt$$

(4.37b)

$$= \int_{-\infty}^{\ell} \frac{e^{-[v_0 - s_{02}(T)]^2/2\sigma_0^2}}{\sqrt{2\pi\sigma_0^2}} \, dv_0$$

These probabilities are shown as shaded surfaces in Fig. 4.10. Combining (4.36) and (4.37), we obtain

$$P_e = \frac{1}{2} \int_{\ell}^{\infty} \frac{1}{\sqrt{2\pi N_T}} \exp\left[\frac{-(v_0 - s_{01})^2}{2N_T}\right] dv_0$$

$$+ \frac{1}{2} \int_{-\infty}^{\ell} \frac{1}{\sqrt{2\pi N_T}} \exp\left[\frac{-(v_0 - s_{02})^2}{2N_T}\right] dv_0$$

(4.38)

The optimum threshold level ℓ_{opt} can be derived by finding the first differential with respect to ℓ in expression (4.38). If we set the first differential to zero, we obtain the optimum setting of the threshold level ℓ:

$$\ell_{opt} = 0.5[s_{01}(T_b) + s_{02}(T_b)]$$

(4.39)

This level is at the intersection of the two conditional pdf's illustrated in Fig. 4.10.

By insertion of the best choice of the threshold level ℓ_{opt} into equation (4.38), we obtain

$$P_e = 0.5 \, \text{erfc} \left[\frac{s_{02}(T_b) - s_{01}(T_b)}{2\sqrt{2}\sigma_0}\right]$$

(4.40a)

where erfc (y) is defined by

$$\text{erfc} (y) \triangleq \frac{2}{\sqrt{\pi}} \int_{y}^{\infty} e^{-z^2} \, dz = 2Q(\sqrt{2}y) \qquad y > 0$$

(4.40b)

For numerical values, see Table 3.1 and Fig. 3.25. [*Note:* The application of equation (4.40a) is limited to the additive Gaussian noise case.]

Transfer Function of the Optimal Receiver (For White Gaussian Noise). From equation (4.40) we conclude that the probability of error is a function of the *difference* (distance) between the two output signals in the sampling instant, $t = T$, and of the rms noise voltage σ_o. The error function erfc (y) decreases monotonically with y; thus the probability of error decreases with the increasing of the difference between the received sampled signals. Thus the *matched filter* transfer function $H(f)$, or equivalently its impulse response, $h(t)$, leads to the smallest P_e, and therefore best performance, by maximization of the ratio

$$\gamma \triangleq \frac{s_{02}(T_b) - s_{01}(T_b)}{\sigma_o} \qquad (4.41)$$

If we maximize γ^2 instead, then the derivation can be simplified by eliminating the requirement $s_{01}(T_b) < s_{02}(T_b)$. Let us define the difference in the filtered outputs (Fig. 4.9) by

$$S_{od}(t) = s_{02}(t) - s_{01}(t) \qquad (4.42)$$

and the difference in the sampling instants by

$$S_{od}(T_b) \triangleq s_{02}(T_b) - s_{01}(T_b) \qquad (4.43)$$

Thus

$$\gamma^2 = \frac{s_{od}^2(T_b)}{\sigma_o^2} = \frac{s_{od}^2(t)}{E\{n_o^2(t)\}}\bigg|_{t=T_b} \qquad (4.44)$$

The difference of the filtered outputs may be obtained by inverse Fourier transform,

$$S_{od}(t) = F^{-1}[S_{od}(f)H(f)] = \int_{-\infty}^{\infty} H(f)S_{od}(f)e^{j2\pi ft}\,df \qquad (4.45)$$

The noise power (or variance) of the stationary noise process at the filter output is given by

$$\sigma_o^2 = E\{n_o^2(t)\} = E\{n_o^2(T_b)\} = \int_{-\infty}^{\infty} G_n(f)|H(f)|^2\,df \qquad (4.46)$$

$$= \frac{1}{2}N_o \int_{-\infty}^{\infty} |H(f)|^2\,df$$

If we set $t = T_b$ in (4.45) and insert this equation, together with equation (4.46) into (4.44), we obtain

$$\gamma^2 = \frac{\left|\int_{-\infty}^{\infty} H(f)S_{od}(f)e^{j2\pi fT_b}\,df\right|^2}{\frac{1}{2}N_o \int_{-\infty}^{\infty} |H(f)|^2\,df} \qquad (4.47)$$

To maximize γ^2 we employ *Schwarz's inequality*. This inequality states that

$$\left|\int_{-\infty}^{\infty} X(f)Y^*(f)\,df\right|^2 \le \int_{-\infty}^{\infty} |X(f)|^2\,df \int_{-\infty}^{\infty} |Y(f)|^2\,df \qquad (4.48)$$

where $X(f)$ and $Y(f)$ are arbitrary complex functions of a common variable f. Equality holds if

$$X(f) = kY^*(f) \tag{4.49}$$

where k is an arbitrary constant and $Y^*(f)$ is the complex conjugate of $Y(f)$. If you are interested in the derivation of Schwarz's inequality, you should read Chapter 8 of [Ziemer and Tranter, 1976] or other classical communication theory textbooks.

In equation (4.48) we perform the following substitution:

$$X(f) = H(f) \tag{4.50}$$

$$Y^*(f) = S_{od}(f)e^{j2\pi f T_b} \tag{4.51}$$

and we obtain

$$\gamma^2 = \frac{1}{G_n(f)} \frac{\left| \int_{-\infty}^{\infty} X(f)Y^*(f)\, df \right|^2}{\int_{-\infty}^{\infty} |H(f)|^2\, df} \leq \frac{2}{N_0} \frac{\left| \int_{-\infty}^{\infty} |H(f)|^2\, df \int_{-\infty}^{\infty} |S_{od}(f)|^2\, df \right|}{\int_{-\infty}^{\infty} |H(f)|^2\, df} \tag{4.52}$$

$$\gamma^2 \leq \frac{2}{N_0} \int_{-\infty}^{\infty} |S_{od}(f)|^2\, df \tag{4.53}$$

Based on Schwarz's inequality and the additive white Gaussian noise assumption, $G_n(f) = N_0/2$, we conclude that equality holds; that is, the signal-to-noise ratio is maximized if

$$H(f) = mS_{od}^*(f)e^{-j2\pi T_b f} \tag{4.54}$$

In (4.54) m is a constant that represents the receive filter gain. The gain constant of linear filters has evidently no influence on the output signal-to-noise ratio. Thus, without loss of generality, we may set $m = 1$. The *optimum receiver transfer function* $H_0(f)$ is thus given by

$$\boxed{H_0(f) = S_{od}^*(f)e^{-j2\pi T_b f}} \tag{4.55}$$

By means of the inverse Fourier transform, we can establish the receiver impulse response $h_0(t)$, which corresponds to the optimum choice of $H_0(f)$.

$$h_0(t) = \mathscr{F}^{-1}[H_0(f)] = \int_{-\infty}^{\infty} S_{od}^*(f)e^{-j2\pi T_b f}e^{j2\pi f t}\, df \tag{4.56}$$

$$= \int_{-\infty}^{\infty} S_{od}(f')e^{j2\pi f'(T_b - t)}\, df'$$

Expression (4.56) represents the inverse Fourier transform of $s_{od}(t)$ with t replaced by $T_b - t$. Thus we have

$$h_0(t) = s_2(T_b - t) - s_1(T_b - t) \tag{4.57}$$

If we define

$$s(T_b - t) \triangleq s_2(T_b - t) - s_1(T_b - t) \tag{4.58}$$

then we have

$$\boxed{h_o(t) = s(T_b - t)} \tag{4.59}$$

We recall from the system model illustrated in Fig. 4.9 that $s_1(t)$ and $s_2(t)$ represent the transmitted signals at the modulator output. The result, obtained in (4.57), states that the impulse response $h_o(t)$ of the optimal receiver equals the difference of the T_b-second shifted time reverse of $s_2(t)$ and $s_1(t)$. In other words, the receiver is matched to the *difference* of the transmitted waveforms and therefore is known as a *matched filter receiver*. This difference is sampled and compared to the optimal threshold l_{opt} [equation (4.39); see Fig. 4.11(a)].

An alternative optimal receiver structure is illustrated in Fig. 4.11(b). This receiver is known as a *correlation receiver* or *correlation detector*. In the following we show that *the matched filter receiver and the correlation receiver are identical,* and hence have the same performance characteristics.

The output of the matched filter receiver is

$$
\begin{aligned}
v(t) = h(t) * y(t) &= \int_0^\infty s(T_b - \tau)y(t - \tau) \, d\tau \\
&= \int_0^{T_b} s(T_b - \tau)y(t - \tau) \, d\tau
\end{aligned}
\tag{4.60}
$$

since

$$h(t) = \begin{cases} s(T_b - t) & 0 \le t \le T_b \\ 0 & \text{elsewhere} \end{cases} \tag{4.61}$$

If we let $t = T_b$ and change variables to $\alpha = T_b - \tau$, we obtain

$$v(T_b) = \int_0^{T_b} s(\alpha)y(\alpha) \, d\alpha \tag{4.62}$$

The **output of the correlator** at the sampling instant, $t = T_b$, is obtained by inspection of Fig. 4.11(b). It is

$$v'(T_b) = \int_0^{T_b} s(t)y(t) \, dt \tag{4.63}$$

Observe that $v(T_b) = v'(T_b)$; that is, the output of the matched filter receiver and of the correlation receiver are *identical at the sampling instant,* $t = T_b$. The value of the *sampled output* is compared to the threshold level; thus for the equivalence condition it is sufficient to have equal output values at the sampling instant.

The correlation receiver [Fig. 4.11(b)] consists of a multiplier followed by an integrator and a sampler. Note that the $s(t)$ input to the multiplier [equation (4.63)]

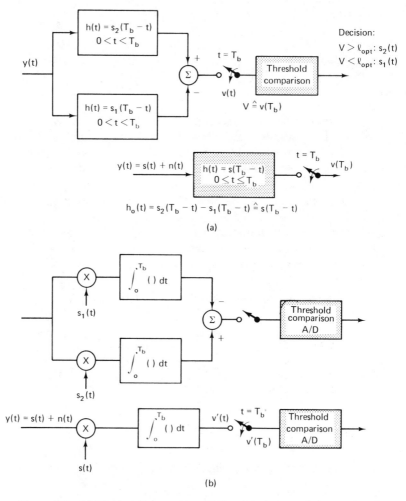

Figure 4.11 (a) Matched filter receiver. (b) Its equivalent correlation receiver for binary signaling in white Gaussian noise. Coherent PSK demodulators are essentially correlation receivers. Note that $s(t)$ is a sinusoidal signal only for infinite-bandwidth systems.

equals the difference of the transmitted, bandlimited signals. Thus $s(t)$ *is not necessarily a simple sinusoidal wave. Solve Problem 4.5.*

$P_e = f(E_b/N_o)$ ***Performance of the Optimal Receiver.*** The probability of error performance, P_e, of the modulated/coherently demodulated binary signal, illustrated in Fig. 4.9, is described by equation (4.40). It is given in the presence of white Gaussian noise by

$$P_e = 0.5 \; \text{erfc} \left[\frac{s_{02}(T_b) - s_{01}(T_b)}{2\sqrt{2}\, \sigma_o} \right]$$

Figure 4.11(a) represents the matched filter implementation of a binary system. The P_e for the matched filter receiver is

$$P_e = \frac{1}{2} \text{erfc} \left(\frac{\gamma}{2\sqrt{2}} \right) \tag{4.64}$$

where γ is the maximum value given by (4.53). It is

$$\gamma = \sqrt{\frac{2}{N_0}} \int_{-\infty}^{\infty} |S_{od}(f)|^2 \, df \tag{4.65}$$

From equation (4.42) we recall that

$$S_{od}(t) = s_2(t) - s_1(t)$$

Thus, using *Parseval's theorem*, we write

$$\gamma^2 = \frac{2}{N_0} \int_{-\infty}^{\infty} [s_2(t) - s_1(t)]^2 \, dt$$

$$= \frac{2}{N_0} \left[\int_{-\infty}^{\infty} s_2^2(t) \, dt + \int_{-\infty}^{\infty} s_1^2(t) \, dt - 2 \int_{-\infty}^{\infty} s_1(t) s_2(t) \, dt \right] \tag{4.66}$$

$$= \frac{2}{N_0} (E_{2b} + E_{1b} - 2\sqrt{E_{1b}E_{2b}} \, \rho_{12})$$

$$= \frac{2}{N_0} (2E_b - 2\sqrt{E_{1b}E_{2b}} \, \rho_{12})$$

where E_b is the average signal energy of the received signal, $E_b = \frac{1}{2}(E_{1b} + E_{2b})$. [*Note:* The a priori probabilities of the $s_1(t)$ and $s_2(t)$ signals are assumed to be equal.] and

$$\rho_{12} \triangleq \frac{1}{\sqrt{E_1 E_2}} \int_{-\infty}^{\infty} s_1(t) s_2(t) \, dt \qquad -1 \le \rho_{12} \le 1 \tag{4.67}$$

is the normalized finite time *correlation coefficient* of the two waveforms $s_1(t)$ and $s_2(t)$. E_1 and E_2 represent the energy of the transmitted $s_1(t)$ and $s_2(t)$ signals measured in a T_b duration interval [see equation (4.21)]. The correlation coefficient ρ_{12} represents a measure of the *similarity* between the two signals $s_1(t)$ and $s_2(t)$. It is such that

$$-1 \le \rho_{12} \le 1 \tag{4.68}$$

From (4.64) and (4.66), we obtain the P_e. It is

$$P_e = \frac{1}{2} \text{erfc} \left(\frac{1}{2} \sqrt{\frac{E_{1b} + E_{2b} - 2\sqrt{E_{1b}\bar{E}_{2b}\rho_{12}}}{N_0}} \right) \tag{4.69}$$

From equation (4.69) we conclude that the P_e performance is a function of the signal energies E_{1b} and E_{2b}, the noise density N_0, and the correlation coefficient ρ_{12}. For the best possible performance, P_e should be *minimized*. This is done by maximizing the term within the parentheses in (4.69). Assuming that E_{1b}, E_{2b}, and N_o are specified, we conclude that this term will have its maximum for $\rho_{12} = -1$. For this case we have

$$P_e = \frac{1}{2} \, \text{erfc} \left(\frac{1}{2} \sqrt{\frac{E_{1b} + E_{2b} + 2\sqrt{E_{1b}E_{2b}}}{N_o}} \right) \tag{4.70}$$

$$= \frac{1}{2} \, \text{erfc} \left(\frac{1}{2} \frac{\sqrt{E_{1b}} + \sqrt{E_{2b}}}{\sqrt{N_o}} \right) \qquad \text{for } \rho_{12} = -1$$

Assuming that the transmitted signals are *equiprobable*, the *average* received signal energy (measured in a T_b-second interval) is

$$E_{av} = \tfrac{1}{2}(E_{1b} + E_{2b}) \tag{4.71}$$

Equation (4.69) may be now written as

$$P_E = \frac{1}{2} \, \text{erfc} \left[\sqrt{\frac{1}{2} \frac{E_{av}}{N_o} (1 - R_{12})} \right] \tag{4.72}$$

where

$$R_{12} \triangleq \frac{2\sqrt{E_{1b}E_{2b}}}{E_{1b} + E_{2b}} \rho_{12} \tag{4.73}$$

The smallest value of R_{12} yields the best P_e performance. *Why?* The minimum value R_{12min} of -1 is attained for $\rho_{12} = -1$ and $E_{1b} = E_{2b}$. For the best possible choice we have

$$P_E = \frac{1}{2} \, \text{erfc} \sqrt{\frac{E_{avb}}{N_o}} \tag{4.74}$$

We conclude that the best performance in an energy (i.e., power)-limited additive white Gaussian channel is obtained if the equiprobable transmit signals are chosen as dissimilar as possible. This is achieved by generation of *antipodal signals* having $R_{12} = -1$. (*Note:* The $+ A \sin \omega_c t$ and $-A \sin \omega_c t$ signals generated in BPSK modulation schemes are examples of antipodal signals that are frequently employed.)

References Related to the Matched Filter Derivation.   The matched filter derivation receives extensive treatment in the readily available literature. Almost all communications textbooks contain derivations similar to the one presented in this section. These references include [Haykin, 1978; Gregg, 1977; Panter, 1965; Taub, and Schilling 1971; Stremler, 1979; Ziemer and Tranter, 1976; Shanmugam, 1979; and Schwartz et al., 1966]. Papers on this subject include [Lawton, 1958, and Turin, 1960].

In [Bennett and Davey, 1965, and Lucky et al., 1968] a somewhat different philosophy of derivation is adopted: the emphasis is on bandlimited Nyquist channels.

However, in this section we present the classical matched filter derivation, in order to gain an insight into the mathematics of optimal receivers and also to obtain the required background to understand the necessary *conditions* required for the *equivalence* of the *Nyquist channel model* (i.e., receiver) and the *matched filter receiver.*

4.5 ARE THE NYQUIST AND THE MATCHED FILTER (CORRELATION) RECEIVERS EQUIVALENT?

If the channel filtering, including the receive filter, satisfies the Nyquist criteria, as presented in Section 3.4, then the sampling and threshold comparison receiver is defined as a *Nyquist receiver.* The differences between the Nyquist receiver and the matched filter receiver, or their equivalence, must now be established and clearly defined. We tackle this problem by asking very specific questions and by presenting detailed answers.

1. Is the probability of error performance of the matched filter, $P_{e(\text{MF})}$, as derived in the preceding section, identical to the probability of error performance of the Nyquist channel model, $P_{e(\text{NYQ})}$, derived in Section 3.6?

2. Is the Nyquist channel model (receiver) identical with the matched filter receiver?

3. Is the matched filter receiver (correlation receiver) suitable for the reception of synchronous bandlimited data?

These questions have been asked by many design engineers who are required to apply their theoretical knowledge to the design of operational systems.

A detailed analysis of these three questions now follows:

1.   Is the Probability of Error $P_{e(\text{MF})} = P_{e(\text{NYQ})}$?   The probability of error of the matched filter receiver $P_{e(\text{MF})}$, derived in the preceding section, is

$$P_{e(\text{MF})} = \frac{1}{2} \operatorname{erfc} \sqrt{\frac{E_{\text{av}}}{N_o}} \tag{4.75}$$

where

E_{av} = average received signal energy at the input of the matched filter (correlation) receiver; in binary systems, $E_{\text{av}} = E_b$, where E_b is the bit energy

N_o = single-sided noise density

The probability of error of the Nyquist channel, $P_{E(\text{NYQ})}$ derived in Section 3.6. is

$$P_{e(\text{NYQ})} = Q\left(\frac{A}{\sqrt{N_T}}\right) = Q\left(\frac{A}{\sigma}\right) \tag{4.76}$$

where

A = peak voltage of the bandlimited (raised-cosine channel) receiver output at the sampling instant

N_T = total noise power at the output of the receive filter

σ = rms noise voltage at the output of the receive filter

$$P_{e\,(MF)} = \frac{1}{2}\, \text{erfc}\, \sqrt{\frac{E_{av}}{N_o}} = \frac{2}{2}\, Q\left(\sqrt{\frac{2E_{av}}{N_o}}\right)$$

$$= Q\left(\sqrt{\frac{2A^2 T_b}{N_o}}\right) = Q\left(\sqrt{\frac{2A^2 T_b}{2 N_T T_b}}\right) \tag{4.77}$$

$$= Q\left(\sqrt{\frac{A^2}{N_T}}\right) = Q\left(\frac{A}{\sigma}\right) = P_{E\,(NYQ)}$$

$$\boxed{P_{e\,(MF)} = P_{e\,(NYQ)}} \tag{4.78}$$

Thus question 1 has been answered.

In equation (4.77) we start with the left-hand expression of the probability of error of the matched filter receiver, $P_{e\,(MF)}$, perform a number of equivalent *substitutions*, and obtain the probability of error expression for the Nyquist channel, $P_{e\,(NYQ)}$. The following substitutions are used:

$$\text{erfc}\,(y) = 2Q(\sqrt{2}\,y) \qquad y > 0 \tag{4.79}$$

See Table 3.1, expression (4), and Fig. 3.29. Hence

$$E_{av} = A^2 T_b \tag{4.80}$$

The average bit energy, at the input of the Nyquist receiver (Fig. 3.24) equals the product of the signal power (A^2) and the bit duration T_b. At the threshold comparator input (receive filter output) the mean bit power is $A^2[(4 - \alpha)/4]$.

At this time, it is appropriate to review *Problem 3.10*, as its solution assists you in the following text. It is assumed that the received voltage is $+A$ or $-A$ volts, measured across a 1-Ω normalized impedance.

$$N_o = \frac{N_T}{\text{Nyquist bandwidth}} = \frac{N_T}{\frac{1}{2}(1/T_b)} \tag{4.81}$$

$$= N_T 2 T_b = \sigma^2 2 T_b$$

The single-sided density, N_o, is obtained by dividing the total noise power by the receiver noise bandwidth. This bandwidth, for the case of the Nyquist channel, equals the Nyquist bandwidth, $f_N = 1/(2T_b)$.

Equations (4.77) and (4.78) indicate that the P_e performance of the matched filter receiver $P_{e\,(MF)}$ and of the Nyquist channel $P_{e\,(NYQ)}$ are identical. However, we should remember that in the P_e derivation of matched filters there is **no discussion** of intersymbol-interference (i.e., a single-shot receiver is assumed). Furthermore, it is assumed that the cascading of the transmit and receive Nyquist filters is real.

   2.   Are the Matched Filter and the Nyquist Channel Receivers Equivalent?   In most data transmission systems a *sequence* of synchronous pulses (not a single pulse) is transmitted. Matched filter receivers do not assure intersymbol-interference (ISI)-free data reception. To avoid performance degradation, the matched filter receiver and the overall bandlimited channel have to be designed so as to satisfy the Nyquist theorem. Thus for synchronous spectral efficient data transmission, the matched filter receiver performance is *optimal,* per equation (4.77), if it satisfies the Nyquist ISI-free transmission theorem.

   The Nyquist channel model assures intersymbol-interference-free transmission (assuming that linear amplifiers are used). The transmit and receive filters can have an arbitrary split, as long as the *cascaded* transfer function meets the Nyquist criteria (e.g., the raised-cosine channel model). *For optimal P_e performance, the Nyquist channel filter partitioning must satisfy the matched filter criteria.* For the real Nyquist channel this means that

$$H_T(f) = H_R(f) \tag{4.82}$$

That is, the transmit and receive filters are identical.

   The equivalence of matched filter and Nyquist receivers and the required conditions for optimal performance are summarized in Fig. 4.12.

   Thus question 2 has been answered.

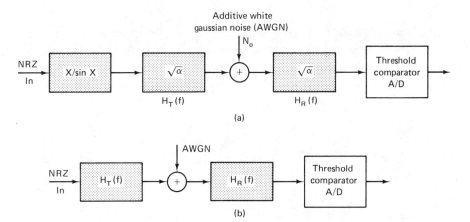

Figure 4.12   Equivalence of the Nyquist channel and matched filter. Required conditions for optimal performance of synchronous NRZ data. (a) Nyquist channel. Optimal P_e performance is achieved by having the $x/\sin x$ shaping in the transmitter; the raised-cosine channel characteristics is equally split between transmitter and receiver. Thus the receiver is matched to the transmitter. For real-valued transfer function, $H_T(f) = H_R(f)$. $\sqrt{\alpha}$ indicates the square root of the raised-cosine Nyquist equation (3.48a). (b) Matched filter channel. The receiver is matched to the transmitter; that is, $H_R(f) = G^*(f)e^{-j2\pi Tf}$; see equation (4.55). In bandlimited systems the matched filter has optimal performance, that is, ISI-free transmission if it satisfies the Nyquist transmission criteria.

3. Is the Matched Filter (Correlation Receiver) Suitable for the Reception of Synchronous Bandlimited Data?   The derivation of the matched filter receiver presents a mathematical insight into both the P_e performance and the concepts of optimal binary systems. This receiver is optimal for *single-shot* transmission, that is, for systems in which only one symbol is transmitted, and in wideband systems where each pulse is confined to its bit interval (i.e., intersymbol-interference is negligible). For optimal spectral efficiency (i.e., a system which has a bandwidth narrower than the bit rate), the Nyquist and the matched receiver criteria must be *simultaneously* satisfied.

Thus the answer to question 3 is **no;** the matched filter receiver, in itself, is *insufficient* for optimal efficiency in the reception of synchronous bandlimited data.

4.6 P_e PERFORMANCE OF BINARY COHERENT PSK SYSTEMS

In this section the correlation receiver (matched filter) theory is applied to the P_e derivation of optimal coherent binary phase-shift-keyed (BPSK) systems. The carrier and symbol timing recovery circuits are assumed to provide ideal synchronization signals that are free from phase noise. The principles, design, and performance degradations caused by imperfect synchronization circuits are presented in Chapter 7.

The BPSK signals are

$$s_1(t) = +C \cos \omega_c t \tag{4.83}$$

$$s_2(t) = -C \cos \omega_c t \tag{4.84}$$

To simplify the P_e derivation, we assume that the frequency of the carrier generator, shown in Fig. 4.13, is an integer multiple of the bit rate; that is,

$$f_c = n f_b = \frac{n}{T_b} \tag{4.85}$$

In numerous applications the data rate fluctuates (e.g., the source rate could change ± 50 b/s of its nominal 10-Mb/s rate). On the other hand, the carrier frequency and the radiated spectra must be tightly controlled. Thus, to satisfy equation (4.85), a fairly complex modem including a *bit stuffing* device may be required [Intelsat, 1980]. Fortunately, even though equation (4.85) is **not satisfied,** the P_e performance degradation is **practically negligible** if the carrier frequency is considerably higher than the bit rate. For $f_c \geq 10/T_b$, this degradation is in the order of 0.1 dB [Spilker, 1977].

To derive the P_e of the infinite bandwidth correlation receiver, we compute E_1, E_2, $\sqrt{E_1 E_2}\, \rho_{12}$, and R_{12} [see equation (4.73)]. Afterward, these computed values can be substituted into (4.72), the P_e equation.

$$\rho_{12} \sqrt{E_1 E_2} = \int_0^{T_b} s_1(t) s_2(t) \, dt = \tfrac{1}{2} C^2 T_b \tag{4.86}$$

$$R_{12} = \frac{2\sqrt{E_1 E_2}}{E_1 + E_2} \rho_{12} = -1 \tag{4.87}$$

$$P_e = \tfrac{1}{2} \operatorname{erfc}\left[\sqrt{\frac{1}{2}\frac{E_{av}}{N_o}\,(1 - R_{12})} \right] \qquad (4.88)$$

$$\boxed{P_e = \tfrac{1}{2}\operatorname{erfc}\sqrt{\frac{E_{av}}{N_o}} = \frac{1}{2}\operatorname{erfc}\sqrt{\frac{E_b}{N_o}}} \qquad \text{coherent BPSK} \qquad (4.89)$$

where

$$\operatorname{erfc}(x) \triangleq \frac{2}{\sqrt{\pi}}\int_x^\infty e^{-y^2}\,dy,\ x > 0$$

(numerical values are presented in Table 3.1)

E_{Av} = average bit energy E_b at the receiver input $E_{av} = \dfrac{C^2 T_b}{2}$
(in PSK systems, $E_{av} = E_b$)

N_o = noise density at the receiver input

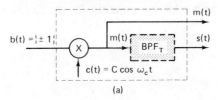

(a)

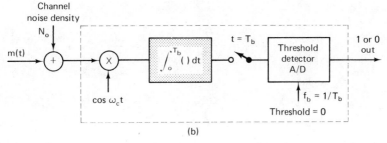

(b)

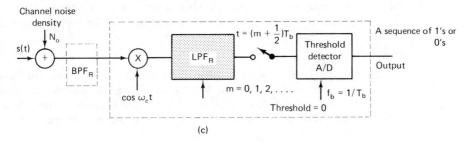

(c)

Figure 4.13   Coherent BPSK modulator, corresponding correlation receiver, and Nyquist receiver. (a) BPSK modulator. $m(t)$, infinite bandwidth signal. $s(t)$, bandlimited signal. (b) Correlation receiver. This correlation (matched filter) receiver is for the infinite bandwidth transmitter and channel model. (c) Nyquist receiver. LPF_R is the equivalent LPF of the bandpass channel (in some applications replaces the BPF_R filter). Its transfer function equals the square root of the raised-cosine channel.

Equation (4.89) represents the P_e performance of a BPSK matched filter correlation receiver. It is identical with equations (3.69) and (4.77) of the baseband Nyquist channel. If a BPSK system is bandlimited in accordance with the Nyquist criteria, it has a performance identical with that of the matched filter receiver.

4.6.1 Relationship Between E_b/N_o and C/N

The required bit energy, E_b, to noise density, N_o, ratio E_b/N_o is a convenient quantity for system calculations and performance comparisons. However, in *practical* measurements it is more convenient to measure the average carrier-to-average noise (C/N) power ratio. Most tests are performed using power and root-mean-square (rms) voltage meters, which are readily available; bit energy meters are not commercially manufactured. The following simple relations are useful for the E_b/N_o to C/N transformations:

$$E_b = CT_b = C\left(\frac{1}{f_b}\right) \tag{4.90}$$

$$N_o = \frac{N}{B_w} \tag{4.91}$$

$$\frac{E_b}{N_o} = \frac{CT_b}{N/B_w} = \frac{C/f_b}{N/B_w} = \frac{CB_w}{Nf_b} \tag{4.92}$$

$$\boxed{\frac{E_b}{N_o} = \frac{C}{N} \cdot \frac{B_w}{f_b}} \tag{4.93}$$

The E_b/N_o ratio equals the product of the C/N ratio and of the receiver **noise bandwidth**-to-bit rate ratio (B_w/f_b). It should be noted, of course, that any C/N measuring instrument can be recalibrated to read E_b/N_o directly, if required.

Example 4.2

A binary coherent BPSK modem operates at a rate of $f_b = 10$ Mb/s. The unmodulated carrier frequency is $f_c = 60$ MHz. What is the P_e performance of this system if the available E_b/N_o ratio is 8.4 dB? Assume that Nyquist channel shaping is employed. How much is the corresponding C/N ratio if it is measured at the receive bandpass filter output?

Solution. The $P_e = f(E_b/N_o)$ curve [equation (4.89)] is shown in Fig. 4.14. This curve is obtained from Table 3.1 and Fig. 3.25 From the curve we conclude that an $E_b/N_o = 8.4$ dB will yield a $P_e = 10^{-4}$. The double-sideband radio-frequency noise

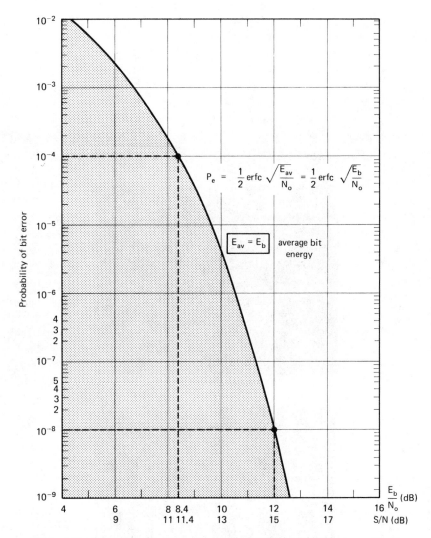

Figure 4.14 Probability of error performance of coherent BPSK modems. T_b, bit duration; C, average modulated carrier power; N_o, noise density (noise power in 1 Hz); N, total noise power measured in baseband Nyquist bandwidth; S, baseband E_b, average bit energy.

159

bandwidth of the Nyquist channel equals the bit rate (i.e., 10 MHz). From (4.93) we have

$$\frac{C}{N} = \frac{E_b}{N_0}\frac{f_b}{B_w}$$

or, expressed in dB-s,

$$\frac{C}{N} = \frac{E_b}{N_0} + 10\log\frac{f_b}{B_w}$$

$$= 8.4 + 10\log\frac{10\text{ Mb/s}}{10\text{ MHz}} = 8.4\text{ dB}$$

In Figs. 4.15 and 4.16, the measured eye diagrams of both wideband and narrowband BPSK systems are illustrated. The photographs of the demodulated eye diagrams

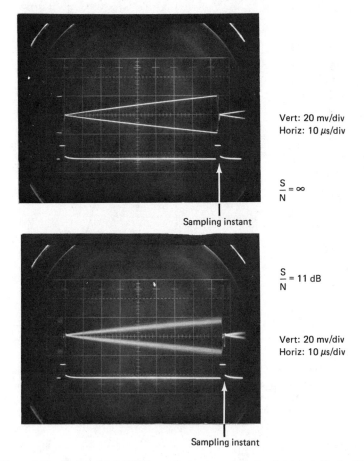

Vert: 20 mv/div
Horiz: 10 μs/div

$$\frac{S}{N} = \infty$$

Sampling instant

$$\frac{S}{N} = 11\text{ dB}$$

Vert: 20 mv/div
Horiz: 10 μs/div

Sampling instant

Figure 4.15 Demodulated BPSK eye diagrams—correlation (matched filter) wideband receiver implemented as an integrate–sample–dump detector.

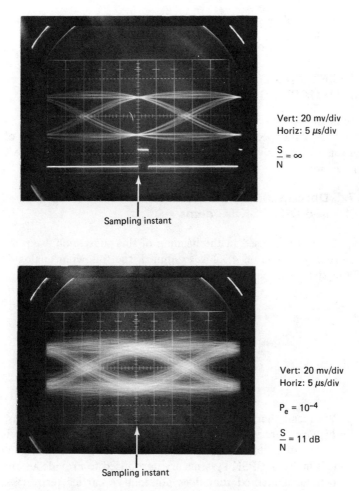

Vert: 20 mv/div
Horiz: 5 μs/div

$$\frac{S}{N} = \infty$$

Sampling instant

Vert: 20 mv/div
Horiz: 5 μs/div

$P_e = 10^{-4}$

$$\frac{S}{N} = 11 \text{ dB}$$

Sampling instant

Figure 4.16 Demodulated BPSK eye diagrams—Nyquist filtered channel. Fourth-order phase-equalized Butterworth filter having a 3-dB cutoff frequency at 17 kHz. For a bit rate $f_b = 32$ kb/s this filter approximates an $\alpha = 0.5$ roll-off Nyquist channel.

were taken at the output of the integrator and of the receive low-pass filter. For optimal performance, the integrator used in the correlation receiver of Fig. 4.13 is sampled at the end of the bit interval. Immediately after sampling, the integrator is instantaneously discharged (dumped). In high-speed wideband systems, operating above 10 Mb/s, it is customary to replace the "ideal integrator" with a simple RC time constant, or first-order filter. In this case the performance degradation is in the order of 1 dB. *Solve Problems 4.6 and 4.7.*

In spectrally efficient systems, Nyquist filtering is used. A demodulated eye diagram of a BPSK system, having $\alpha = 0.5$ raised-cosine channel characteristics is illustrated in Fig. 4.16. From the photograph it is evident that the intersymbol-interference-free sampling instant is at the center of the bit interval; sampling at any other time could degrade the system performance.

4.7 COHERENT AND DIFFERENTIALLY COHERENT-QUADRIPHASE PSK SYSTEMS (QPSK)

A functional description of the most frequently employed QPSK modems is followed by a study of the spectral efficiency of these systems. A probability of error performance analysis in ideal Nyquist shaped additive white Gaussian noise channels concludes this section.

4.7.1 Description of QPSK, DEQPSK, DQPSK, and OKQPSK Modems

The abbreviations used in the heading of this subsection are not completely new. If the heading of Section 4.3.1 is examined, the following analogies with quadriphase PSK systems can be noted.

Binary PSK	Quadriphase PSK
BPSK	QPSK
DEBPSK	DEQPSK
DBPSK	DQPSK
—	OKQPSK

We recall that BPSK systems are *binary coherent* systems, which require an *unambiguous* carrier recovery circuit. Differential encoding is introduced into DEBPSK systems to enable operation with 180° phase ambiguous carrier recovery circuits. Finally, DBPSK systems are introduced to provide an alternative demodulation method, a method that does not require carrier recovery circuitry (see Figs. 4.4 and 4.6).

Quadriphase modems (QPSK) are used in systems applications where the 1-b/s/Hz theoretical spectral efficiency of BPSK modems is insufficient for the bandwidth available. The various demodulation techniques used in binary phase-shift-keyed systems also apply to quadriphase PSK systems. In addition to the straightforward extensions of the binary modem techniques, a technique known as *offset-keyed or staggered* quadriphase modulation (OKQPSK) is also in use.

In QPSK systems the modulated signal has four distinct phase states. These states are generated by a unique mapping scheme of consecutive *dibits* (pairs of bits) into symbols. The corresponding phase states are maintained during the signaling interval T_s. This interval has a two-bit duration (i.e., $T_s = 2T_b$). The four possible dibits are frequently mapped in accordance with the *Gray code*. An important property of this code is that *adjacent symbols* (phase states) *differ by only one bit* [see Fig. 4.17(c)]. In transmission systems corrupted by noise and interference, the most frequent

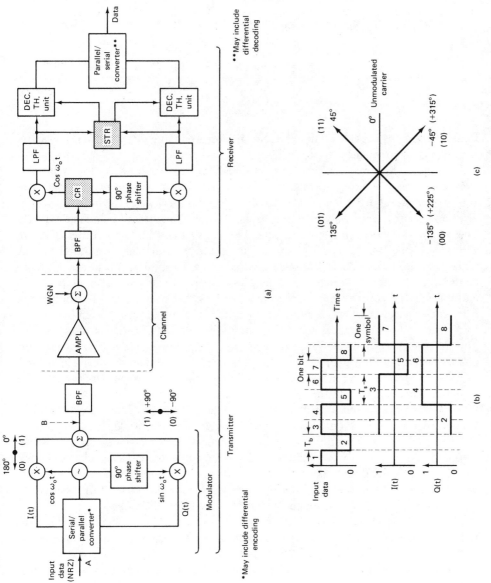

Figure 4.17 QPSK system representation. (a) Block diagram. CR, carrier-recovery circuit; STR, symbol timing-recovery circuit. (b) Modulator data streams. Data in bit number n is converted to I/Q baud number n. Bit rate $= 1/T_b$; baud or symbol rate $= 1/T_s$; $T_b = T_s/2$. (c) Signal-space diagram. Gray-coded vector presentation of signal states. (With permission from Dr. D. Morais, Farinon, Canada.)

errors are introduced by making decision errors between adjacent states. In this case the Gray code assures that a single symbol error corresponds to a single bit error. The Gray code is compared to the binary-coded-decimal code in *Problem 4.8.* Gray codes are advantageous, particularly when QPSK systems are followed by single-error-correcting decoders. In Chapters 6 and 10 we examine the performance of coded and modulated satellite systems.

The four Gray-coded signaling states may be described as follows:

$$
\begin{aligned}
s_{11}(t) &= A \cos\left(2\pi f_0 t + 45°\right) \\
s_{01}(t) &= A \cos\left(2\pi f_0 t + 135°\right) \\
s_{00}(t) &= A \cos\left(2\pi f_0 t + 225°\right) \\
s_{10}(t) &= A \cos\left(2\pi f_0 t + 315°\right)
\end{aligned}
\tag{4.94}
$$

In these equations the subscripts represent the corresponding Gray-coded signal states and the carrier frequency respectively.

A block diagram of a conventional QPSK modem is shown in Fig. 4.17. The NRZ data stream entering the modulator is converted by a serial-to-parallel converter into two separate NRZ streams. One stream is in phase, $I(t)$, and the other is quadrature phase, $Q(t)$, with a symbol rate equal to *half* that of the incoming bit rate. The relationship between the input data stream and the I and Q streams is shown in Fig. 4.17(b). Both I and Q streams are separately applied to multipliers (the terms "balanced mixer" and "product modulator" are also in use). The second input to the I multiplier is the carrier signal, $\cos \omega_0 t$, and the second input to the Q multiplier is the carrier signal shifted by exactly 90° (i.e., $\sin \omega_0 t$). The outputs of both multipliers are BPSK signals. The I multiplier output signal has phase 0° or 180° relative to the carrier, and the Q multiplier output signal has phase 90° or 270° relative to the carrier. The multiplier outputs are then summed to give a four-phase signal. Thus *QPSK can be regarded as two BPSK systems operating in quadrature.*

The four possible outputs of the modulator and their corresponding IQ digit combinations are shown in the signal space diagram of Fig. 4.17(c). Note that either 90° or 180° phase transitions are possible. As an example, a 180° phase transition would occur when the IQ digit combination changed from 11 to 00. For an unfiltered QPSK signal, phase transitions occur instantaneously and the signal has a constant-amplitude envelope. However, phase changes for *filtered* QPSK signals result in a *varying envelope amplitude.* In particular, a 180° phase change results in a momentary change to zero in envelope amplitude. In later sections we study the effects of envelope fluctuations on the P_E performance in nonlinear satellite channels.

The QPSK signal at the modulator output is normally filtered to limit the radiated spectrum, amplified, then transmitted over the transmission channel to the receiver input. Because the I and Q modulated signals are in quadrature (orthogonal), the receiver is able to demodulate and regenerate them independently of each other, operating effectively as two BPSK receivers.

A comprehensive treatment of orthogonal functions and their applications to communications systems can be found in [Stremler, 1979]. The regenerated I and Q streams are then recombined in a parallel-to-serial converter to form the original input data stream; however, this stream is of course subject to error because of the effects of noise and filtering.

A block diagram of an *offset-keyed quatenary phase-shift-keyed* (OK-QPSK) system is shown in Fig. 4.18(a). The block diagram of this system is very similar to conventional QPSK. The difference lies in the data transitions between the I and Q streams as they enter the multipliers. The incoming data stream is applied to a serial-to-parallel converter. One of the converter output streams, the Q stream in the case shown in Fig. 4.18(a), is then *offset* with respect to the other by delaying it by an amount equal to the incoming signal bit duration, $T_b = T_s/2$. This results in the relationship between the I and Q streams and the input data stream shown in Fig. 4.18(b). The resulting instantaneous phase states at the modulator output are the same as for QPSK. However, because both data streams applied to the multipliers can never be in transition simultaneously, only one of the vectors that comprise the offset-keyed quadriphase modulator output signal can change at any one time. The result is that only 90° phase transitions occur in the modulator output signals. Like QPSK, an *unfiltered* offset QPSK signal has a *constant-amplitude envelope*. However, for filtered offset QPSK signals, the result is a maximum amplitude envelope variation of 3 dB (30%) compared to the 100% amplitude envelope variation for conventional QPSK systems. In Section 4.10 we demonstrate that this lower-amplitude envelope variation imparts certain advantages to offset QPSK as compared to QPSK in both nonlinear satellite and also line-of-sight microwave systems. For example, when a bandlimited offset QPSK signal is transmitted through an amplitude-limiting device, there is only partial regeneration of the spectrum amplitude back to the unfiltered level. For QPSK under the same circumstances, however, there is an almost complete regeneration to the unfiltered level.

The receiver shown in Fig. 4.18(a) is identical to that shown in Fig. 4.17(a) for QPSK, with the exception that the regenerated I data stream is delayed by a unit bit duration $T_b = T_s/2$, so that when combined with the regenerated Q stream, the original "input data" stream is recreated. Of course, this is subject to error because of the effects of noise and filtering.

In Fig. 4.19(a) the block diagram of a *synchronous* QPSK/OKQPSK modulator is presented. The serial NRZ data stream $\{s_k\}$ is converted serial-to-parallel (S/P) and becomes two parallel data streams, $\{a_k\}$ and $\{b_k\}$. A *differential encoder* (DIFF ENC) may be inserted into the modulator. This encoder, with a complementary differential decoder of the receiver, is required if the carrier recovery circuitry introduces a phase ambiguity, or if a DQPSK demodulation scheme without carrier recovery circuitry is used. The serial-to-parallel converter assures that the I and Q data streams are synchronous. The $T_b = T_s/2$ delay element is inserted if OKQPSK modulation is required. This delay line does not change the synchronous relationship needed between the I and Q channels. In a similar manner to the binary PSK case, the local oscillator (LO) may be locked to an integer multiple of the data rate. To achieve this lock a voltage-controlled oscillator (VCO) is required. The VCO is normally part of a phase-locked-loop circuit [Gardner, 1979]. Even if the local oscillator is not locked (i.e., in synchronism) to an integer multiple of the data rate, the system

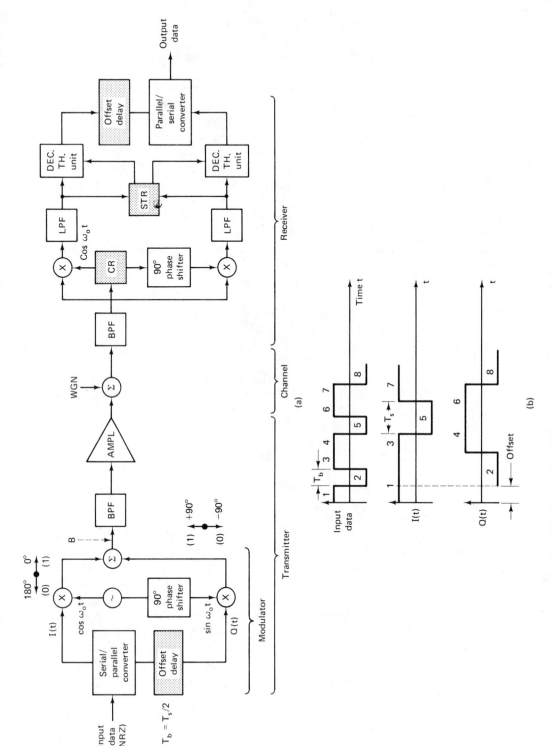

Figure 4.18 Offset (staggered) QPSK system representation. (a) Block diagram. (b) Modulator data streams. Input data bit number n is converted to I/Q baud number n. $T_b = T_s/2$. (With permission from Dr. D. Morais, Farinon, Canada.)

166

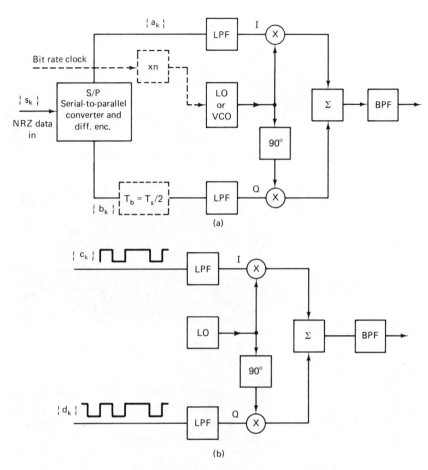

Figure 4.19 Differentially encoded and asynchronous QPSK (2L-QAM) modulators. (a) Synchronous. Differential encoding in the modulator and differential decoding in the demodulator is required if the carrier recovery introduces a phase ambiguity. (b). Asynchronous. 2L-QAM = two-level quadrature AM.

performance degradation in most satellite systems applications is very small (of the order of 0.1 dB or less). Spectral shaping may be achieved by means of premodulation low-pass filters (LPF) or postmodulation bandpass filters (BPF). In a number of applications a combination (or cascade) of these filters can be used.

If the incoming data rates are *asynchronous,* as illustrated in Fig. 4.19(b), it is frequently advantageous to modulate directly the *I* and *Q* data streams. Here we assume that the $\{c_k\}$ and $\{d_k\}$ data *rates* are independent of each other and are fluctuating around specified bit rates. In this case the QPSK modulator provides, in effect, an *asynchronous multiplexing* scheme. If the data streams are bandlimited with a low-pass filter before modulation, this scheme is known as a *two-level quadrature*

Figure 4.20   44.7-Mb/s offset QPSK modulator and demodulator. (Courtesy of Dr. D. H. Morais, Farinon, Canada.)

AM (2L-QAM) modulator. The spectral efficiency and P_e performance of the QAM system is identical with that of QPSK systems. *Solve Problem 4.9.*

The block diagrams of coherent QPSK and offset-keyed QPSK demodulators are shown in Figs. 4.17 and 4.18. Most carrier recovery circuits introduce phase ambiguities, and for this reason differential encoding (DE) and decoding is required; thus the terms DEQPSK and DEOK-QPSK. A photograph of a differentially encoded 44.7-Mb/s offset-QPSK modulator and demodulator board is shown in Fig. 4.20.

Differential coding/decoding—QPSK/OK-QPSK.   A frequently used carrier recovery circuit is the *quadrupler*. The phase-locked loop of this circuit, described in Chapter 7, locks onto the fourth harmonic of the unmodulated carrier frequency. A four-phase ambiguity results in that any of the transmitted phases cos $(\omega t + n\pi/2)$, where $n = 0, 1, 2, 3$, gives a cos $4\omega t$ term for the phase-locked loop to lock onto. Thus the transmitted *prime* phase is not available as a reference phase. Depending on the phase of the recovered carrier, the (P, Q) data pair at the modulator input can be either (P, Q), $(\bar{P}, \bar{Q})$, $(Q, \bar{P})$, or $(\bar{Q}, P)$ at the demodulator output, which except for the first would not give the same output as the input. This dilemma may be resolved by utilizing encoding in the modulator and differential decoding in the demodulator (see Fig. 4.21).

Differential coding encodes the pairs so that only changes in the phase of the QPSK represent the transmitted information, not the absolute phase, thus eliminating

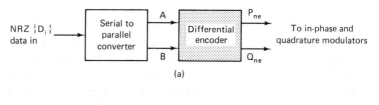

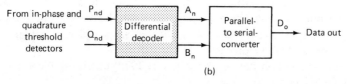

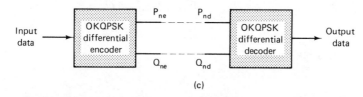

Figure 4.21 Differential encoder/decoder for QPSK and OK-QPSK modems. (a) Encoder (transmit side); (b) decoder (receive side); (c) OKQPSK encoder/decoder.

the need for a reference phase [Clewer, 1979]. The equations for encoding and decoding of QPSK signals are:

QPSK encoder:

$$P_{ne} = \overline{(A_n \oplus B_n)}\,(A_n \oplus P_{(n-1)e}) + (A_n \oplus B_n)\,(B_n \oplus Q_{(n-1)e})$$
$$Q_{ne} = \overline{(A_n \oplus B_n)}\,(B_n \oplus Q_{(n-1)e}) + (A_n \oplus B_n)\,(A_n \oplus P_{(n-1)e})$$

$$(4.95)$$

where $\oplus$ denotes exclusive-OR addition.

QPSK decoder:

$$A = \overline{(P_{nd} \oplus Q_{nd})}\,(P_{nd} \oplus P_{n-1d}) + (P_{nd} \oplus Q_{nd})\,(Q_{nd} \oplus Q_{n-1d})$$
$$B = \overline{(P_{nd} \oplus Q_{nd})}\,(Q_{nd} \oplus Q_{(n-1)d}) + (P_{nd} \oplus Q_{nd})\,(P_{nd} \oplus P_{(n-1)d})$$

$$(4.96)$$

where for the encoder (P_{ne}, Q_{ne}) and $(P_{(n-1)e}, Q_{(n-1)e})$ represent the present and previous outputs and (A_n, B_n) represents the present input, and for the decoder (A_n, B_n) represents the decoded output and (P_{nd}, Q_{nd}) and $(P_{(n-1)d}, Q_{(n-1)d})$ represent the present and previous inputs from the regenerators. In a noise-free environment the input data sequence $\{D_i\}$ equals the output data sequence $\{D_o\}$, but the $\{P_{ne}, Q_{ne}\}$ sequence is not necessarily the same as the $\{P_{nd}, Q_{nd}\}$ sequence. *Why?*

Input data are split into two paths by the serial to parallel converter, with even bits going to the A channel and odd bits to the B channel (see Fig. 4.21). The differential encoder outputs the P_{ne} and Q_{ne} bits according to the present input pair and the last output pair. The decoder looks at the last and present demodulator outputs and reconstructs the initial (A, B) pair. This pair is passed on to the parallel-to-serial converter, where the input data stream is reconstructed.

Offset QPSK presents a slightly different problem. Let us assume that OQPSK is generated by simply delaying one channel of a differentially encoded (P, Q) pair by one-half of a symbol period prior to modulation and realigning after demodulation. However, when the recovered carrier is out of phase by $\pi/2$ or $3\pi/2$, P and Q are interchanged to become either $(\overline{Q}, P)$ or $(Q, \overline{P})$ at the demodulator output, and the delayed channel is inadvertently realigned with the wrong bit. Despite differential coding, the output in such a case is invalid, so a different scheme for coding to solve the time ambiguity problem is required.

A simple solution derived by [Clewer, 1979, based on Weber, 1978] is:
OKQPSK encoder:

$$P_{\mathrm{ne}} = D_{\mathrm{EVEN}} \oplus Q_{ne}$$
$$Q_{\mathrm{ne}} = D_{\mathrm{ODD}} \oplus \overline{P}_{ne}$$
(4.97)

OKQPSK decoder:

$$D_{\mathrm{EVEN}} = P_n \oplus Q_n$$
$$D_{\mathrm{ODD}} = \overline{P}_n \oplus Q_n$$
(4.98)

where D_{EVEN} and D_{ODD} represent even and odd bits in the input/output serial data streams, and P_n and Q_n represent inputs and outputs of the encoder/decoder at the same point in time.

The encoder/decoder for OKQPSK is considerably less complex than the one for QPSK, which is an unanticipated advantage of OK-QPSK. An in-depth treatment of this subject is given in [Cacciami and Wolejsza, 1971].

DQPSK demodulation. The design of carrier recovery circuits is a difficult task, particularly if fast modem synchronization is required. To avoid the need of a complex carrier recovery circuit and to improve the synchronization speed of the demodulator, *differential demodulation* may be employed instead of coherent demodulation.

A number of planned satellites, with digital regeneration transponders which will operate in a multiple-access burst mode, are also expected to use differential demodulators, directly at the received radio frequency in the range 10 to 40 GHz [European Space Center, 1979].

A typical quadriphase demodulator which employs the differential demodulation principle is shown in Fig. 4.22. Following the receive BPF the modulated carrier is split and sent to two differential DPSK demodulators. These demodulators differ from the one described in Fig. 4.6 in that the value of the delay element is different. (*Note:* Here we have $T_s = 2T_b$ whereas in Fig. 4.6 we had $T_s = T_b$. Why?)

To assure quadrature demodulation, a 90° wideband phase shifter has to be added into the quadrature channel. The symbol timing recovery circuit (STR) required for sampling of the threshold comparator (analog-to-digital converter), may be con-

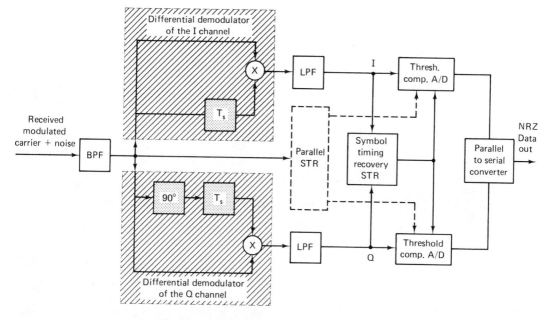

Figure 4.22 DQPSK demodulator block diagram.

nected to the demodulated I and/or Q channels or to the output of the bandpass filter. This symbol timing recovery system, known as *parallel STR,* is employed when fast synchronization speed is required [Feher and Takhar, 1978, and Chapter 7 of this book].

4.7.2 Spectrum and Spectral Efficiency of QPSK Modems

The block diagram of a QPSK modulator, Fig. 4.17, indicates that a QPSK signal can be generated by linear addition of two quadrature (90° shifted) binary phase-shift-keyed (BPSK) signals. The inphase and quadrature baseband drive signals $I(t)$ and $Q(t)$ are independent synchronous data streams; that is, the signal polarity of channel I is independent of the signal polarity of channel Q. The symbol rates of the I and Q channels equal one-half of the bit rate (i.e., $f_s = f_b/2$).

If the input data are random and equiprobable, the power spectrum of an unfiltered (i.e., infinite bandwidth) BPSK carrier, as derived in Section 4.3.2 is

$$S_{\text{BPSK}} = KA^2 T_b \left[\frac{\sin \pi (f - f_c) T_b}{\pi (f - f_c) T_b} \right]^2 \tag{4.99}$$

The QPSK spectrum is obtained by superposition of the two independent BPSK spectra; thus the shape of the power spectra, described in (4.99), does not change. However, note that the non-return-to-zero I and Q polar baseband signals have an $f_s = f_b/2$ rate. Thus the resultant QPSK spectrum is

$$S_{\text{QPSK}}(f) = CA^2 T_s \left[\frac{\sin \pi (f - f_c) T_s}{\pi (f - f_c) T_s} \right]^2 \qquad (4.100a)$$

An equivalent form of this equation is

$$S_{\text{QPSK}}(f) = CA^2 T_b \left[\frac{\sin 2\pi (f - f_c) T_b}{2\pi (f - f_c) T_b} \right]^2 \qquad (4.100b)$$

where

$CA^2 =$ total infinite bandwidth signal power normalized across 1-Ω resistance

$T_b = 1/f_b =$ bit duration

$T_s = 1/f_s =$ symbol duration of the I and Q channels

The differential encoding and offset keying operations (delaying the Q channel by T_b seconds) do not modify the power spectral density. Thus for random equiprobable

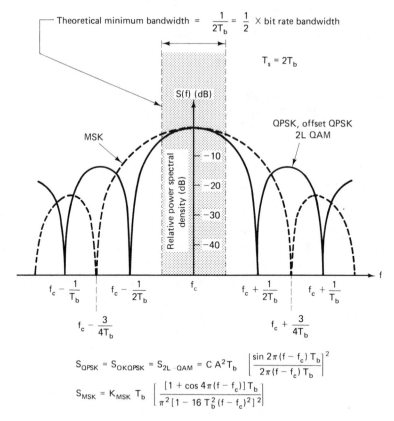

Figure 4.23 Power spectral density of QPSK offset QPSK, 2L-QAM, and MSK systems. (With permission from Dr. D. Morais, Farinon, Canada.)

input data, equation (4.100) represents the power spectrum of unfiltered QPSK, OKQPSK, and DEQPSK modulated systems. Similarly, this equation applies to two-level QAM (2L-QAM) systems.

The unfiltered power spectral density of QPSK, OKQPSK, and 2L-QAM modulated systems is illustrated in Fig. 4.23. Note that the first spectral nulls are at $f_c \pm 1/(2T_b)$, and that the minimum double-sided radio-frequency bandwidth requirement is $1/2T_b$. For reference, the power spectral density of minimum-shift-keyed (MSK) systems is also illustrated.

The intersymbol-interference-free channel filtering requirements are obtained from the Nyquist baseband transmission theorems, the equivalence of low-pass and of bandpass channel models, and the superposition theorem. The 1-b/s/Hz theoretical spectral efficiency of bandlimited BPSK systems becomes 2 b/s/Hz for the case of QPSK systems. The spectral shaping filters and bandlimited power spectra for the BPSK modulator, illustrated in Fig. 4.8, apply also to the QPSK case. In the BPSK case the symbol rate equals the bit rate, $f_s = f_b$, whereas in the QPSK case the symbol rate equals one-half of the bit rate, $f_s = f_b/2$. *Why?*

Example 4.3

Illustrate the unfiltered spectral density of an $f_b = 64$ kb/s rate QPSK signal. Assume that the source is an equiprobable random generator and that the carrier frequency is $f_c = 512$ kHz. Determine the transmitted power spectrum if Nyquist shaped filters having the square root of $\alpha = 0.3$ roll-off factor are employed.

Solution. The measured unfiltered and filtered power spectra are illustrated in Fig. 4.24. The unfiltered spectrum has the expected, $(\sin x/x)^2$ shape. The spectral nulls are located at $f_c \pm nf_s$, where $n = 1, 2, 3, \ldots$. The filtered spectrum shows an almost flat segment in the $f_c \pm \frac{1}{2}f_s$ frequency range. This flat segment is due to the $(x/\sin x)$-shaped equalizer of the transmit filter.

■

4.7.3 Error Probability Performance of QPSK Receivers

The probability of error, P_e, performance of coherent QPSK, offset-keyed QPSK, DEQPSK, and DQPSK systems is derived. Due to the **equivalence** of coherent QPSK systems with that of two-level coherent QAM, the results apply also to QAM systems.

$P_e = f(E_b/N_o)$ **derivation for coherent QPSK systems.**   In this section two separate P_e performance derivations are presented. Both of them are based on the theoretical coherent demodulator model, shown in Figs. 4.17 and 4.27. In the first, somewhat heuristic derivation, we use the result obtained for BPSK modems, presented in Section 4.6. Here a relationship between the *probability of bit error* P_e and probability of *symbol error* P_E, is also derived. The second derivation is somewhat more complex. It illustrates an application of the orthogonality principle and provides an insight into the mathematical theory of operation of quadrature demodulators. The books [Haykin, 1978 and Shanmugam, 1979] provide background material and present a good treatment of this subject.

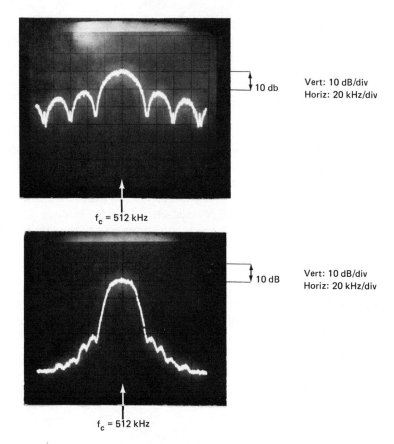

10 db

Vert: 10 dB/div
Horiz: 20 kHz/div

f_c = 512 kHz

10 dB

Vert: 10 dB/div
Horiz: 20 kHz/div

f_c = 512 kHz

Figure 4.24 (a) Unfiltered (infinite bandwidth) and (b) filtered power spectra of 64-kb/s QPSK (2L-QAM) modulated signals. Fourth-order Butterworth premodulation low-pass filters having 3-dB attenuation at 17.5 kHz bandlimit the transmitted spectrum. This modem satisfies the INTELSAT single-channel-per-carrier (SCPC) satellite system specifications. The channel filtering approximates an $\alpha = 0.5$ Nyquist channel. (Courtesy of H. Pham Van, Spar Aerospace Limited.)

1. Heuristic Derivation for Coherent QPSK Systems. From an examination of the block diagram of the coherent QPSK demodulator (Figs. 4.17 and 4.27) we may conclude that the P_e performance of the in-phase, I, and quadrature, Q, channels is independent and that in a symmetrical bandpass channel the performance of these orthogonal channels is identical.

We designate the probability of error of the in-phase BPSK demodulator (upper path) as P_{eI} and of the quadrature (lower path) as P_{eQ}. The average probability of *correct symbol* reception, P_C, at the QPSK receiver output equals the product of P_{cI}, P_{cQ}, where P_{cI} and P_{cQ} represent the independent correct decision probabilities of the I and Q binary PSK demodulators, respectively. Thus we have

$$P_C = (1 - P_{eI})(1 - P_{eQ}) \qquad (4.101)$$

The corresponding QPSK *symbol error* rate is

$$P_e = 1 - P_C = P_{eI} + P_{eQ} - P_{eI}P_{eQ} \qquad (4.102)$$

Since $P_{eI} = P_{eQ}$ and P_{eI} and P_{eQ} are typically small quantities ($<10^{-2}$), we obtain a good approximation for the overall probability of *symbol* error of the QPSK demodulator. It is given by

$$P_E \approx P_{eI} + P_{eQ} = 2P_{eI} = 2P_{eQ} \qquad (4.103)$$

The probability of error of the individual BPSK demodulators is

$$P_{eI} = P_{eQ} = \frac{1}{2} \operatorname{erfc} \sqrt{\frac{E_b}{N_o}} = P_{e\,\text{BPSK}} \qquad (4.104)$$

Finally, the probability of *symbol error*, $P_{E,\,\text{QPSK}}$ of the coherent QPSK receiver is

$$P_{E\,\text{QPSK}} = 2P_{eI} = 2 \cdot \frac{1}{2} \operatorname{erfc} \sqrt{\frac{E_b}{N_o}}$$

$$\boxed{P_{E\,\text{QPSK}} = \operatorname{erfc} \sqrt{\frac{E_b}{N_o}}} \qquad (4.105)$$

where

$$\operatorname{erfc}(y) \triangleq \frac{2}{\sqrt{\pi}} \int_y^\infty e^{-z^2}\, dz \qquad for\ y > 0 \qquad (4.106)$$

Numerical values of the error function are listed in Table 3.1 and an illustration is given in Fig. 3.25.

Assuming that the system parameter E_b/N_o is specified, the probability of symbol error of the QPSK system is twice that of the BPSK system. This corresponds to an increased E_b/N_o requirement of approximately 0.3 dB if the same symbol error rate as in BPSK systems is required.

From a conceptual point of view the symbol error rate, $P_{E\text{QPSK}}$, is an important system parameter. It is also important in the analysis of systems in which the QPSK modem operates in conjunction with certain forward error-correcting decoding (codec) devices. However, in many applications it is required to know the *bit error rate* of the QPSK system, $P_{e\,\text{QPSK}}$. This error rate is related to the symbol error rate, $P_{E\,\text{QPSK}}$, as stated in the following paragraphs.

Relation of the Bit Error Rate, $P_{e\,\text{QPSK}}$, to the Symbol Error Rate, $P_{E\,\text{QPSK}}$.   The Gray-coded and non-Gray-coded cases are illustrated in Fig. 4.25. In the Gray-coded "constellation," adjacent phase states or symbols differ by only one bit, whereas in the non-Gray-coded case the adjacent states differ by one or by two bits. We focus attention on decisions in which *adjacent* phase states are erroneously interchanged. The probability of interchanging 180° phase-shifted signals is negligible when compared to the probability of error of adjacent, 90° shifted, phase states. Thus, in the

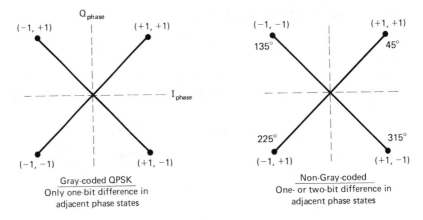

First no. represents I channel
Second no. represents Q channel

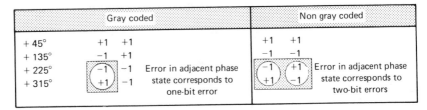

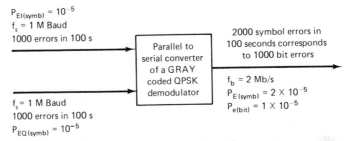

Figure 4.25 Bit error rate, P_e, and symbol error rate, P_E, relation in QPSK systems.

Gray-coded signal constellation case, *one symbol error corresponds to one bit error.* In the non-Gray-coded case illustrated, an interchange of adjacent phase states corresponds, 50% of the time, to two bit errors.

The **bit error rate,** P_e, and **symbol error rate** P_E, relationship is explained with the aid of an example illustrated in Fig. 4.25. Assume that the QPSK modem bit rate is $f_b = 2$ Mb/s; thus the corresponding symbol rate is $f_s = 1$ Mbaud (1M symbols/second). The symbol error rates of the I and Q channels are $P_{EI} = P_{EQ} = 2 \times 10^{-5}$. In this case at the output of the parallel-to-serial converter there are on an average 2000 erroneous bits in a 100-s interval (Gray coding is assumed). The bit error rate of this modem is calculated with the aid of the following well-known relationship

$$P_e = \frac{\text{number of bits in error}}{\text{total number of bits}} = \frac{2000 \text{ errors}}{2 \text{ Mb/s} \cdot 100 \text{ s}} = 10^{-5}$$

Thus *the bit error rate of a Gray-coded QPSK modem equals one-half of its symbol error rate.* A more rigorous analysis of the P_e and P_E relationship is presented in [Lindsay and Simon, 1973]. The *bit error rate* of the Gray-coded QPSK modem is, therefore,

$$P_{e\,\text{QPSK}} = \tfrac{1}{2} P_{E\,\text{QPSK}} = \tfrac{1}{2} \text{erfc} \sqrt{\frac{E_b}{N_o}} \qquad (4.107)$$

The theoretical bit error rate performance of coherent QPSK systems, $P_{e\,\text{QPSK}} = f(E_b/N_o)$, is illustrated in Fig. 4.26. The performance of other modems is also illustrated in this figure. Note that the bit error rate performance of the illustrated *Gray-coded QPSK* and of the Gray-coded offset-keyed four-phase OK-QPSK modems is *identical* with that of coherent BPSK modems.

   2. **Detailed $P_e = f(E_b/N_o)$ Derivation for Coherent QPSK Systems.**   In the following derivation we assume that the transmitted QPSK signal is unfiltered and that the channel has infinite bandwidth (Fig. 4.27). A channel which has a bandwidth three times larger than the bit rate is for all practical purposes an *infinite bandwidth channel.* Based on the equivalence conditions of matched filter (correlation receivers) and of Nyquist shaped channels, as described in Section 4.5, we conclude that this derivation also applies to bandlimited ISI free QPSK systems.

   Prior to filtering, the received QPSK signal, corrupted by zero mean additive white Gaussian noise (AWGN) may be expressed as

$$y(t) = A\sqrt{2} \, \cos(2\pi f_c t + \theta_m) + n(t) \qquad (4.108a)$$

where $\theta_m = 45°$, $135°$, $225°$, or $315°$ [see equation (4.94)] or, equivalently,

$$y(t) = \pm A \, \cos(2\pi f_c t) \pm A \, \sin(2\pi f_c t) + n(t) \qquad \text{for } 0 \leq t \leq T_s \qquad (4.108b)$$

The polarity of the cos $(\cdot)$ and sin $(\cdot)$ coefficients depends on which particular symbol is transmitted. We recall that in QPSK four different symbols exist and each symbol is characterized by its phase. The transmitted phase, as shown in the signal state space diagram of Fig. 4.27, depends on which particular pair of bits has been transmitted. At the end of a symbol interval, T_s (decision or sampling instant of the infinite bandwidth receiver), the demodulated baseband output of the in-phase correlator is

$$b_{oI} = \int_0^{T_s} [A\sqrt{2} \, \cos(2\pi f_c t + \theta_m) \cos 2\pi f_c t + n(t) \cos 2\pi f_c t] \, dt$$

$$= \int_0^{T_s} \{[\pm A \, \cos(2\pi f_c t) \pm A \, \sin(2\pi f_c t) + n(t)] \cos 2\pi f_c t\} \, dt \qquad (4.109)$$

$$= \pm \frac{1}{2} A T_s + 0 + \int_0^{T_s} n(t) \cos(2\pi f_c t) \, dt$$

The integral of the second term equals zero as the cos $(\cdot)$ and sin $(\cdot)$ terms are **orthogonal over the 0 to T_s symbol interval.** This *orthogonal condition holds only if*

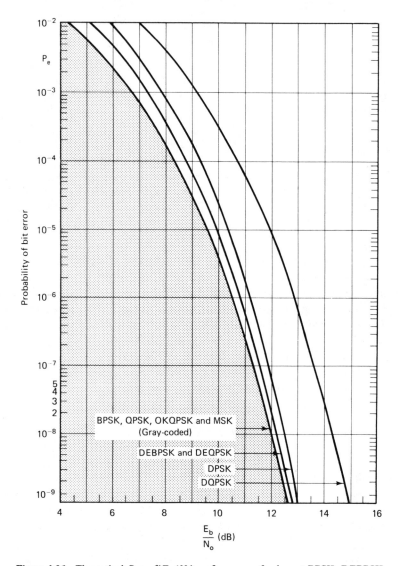

Figure 4.26 Theoretical $P_e = f(E_b/N_o)$ performance of coherent BPSK, DEBPSK, coherent QPSK, and DQPSK modems (Gray encoded). Additive white Gaussian noise and intersymbol-interference-free model.

the carrier frequency is an integer multiple of the symbol rate [i.e., $f_c = kf_s$ (k an integer > 0)].

Similarly, we conclude that the demodulated baseband output of the quadrature correlator is [Haykin, 1978]

$$b_{oQ} = \pm \frac{1}{2} AT_s + \int_0^T n(t) \sin(2\pi f_c t) \, dt \qquad (4.110)$$

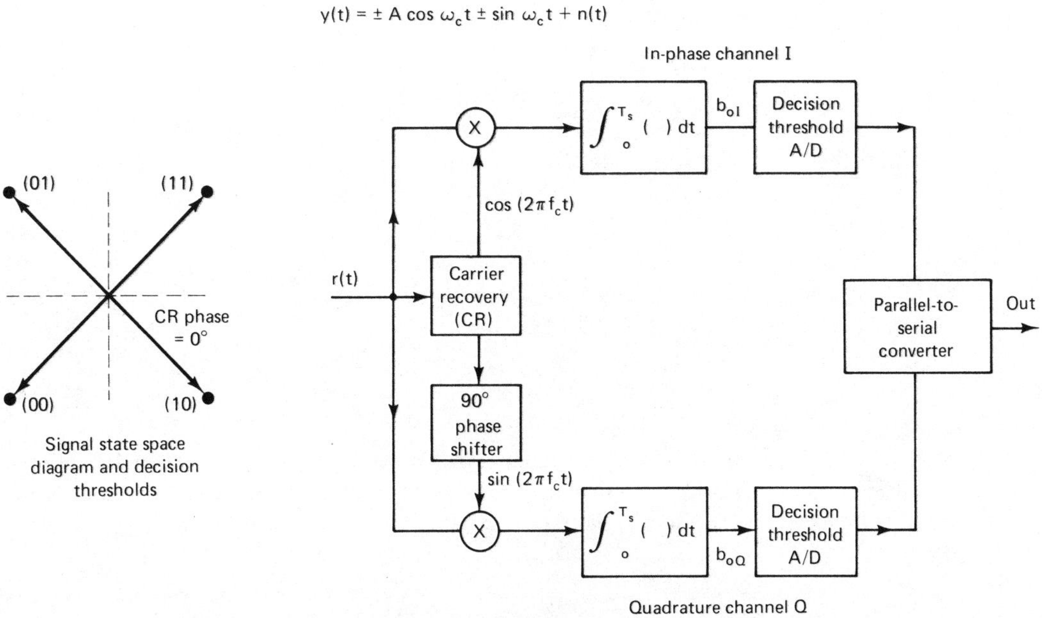

$$y(t) = \pm A \cos \omega_c t \pm \sin \omega_c t + n(t)$$

Figure 4.27   Coherent QPSK demodulator for unfiltered "infinite-bandwidth" signals. For band-limited signals replace the integrators by low-pass filters, as illustrated in Fig. 4.17. Solid line, signal vectors; dashed line, decision boundary.

The corresponding random variables B_{oI} and B_{oQ} are uncorrelated variables. If the $f_c = kf_s$ relationship is satisfied, these Gaussian variables are statistically indepen-dent.

The mean value (expected value) of B_{oI} and B_{oQ} is

$$E[B_{oI}] = E[B_{oQ}] = \pm \frac{AT_s}{2} \tag{4.111}$$

The variance of b_{oI} and b_{oQ} represents the ac noise power. This power is given by

$$\begin{aligned}
\text{Var}\,[b_{oI}] &= E\left\{\left[\int_0^{T_s} n(t) \cos(2\pi f_c t)\, dt\right]^2\right\} \\
&= E\left\{\int_0^{T_s}\int_0^{T_s} n(t)n(\sigma) \cos(2\pi f_c t) \cos(2\pi f_c \sigma)\, dt\, d\sigma\right\} \\
&= \int_0^{T_s}\int_0^{T_s} \frac{N_o}{2}\, \delta(t - \sigma) \cos(2\pi f_c t) \cos(2\pi f_c \sigma)\, dt\, d\sigma \\
&= \frac{N_o}{2}\int_0^{T_s} \cos^2(2\pi f_c t)\, dt = \frac{N_o T_s}{4}
\end{aligned} \tag{4.112}$$

(*Note*: $N_o/2$ represents the spectral density of the additive white Gaussian noise.)

Similarly, the variance of the lower correlator is

$$\text{Var}\,[b_{oQ}] = \frac{N_o T_s}{4} \tag{4.113}$$

From these equations we obtain

$$P_{eI} = P_{eQ} = \tfrac{1}{2}\,\text{erfc}\,\sqrt{\frac{A^2 T_s}{2N_o}} \tag{4.114a}$$

The symbol duration is $T_s = 2T_b$, the signal energy is $E_s = A^2 T_s$, and the bit energy is $E_b = A^2 T_b$. Thus we have

$$P_{eI} = P_{eQ} = \tfrac{1}{2}\,\text{erfc}\,\sqrt{\frac{E_s}{2N_o}} = \frac{1}{2}\,\text{erfc}\,\sqrt{\frac{E_b}{N_o}} \tag{4.114b}$$

This expression is the same as equation (4.89), the probability of error of coherent BPSK systems. The probability of *symbol* error of the QPSK receiver is double that of the individual BPSK receivers [see equation (4.103)]. Thus we have

$$\boxed{P_{E(\text{QPSK})} = \text{erfc}\,\sqrt{\frac{E_b}{N_o}}} \tag{4.115}$$

The probability of bit error of **Gray-coded** QPSK systems equals one-half of the probability of symbol error [see equation (4.107)]; thus the probability of *bit error* is

$$\boxed{P_{e\,(\text{QPSK})} = \frac{1}{2}\,\text{erfc}\,\sqrt{\frac{E_b}{N_o}}} \tag{4.116}$$

Example 4.4

In the INTELSAT-V time-division multiple-access (TDMA) satellite system, coherent four-phase phase-shift keying modems are used. The back-to-back modem performance is within 2.5 dB of the theoretical QPSK probability of error curve. Due to earth station–satellite power amplifier nonlinearities and other system impairments, an additional 1.6-dB degradation in the overall $P_e = f(E_b/N_o)$ performance exists. The transmitted bit rate, f_b, is 120 Mb/s. The specified channel filters approximate the $\alpha = 0.4$ roll-off Nyquist channel. Sketch the transmit and receive filter characteristics. Determine the required E_b/N_o and C/N ratio if a P_e of 10^{-7} is the objective. What would the required C/N ratio and the associated radio-frequency (RF) bandwidth be if the four-phase QPSK modems were replaced by two-phase BPSK modems?

Solution. The amplitude and group delay masks, as specified by [Intelsat, 1980] are shown in Fig. 4.28. The modulator and demodulator amplitude and group delay characteristics are shown separately. Separate specifications for the transmit and for the receive filter characteristics are required to assure that modulators manufactured by company X and demodulators manufactured by company Y are "matched" (i.e., the cascaded modem filters satisfy the Nyquist transmission criteria).

The $x/\sin x$ amplitude shaping is included in the transmitter in Fig. 4.28(a). The symbol rate of the 120-Mb/s QPSK modem f_s is 60 Mbaud (or 60 Msymbols/s); thus the Nyquist frequency is 30 MHz. To achieve the illustrated amplitude and group delay characteristics, either premodulation low-pass filters or postmodulation bandpass filters must be employed. For the bandpass filters, symmetry around the carrier frequency is assumed.

The theoretical E_b/N_o requirement is obtained from Fig. 4.26 or from equation (4.107). For a $P_e = 10^{-7}$, a theoretical QPSK modem, having no degradations, requires an E_b/N_o of 11.5 dB. Due to the assumed 2.5-dB modem imperfections and additional 1.6-dB system impairment previously noted, the practical INTELSAT-V modem and system E_b/N_o requirements are:

$$(E_b/N_o)_{\text{pract. modem}} = (E_b/N_o)_{\text{theoret}} + (\text{modem degrad})_{\text{dB}}$$
$$= 11.5 \text{ dB} + 2.5 \text{ dB} = 14 \text{ dB} \tag{4.117}$$

$$(E_b/N_o)_{\text{INTELSAT-V}} = (E_b/N_o)_{\text{pract. modem}} + (\text{additional syst. degrad})_{\text{dB}}$$
$$= 14 \text{ dB} + 1.6 \text{ dB} = 15.6 \text{ dB} \tag{4.118}$$

The required C/N ratio is obtained from equation (4.93):

$$\frac{C}{N} = \frac{E_b}{N_o} \frac{f_b}{B_w}$$

$$\left(\frac{C}{N}\right)_{\text{dB}} = 15.6 + 10 \log \frac{120 \text{ Mb/s}}{60 \text{ MHz}} = 18.6 \text{ dB} \tag{4.119}$$

If the 120-Mb/s QPSK modem is replaced by a 120-Mb/s BPSK modem, then the C/N requirement obtained from (4.118) and (4.119) is now

$$(C/N)_{\text{BPSK}} = 18.6 \text{ dB} - 3 \text{ dB} = 15.6 \text{ dB}$$

The 3-dB C/N difference between QPSK and BPSK systems is due to the use of four phases instead of two. Geometrically, we may interpret this as follows: If for both modems we assume the same unmodulated carrier voltage A, then the differential between two signal phasors in BPSK systems is $2A$, whereas in QPSK it is only $\sqrt{2} A$.

The E_b/N_o requirement of the BPSK and QPSK modems is practically the same (within 0.3 dB in the range $P_E = 10^{-5}$ to 10^{-10}).

The radio-frequency bandwidth requirement of the BPSK modem is 120 MHz, whereas in the QPSK case it is only 60 MHz.   ■

P_e **performance of OK-QPSK and DEQPSK systems.**   The fundamental difference between coherent QPSK and OK-QPSK modem hardware lies in the presence or absence of the T_b-second offset delay element (see Figs. 4.17 and 4.18). As the in-phase and quadrature BPSK channels are orthogonal and independent at the sampling instant, the insertion of a delay element into the offset-keyed QPSK system does not affect the performance of the coherent QPSK modem. Thus the P_e performance of the OK-QPSK modem is described by

$$P_{e\,(\text{OK-QPSK})} = \frac{1}{2} \text{erfc} \sqrt{\frac{E_b}{N_o}} \tag{4.120}$$

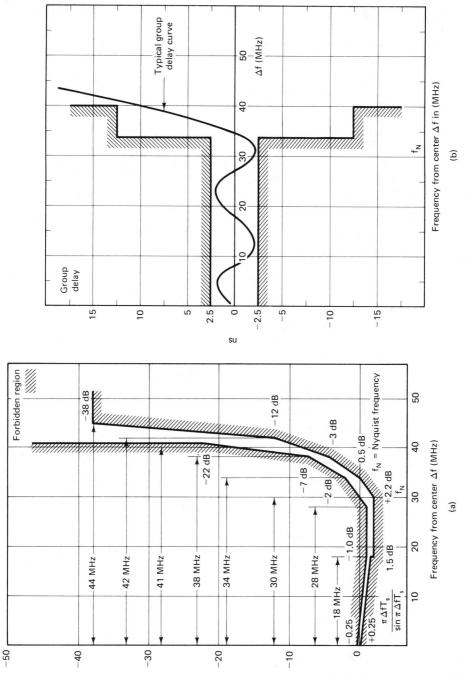

Figure 4.28 (a) Modulator filter amplitude mask (INTELSAT-V = 120-Mb/s QPSK). (b) Modulator filter group delay mask (INTELSAT-V = 120 Mb/s). (c) Demodulator filter amplitude mask (INTELSAT-V = 120-Mb/s QPSK system). (d) Demodulator filter group delay mask (INTELSAT-V = 120-Mb/s QPSK system). (With permission from The International Telecommunications Satellite Organization [Intelsat], TDMA-DSI Specifications BG-42-6SE, June 1980.)

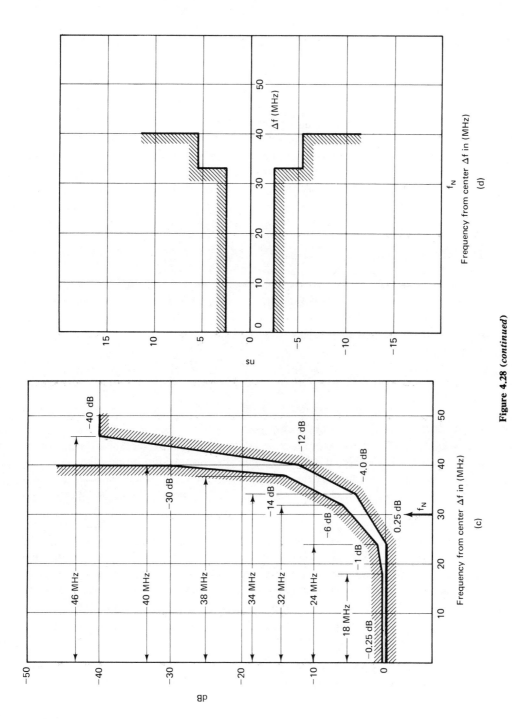

Figure 4.28 (*continued*)

This probability of bit error equation is the same as previously described for QPSK systems [see equation (4.116)]. *Restriction:* The theoretical performance of these intersymbol-interference-free modems is identical only in linear additive white Gaussian noise systems [Simon and Smith, 1974].

In differentially encoded coherent QPSK systems (DEQPSK), the differential decoding process is performed after signal regeneration. During decoding error multiplication by a factor of 2 occurs.

Thus the P_e performance of the Gray-coded DEQPSK modem is

$$P_{e\,(\text{DEQPSK})} = 2P_{e\,(\text{QPSK})} = \text{erfc}\,\sqrt{\frac{E_b}{N_o}} \tag{4.121}$$

The bit error rate performance curves, $P_e = f(E_b/N_o)$, of offset-keyed and differentially encoded coherent QPSK modems are shown in Fig. 4.26.

P_e **performance of DQPSK modems.**   The derivation of the probability of error function for DQPSK modems is somewhat involved; therefore, we will only state the final result. If interested in the derivation, you may consult [Lucky et al., 1968].

The theoretical $P_e = f(E_b/N_o)$ curve of DQPSK systems is shown in Fig. 4.26. The probability of bit error performance is approximated by

$$P_{e\,(\text{DQPSK})} \approx e^{(-A^2/2\sigma^2)(1-1/\sqrt{2})} \tag{4.122}$$

where $A^2/2\sigma^2$ is the carrier-to-noise ratio.

The DQPSK system, illustrated in Fig. 4.22, requires a level some 2 dB higher in E_b/N_o than the coherent QPSK system. In certain applications this increased E_b/N_o requirement is offset by the simpler hardware requirement; that is, no carrier recovery circuitry is needed to construct the DQPSK demodulator [European Space Center, 1979].

The **physical reason** for this higher E_b/N_o requirement of DQPSK modems is explained as follows. In the DQPSK demodulator, the modulated carrier degraded by noise is multiplied by a one-symbol "delayed replica" of both the carrier and noise component. In the coherent QPSK demodulator the delayed replica is replaced by a *clean, noise-free* signal generated by the carrier recovery circuitry. Thus the demodulator (phase comparator) has in the coherent QPSK case a noise-free reference signal, whereas in the DQPSK case the reference signal is degraded by the same amount of noise as in the modulated carrier. *Solve Problems 4.10 to 4.12.*

4.7.4 P_e Performance Degradations in Imperfect Linear Channels

In previous sections we described the performance of various binary and four-phase PSK modulated systems in the presence of Gaussian noise. Idealized channels that satisfy the Nyquist intersymbol-interference-free transmission criteria were assumed. In many applications, a virtually ideal received pulse amplitude spectrum can be achieved with available hardware. The variation from the ideal arises from the equalization of the phase characteristics. This phase distortion, and its corresponding *group delay* distortion (also known as *envelope delay*), along with any amplitude distortions, introduces intersymbol-interference, which requires an increased E_b/N_o requirement.

Here a **computer simulation** is presented, together with measurement results of illustrative system $P_e = f(E_b/N_o)$ degradations. An in-depth theoretical study and computer simulated results of group delay and amplitude distortion effects on linear system P_e performance is reported in [Sunde, 1961, and 1969]. Performance degradation of OK-QPSK and more spectral efficient modulation systems are described by [Morais, et al., 1979]. In this reference it is shown that for terrestrial line-of-

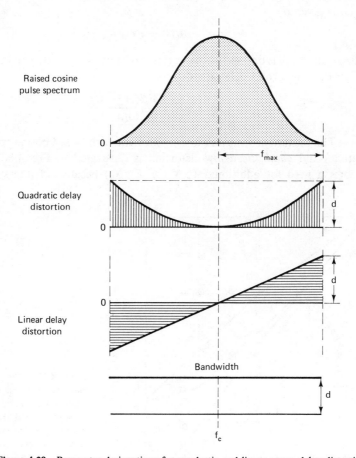

Figure 4.29 Parameter designations for quadratic and linear group delay distortion with raised-cosine (Nyquist) pulse spectrum at demodulator input. The double-sided radio-frequency bandwidth is $2f_{max}$. The ideal group delay in the $f_c \pm f_{max}$ range is constant. d, maximum delay distortion over band f_{max}; f_c, carrier frequency. (After E. D. Sunde, *Bell Syst. Tech. J.*, copyright 1961, American Telephone and Telegraph Company with permission.)

sight microwave systems, the amplitude-created distortion of frequency selective fade is more severe than that of the time-variable group delay. Parameter designations are followed by illustrative examples of performance degradations caused by amplitude and phase distortions.

In Fig. 4.29 the parameter designations for the quadratic and linear group delay distortions are presented. Note that d represents the maximum *group delay at the edge of the band* that is at frequencies $f_c \pm f_{max}$. At these limits the channel filter attenuation is infinity; that is, the raised cosine pulse spectrum at and beyond these frequencies is zero. For practical systems, those frequencies at which the channel filters reach 10 to 15 dB attenuation are considered to be the band limits [Wolejsza and Chakraborty, 1979]. For quadratic group delay distortion, the phase distortion $\beta(\omega)$ is given by

$$\beta(\omega) = C(\omega - \omega_c)^3 \qquad (4.123)$$

The group delay is defined as the negative of the first derivative of the phase characteristics; thus the group delay distortion, τ, is

$$\tau = -\frac{d\beta(\omega - \omega_c)}{d\omega} \qquad (4.124)$$

The maximum transmission impairment with raised cosine spectra (perfect amplitude) and *quadratic delay* distortion is illustrated in Fig. 4.30. The impairment creates a need for additional C/N (or E_b/N_o) because of intersymbol-interference

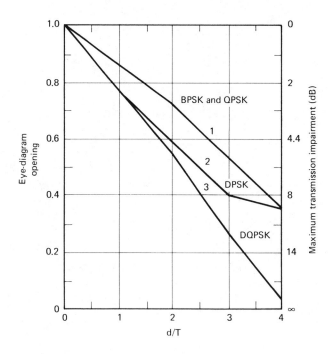

Figure 4.30 Maximum transmission impairments with raised-cosine pulse spectrum and quadratic delay distortion for: 1, two-phase (BPSK) and four-phase (QPSK) systems with synchronous detection; 2, two-phase systems with differential phase detection (DPSK); 3, four-phase systems with differential phase detection (DQPSK). T, symbol duration; d, maximum delay distortion over band f_{max}. (After E. D. Sunde, *Bell Syst. Tech. J.*, March 1961, with permission.)

caused by this quadratic delay. The maximum transmission impairment expressed in dB corresponds to a specified peak eye-diagram opening. Note that the coherent BPSK and QPSK systems have the same impairment; the DPSK and DQPSK systems have larger impairmen.ts

The maximum transmission impairments of BPSK, DPSK, QPSK, and DQPSK systems caused by *linear delay* distortion are shown in computer-derived results in Fig. 4.31. Here the QPSK system suffers more degradation than the BPSK system because the crosstalk between the I and Q channels, introduced by linear delay distortion, degrades the eye diagram. The DQPSK scheme is particularly sensitive to linear delay distortion.

Computer calculated and measured C/N degradations due to amplitude slope and delay slope distortion are shown in Figs. 4.32 and 4.33. For the measurements, a 45-Mb/s bit rate offset-QPSK modem was employed.

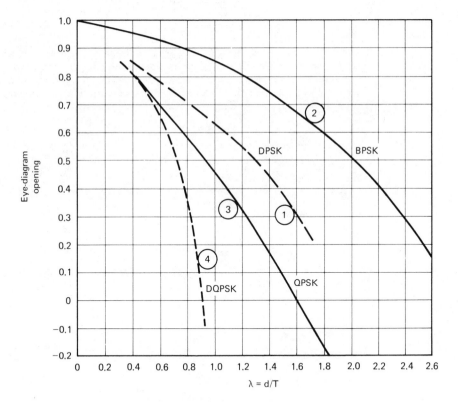

Figure 4.31 Maximum transmission impairments with raised-cosine spectrum and linear delay distortion for: 1, binary PM with differential phase detection (DPSK); 2, binary PM with synchronous detection (BPSK); 3, four-phase modulation with synchronous detection (QPSK); 4, four-phase modulation with differential phase detection (DQPSK). (After E. D. Sunde, *Bell. Syst. Tech. J.* Copyright: March 1961, American Telephone & Telegraph Company, with permission.)

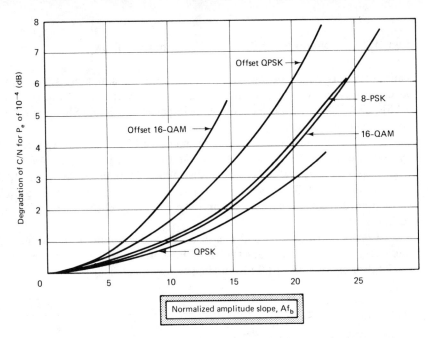

Figure 4.32 Computer-calculated C/N degradations due to amplitude slope distortion. A, amplitude slope (dB/Hz); f_b, bit rate bandwidth (Hz); S, average signal power at receiver filter input; N, average noise power (WGN) in a given bandwidth at receiver filter input. (After [D. Morais, Ph.D. Thesis, University of Ottawa, Canada, 1981], with permission.)

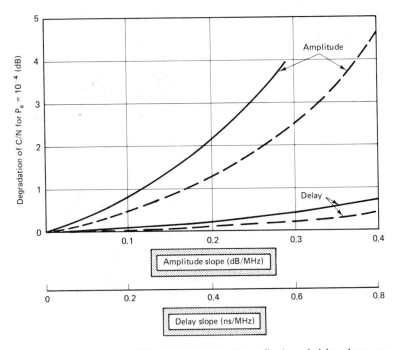

Figure 4.33 Measured C/N degradation due to amplitude and delay slope on a 45-Mb/s offset QPSK, 50% raised-cosine system. Solid line, measured; dashed line, computed. (After [Morais, 1981], with permission.)

Example 4.5

The required transmit and receive amplitude mask of the INTELSAT V-QPSK modem is as shown in Fig. 4.28. We assume that the designed amplitude function approximates a raised-cosine Nyquist channel. Also assume that the group delay of the cascaded transmit and receive filters is quadratic (i.e., parabolic), and that the group delay difference between the edge of the band and at the center frequency is $d = 16.6$ ns. What is the system E_b/N_o requirement if this group delay–created impairment is the only noticeable impairment and if a $P_e = 10^{-8}$ is required? The system bit rate is 120 Mb/s.

Solution.   For an ideal system, $P_e = 10^{-8}$ for $E_b/N_o = 12$ dB (see Fig. 4.26):

$$\text{System degradation due to parabolic group delay:} \qquad d = 16.6 \text{ ns}$$

$$\frac{d}{T_s} = \frac{16.6 \text{ ns}}{1/60 \text{ Mbaud}} = 1$$

From Fig. 4.30 for $d/T_s = 1$ we read an approximately 1.3 dB degradation. The system E_b/N_o requirement is

$$E_b/N_o = 12 + 1.3 \text{ dB} = 13.3 \text{ dB}$$

■

Solve Problems 4.13 and 4.14.

4.8 MINIMUM SHIFT KEYING (MSK) AND NONLINEARLY FILTERED (NLF) OFFSET-KEYED QPSK SYSTEMS (FEHER'S QPSK)

4.8.1 MSK Background

Frequency modulation (FM) is among the most frequently used analog modulation techniques. For data transmission a digital FM technique known as *frequency shift keying* (FSK) was developed. Many data sets operating in the range 50 b/s to 1 Mb/s use noncoherent FSK modems. Noncoherent demodulators are simpler but require a higher E_b/N_o than coherent systems. Since any higher E_b/N_o requirement than the minimum possible is not acceptable for power-constrained satellite systems, *noncoherent* frequency shift keying is not acceptable. For further information on noncoherent FSK systems, see [Lucky et al., 1968]. The *coherent* FSK modulation/ demodulation method, known as *minimum shift keying* (MSK) or *fast frequency shift keying* (FFSK), has a good performance. In this section we describe the principle of operation and the characteristics of MSK systems.

4.8.2 Principle of Operation of MSK Modems

The block diagram of an FFSK modulator is shown in Fig. 4.34(a). The voltage-controlled oscillator (VCO) represents a possible implementation of the modulator.

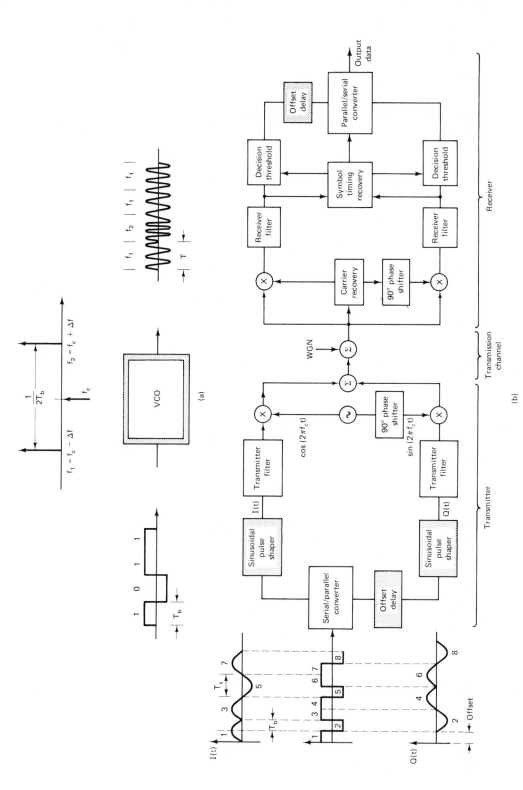

Figure 4.34 Block diagram of FFSK and MSK modems. (a) FFSK representation. (b) Equivalent MSK representation. Data in bit number n is converted to I/Q Baud number n. $T_s = T_b$.

Logic state 1 corresponds to transmit frequency f_2, logic state 0 (-1 V data level) to f_1. The frequency deviation in FFSK is

$$\Delta f = \frac{f_2 - f_1}{2} = \frac{1}{4T_b} \tag{4.125}$$

where T_b is the unit bit duration of the input data stream. Note that a coherent relation between the transmitted frequencies and the bit rate is required.

Now let us demonstrate that the FFSK signal may be generated in a similar manner to premodulation filtered OK-QPSK signals [Huang, 1979].

A frequency shift-keying signal, $s_{FSK}(t)$, can be considered as the transmission of a sinusoid, the frequency of which is shifted between the following two frequencies:

$$f_1 = f_c - \Delta f = f_c - \frac{1}{4T_b}$$
$$f_2 = f_c + \Delta f = f_c + \frac{1}{4T_b} \tag{4.126}$$

It is described by

$$s_{FSK}(t) = A \cos\left[2\pi(f_c \pm \Delta f)t\right]$$
$$= A \cos\left(\pm 2\pi\,\Delta f\, t\right) \cos\left(2\pi f_c t\right) - A \sin\left(\pm 2\pi\,\Delta f t\right) \sin\left(2\pi f_c t\right) \tag{4.127a}$$

In the FFSK case the frequency deviation is $\Delta f = 1/4T_b$; thus the MSK signal $s_{MSK}(t)$ is

$$s_{MSK}(t) = A \cos\left(\pm\pi t/2T_b\right) \cos\left(2\pi f_c t\right) - A \sin\left(\pm\pi t/2T_b\right) \sin 2\pi f_c t \tag{4.127b}$$

Equation (4.127b) is an MSK representation of FFSK. In Fig. 4.34(b) an implementation method of the MSK modulator is shown. The unmodulated carrier of frequency f_c is multiplied by both the in-phase and quadrature baseband signals. The serial-to-parallel converted data are fed into the sinusoidal pulse-shapers. In the baseband I channel the pulse shaper generates a $\cos\left(\pm\pi t/2T_b\right)$ pulse sequence, while in the Q channel the offset delay T_b, in conjunction with the pulse shaper, provides the pulse sequence given by

$$\cos\left[\pm\frac{\pi(t - T_b)}{2T_b}\right] = \sin\left(\pm\frac{\pi t}{2T_b}\right)$$

The sinusoidal pulse shapers could be implemented as nonlinearly switched filters, as described in Chapter 3. Now it is easy to verify that the modulator shown in Fig. 4.34(b) generates the MSK signal described by (4.127b) and that it has a structure which closely resembles an OK-QPSK modulator, shown in Fig. 4.18.

The sinusoidal pulse shaping means that the modulator output has either a positive or negative linear phase-change rate relative to the carrier, depending on the input data [De Buda, 1972]. The amplitude and phase of the output signals of the modulating multipliers is such that the sum of these signals (which is the unfiltered

MSK modulated signal) has a constant-amplitude envelope (i.e., is time invariant). As a result, it can be transmitted through amplitude-limiting devices with minimal additional signal degradation in the same way that conventional FM modulated signals are transmitted.

The MSK demodulator operates in the same manner as the offset QPSK receiver described earlier. However, different filtering is required to ensure ISI-free transmission [Morais and Feher, 1979]. The bit rate and frequency deviation are related by equation (4.125). The application of this phenomenon simplifies the subsystem design for MSK synchronization (i.e., carrier and symbol timing recovery).

Figure 4.35 illustrates typical modulated carrier waveforms for three systems: MSK, offset QPSK, and conventional QPSK. The binary data rate in each case is $1/T_b$. Phase transitions in conventional QPSK occur only every $2T_b$ seconds (i.e., a multiple of the symbol duration), whereas for OKQPSK and MSK, transitions may occur every T_b seconds. In conventional QPSK, abrupt $\pm90°$ phase/transitions and phase reversals of $\pm180°$ are possible, whereas in OKQPSK the abrupt phase transitions are limited to $\pm90°$ over the set of all four signal states. In the MSK case the carrier phase varies linearly over T_b seconds and is continuous. *Solve Problem 4.15.*

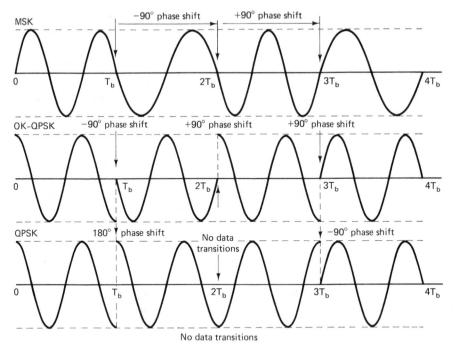

Figure 4.35 Modulated carrier waveforms in MSK, offset QPSK, and conventional QPSK. Note that the amplitude envelope of these *unfiltered* signals is constant. (After [Gronemeyer and McBride, 1976], with permission from the IEEE, © 1976.)

4.8.3 MSK Spectrum and Spectral Efficiency

The power spectrum of sinusoidally shaped baseband signals is presented in Chapter 3 [see Fig. 3.16 and equation (3.55)]. By similar development for the conventional QPSK and OKQPSK cases, we may conclude that the modulated MSK spectrum equals the superimposed and **frequency shifted** power spectra of the I and Q baseband signals. Thus we have

$$G_{\text{MSK}}(f) = \frac{8P_c T_b [1 + \cos 4\pi(f - f_c) T_b]}{\pi^2 [1 - 16 T_b^2 (f - f_c)^2]^2}$$

$$= \frac{4P_c T_s [1 + \cos 2\pi(f - f_c) T_s]}{\pi^2 [1 - 4 T_s^2 (f - f_c)^2]^2}$$

(4.128)

where

f_c = unmodulated carrier frequency

P_c = total power in modulated waveform

$T_b = 1/f_b$ = bit duration

$T_s = 1/f_s = 2T_b$ = symbol duration

A detailed MSK spectral derivation is presented in [Gronemeyer and McBride, 1976].

The normalized power spectral density of conventional QPSK, offset QPSK, and MSK as a function of frequency, and normalized to the binary bit rate $R_b = 1/T_b$, is shown in Fig. 4.36. The QPSK main lobe has a width of $\pm 1/2 T_b$; that of the MSK signal is wider ($\pm 3/4 T_b$). For larger values of $(f - f_c)/R_b$ the unfiltered MSK spectrum falls off at a rate proportional to $(f)^{-4}$, the unfiltered QPSK spectrum at a rate proportional to $(f)^{-2}$. The unfiltered *infinite bandwidth* spectral properties are of particular importance not only in earth station design, where the high power amplifier (HPA) is usually operated in a nonlinear (saturated) mode, but also in satellite systems applications where it is not economic to design Nyquist shaped filters which follow the nonlinear output amplifiers.

As an application of Nyquist's generalized intersymbol-interference-free transmission theorem, we have shown in Example 3.4 that the baseband sinusoidal pulse shapes can be limited to have a spectral efficiency of 2b/s/Hz. Based on the equivalence of bandpass and low-pass channel models (Section 4.2) and on the solution of Example 3.4, we may conclude that *the theoretical spectral efficiency of MSK systems is 2 b/s/Hz*. For a comprehensive study of optimal filter partitioning for spectral efficient MSK systems, see [Morais and Feher, 1979]. *Solve Problem 4.16.*

$P_e = f(E_b/N_o)$ **performance of MSK systems.** It is shown [De Buda, 1972] that the probability of error, P_e, performance of unfiltered FFSK systems is the same as that for ideal matched filter QPSK receivers (recall that MSK is an alternative

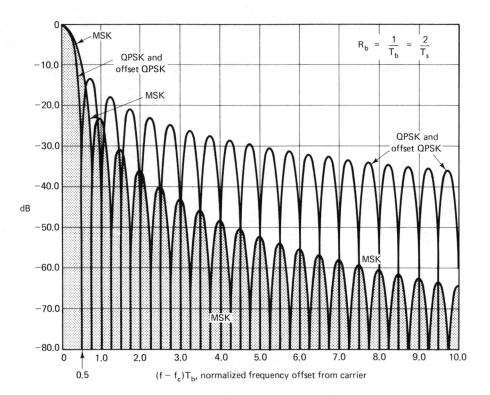

$$G_{MSK}(f) = \frac{8P_c T_b [1 + \cos 4\pi(f - f_c)T_b]}{\pi^2 [1 - 16T_b^2(f - f_c)^2]^2}$$

$$G_{OKQPSK}(f) = 2P_c T \left[\frac{\sin 2\pi(f - f_c)T_b}{2\pi(f - f_c)T_b} \right]^2$$

Figure 4.36 Normalized power spectral densities of unfiltered QPSK, offset QPSK, and MSK systems. The modulated spectrum is symmetrical around the carrier frequency. For this reason only the upper sideband is shown. (After [Gronemeyer and McBride, 1976], with permission from the IEEE, © 1976.)

name for FFSK). More recently, [Morais and Feher, 1979] have shown that the P_e performance of bandlimited Gray-coded MSK systems is identical to that of QPSK systems. Thus if the MSK system is perturbed by additive white Gaussian noise and the modem and channel filtering is in accordance with Nyquist's generalized ISI-free transmission theorem [see equations (3.55) and (4.107)], then

$$P_{e\,(MSK)} = \tfrac{1}{2} \operatorname{erfc} \sqrt{\frac{E_b}{N_o}} = P_{e\,(QPSK)} \tag{4.129}$$

The $P_{e\,(MSK)} = f(E_b/N_o)$ relation is illustrated in Fig. 4.26.

4.8.4 Nonlinearly Filtered Offset QPSK (NLF-OKQPSK) (Feher's QPSK) Modems

A relatively new nonlinear processing (filtering) technique combined with offset QPSK modulation has significant spectral spreading advantages compared to the previously described modulation techniques. This modulator, known as *nonlinearly filtered offset-keyed QPSK* (NLF-OKQPSK) or *Feher's* modulator is shown in Fig. 4.37. In [Feher,

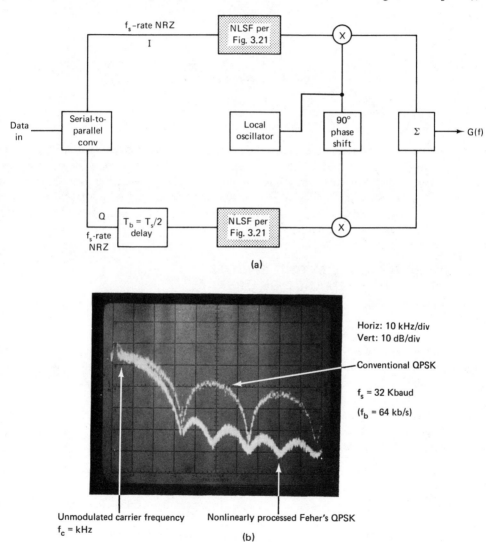

(a)

(b)

Figure 4.37 Nonlinearly filtered-offset-keyed QPSK (NLF-OKQPSK) modulator and corresponding spectrum. This modulator is also known as *Feher's* QPSK. (a) Modulator block diagram. (b) Spectral comparison of conventional with Feher's QPSK.

195

1979; Huang et al., 1979; and Le-Ngoc and Feher, 1982] detailed spectral density derivations, applications, and implementation diagrams are presented. In Section 4.9 we describe the spectral characteristics of this modulator in hard-limited systems.

Referring to Fig. 4.37(a), the serial-to-parallel converted data are fed to the I and Q baseband channels. Prior to modulation, the $f_s = f_b/2$ rate symbol streams are nonlinearly processed by means of Feher's processor, described in Section 3.5.3. The modulated power spectral density is

$$G(f) = C \left\{ \frac{\sin\left[2\pi(f - f_c)T_s\right]}{2\pi(f - f_c)T_s} \frac{1}{1 - 4(f - f_c)^2 T_s^2} \right\}^2 \qquad (4.130)$$

where

$$f_s = 1/T_s \text{ is the symbol rate } (f_s = f_b/2)$$

$$C = \text{proportionality constant}$$

$$f_c = \text{unmodulated carrier frequency}$$

A measured power spectral density plot of both conventional QPSK and Feher's nonlinearly switched QPSK is shown in Fig. 4.37(b). Since the radio-frequency spectrum is symmetrical, only the frequencies *above* the carrier are shown. Note the close resemblance of the filtered modulated spectrum with that of the equivalent baseband shown in Fig. 3.22. The conventional unfiltered QPSK spectrum is also shown.

4.9 FILTERING AND LIMITING EFFECTS ON THE PERFORMANCE OF QPSK, OKQPSK, MSK, AND FEHER'S QPSK

In this section the effects of filtering and limiting on the performance of power-efficient modulation systems are described. Here we focus attention on the performance degradations caused by *filtering and hard-limiting*. More complex nonlinear effects, such as AM-to-PM- and AM-to-AM-conversion-caused P_e degradations are studied in Chapters 8 and 9, where TDMA and regenerative satellite systems are presented. Note that hard-limiting amplifier models approximate both earth station and satellite power amplifiers operating in a saturated mode.

The performance degradations can be classified into two major categories:

1. Performance degradations caused in adjacent channels because of spectral restoration (known as spectral spreading)
2. Performance degradation caused by signal distortions, including I-to-Q crosstalk

These signal impairments are introduced when the time-varying envelope of the modulated bandlimited signal is amplified by a nonlinear power-efficient output amplifier. Returning our attention to Fig. 4.1, we assume that the adjacent channel earth station amplifiers also operate in a saturated mode, for highest power efficiency. For many applications where the radio-carrier frequency is much higher than the data bit rate $(f_c/f_b > 10^5)$, it is very difficult to design spectrally efficient and temperature-stable RF filters. For example, in a 6-GHz uplink using 64-kb/s QPSK, the filters F_{31}, F_{32}, and F_{33} all require a center frequency of 6 GHz and a double-sided bandwidth of 32 kHz. Because of the complexity of this type of filter design, many satellite earth stations do not utilize spectral shaping filters after the higher-power amplifiers. Thus the restored spectrum created by these nonlinear amplifiers falls into the adjacent satellite channels and is the cause of *adjacent channel interference*.

If we could discover a modulation scheme that generates a constant envelope bandlimited signal, these degradations would not exist. However, as this is not the case to date, we have to analyze the envelope fluctuations in order to gain an insight into the distortion mechanisms associated with nonlinear amplification.

4.9.1 Envelope Fluctuations of QPSK, OKQPSK, and MSK Signals

Because of the time coincidence of the two data streams in QPSK, instantaneous phase transitions of 0°, ±90°, and 180° can occur. Figure 4.38(a) shows the RF

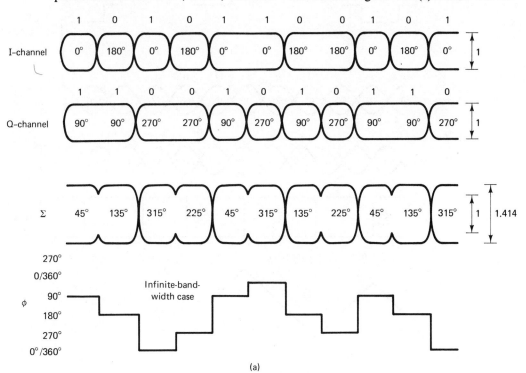

(a)

Figure 4.38 (a) RF amplitude and phase of filtered QPSK signals. (b) RF amplitude and phase of filtered OKQPSK signals. (c) RF amplitude and phase of *unfiltered* MSK signals. (Figures and corresponding descriptions after [Huang, 1979], with permission.)

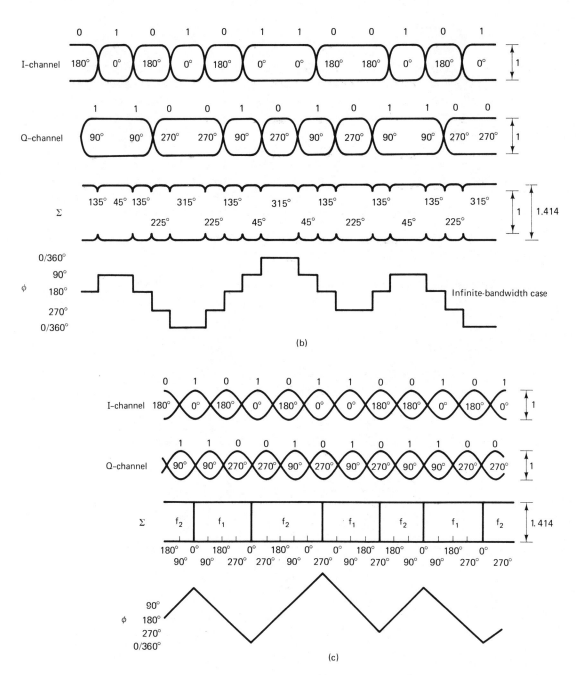

Figure 4.38 (*continued*)

amplitudes and phases of the filtered QPSK signal. At the 180° instantaneous phase transitions, which arise when both the in-phase and the quadrature channel data change phase simultaneously, the envelope goes through zero amplitude. At the ±90° phase transitions, which occur when only one channel data change phase at the keying instant, there is a 3-dB envelope fluctuation.

In OKQPSK, because of the one-bit (half a symbol) duration delay between the two data channels, the 180° phase transitions are prevented. Only ±90° phase transitions arise, with maximum 3-dB envelope fluctuations, as shown in Fig. 4.38(b). The overall envelope fluctuations of OKQPSK are thus smaller than those of QPSK.

In MSK, the baseband pulses are of half cosine shape, as shown in Fig. 4.38(c). From equation (4.127) we may conclude that the modulated signal has a constant envelope. Another important feature of MSK is that the phase transitions are linear and continuous [De Buda, 1976].

Quadrature-modulated signals can be described by

$$z(t) = i(t) \cos \omega_c t + q(t) \sin \omega_c t \qquad (4.131)$$

where

$$i(t), q(t) = \text{in-phase and quadrature baseband signals}$$
$$\omega_c = \text{carrier angular frequency, rad/s } (\omega_c = 2\pi f_c)$$

The *envelope* of this signal is

$$e(t) = \sqrt{i^2(t) + q^2(t)} \qquad (4.132)$$

A convenient method of displaying envelope fluctuations is by using a *signal-space diagram*.

So if we feed $i(t)$ into the horizontal and $q(t)$ into the vertical axis of an oscilloscope, we obtain a *Lissajous* curve which represents the envelope of the quadrature modulated system. The signal-space diagrams of unfiltered modulated systems are shown in Fig. 4.39; measured signal-space diagrams of filtered and nonlinearly processed QPSK modulated systems are shown in Fig. 4.40. Note that the envelope fluctuations measured in the conventional QPSK case are larger than those of the OKQPSK and nonlinearly processed QPSK, which are limited to 3 dB.

4.9.2 Spectral Spreading

Bandlimited systems have finite rise and fall times, so filtering introduces envelope fluctuations into modulated QPSK and MSK systems. These time-varying envelope signals are fed into an ideal infinite-bandwidth (memoryless) limiter, known as a memoryless *hard-limiter,* shown in Fig. 4.41. The output of the limiter has a constant-amplitude envelope; that is, the limiter removes the envelope fluctuations of the input signal. Unfortunately, this nonlinear level-limiting process **restores** (regenerates) the sidebands. The bandlimited spectrum of a 64-kb/s rate QPSK modulator is shown in Fig. 4.41(c) (lower trace); the bandlimited *and* hard-limited spectrum is shown

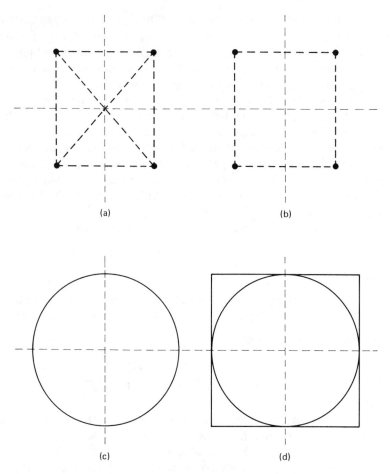

Figure 4.39   Signal-space diagrams for unfiltered (a) QPSK, (b) OKQPSK, (c) MSK (FFSK), and (d) NLF-OKQPSK (Feher's QPSK).

in the upper trace. Note the spectral restoration, particularly for the higher-order sidelobes.

Analytical studies of spectral spreading are complex [Devieux, 1974; Benedetto et al., 1978] and do not provide sufficient insight into the performance degradation problems of adjacent channels. For this reason most systems engineers employ computer simulations to predict the amount of spectral restoration and the degradations caused by this to adjacent channels [Harris, 1978; Lundquist, 1978].

Measured results of bandlimited and hard-limited QPSK, OKQPSK, MSK, and NLF-OKQPSK are shown in Fig. 4.42(a). [Le-Ngoc et al., 1982; Feher, 1979]. The satellite earth station antenna used in these measurements is shown in Fig. 4.42(b). From the measured results it is evident that the NLF-OKQPSK has a fast spectral

Band-pass filtered QPSK

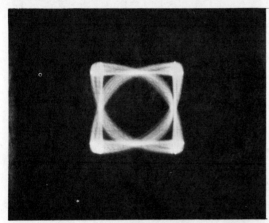

Band-passed filtered OKQPSK

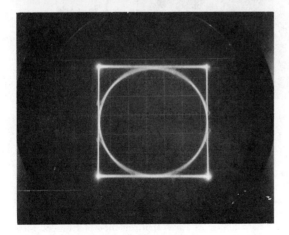

Nonlinearly filtered OKQPSK (Feher's QPSK)

Figure 4.40 Measured signal space diagram of bandlimited QPSK, OKQPSK, and Feher's QPSK.

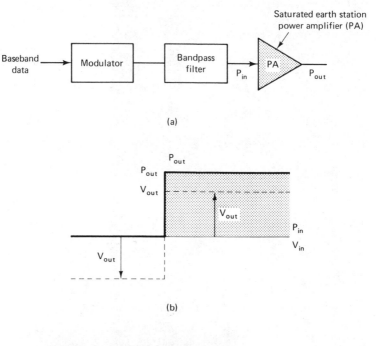

(a)

(b)

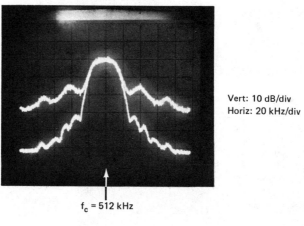

Vert: 10 dB/div
Horiz: 20 kHz/div

$f_c = 512$ kHz

(c)

Figure 4.41 Filtered and hard-limited QPSK spectra. Bit rate, $f_b = 64$ kb/s. Carrier frequency, $f_c = 512$ kHz. Filtered QPSK spectrum (lower trace.) Same modulator and filters as in the spectral measurement of Fig. 4.24. Filtered and hard-limited spectrum (upper trace). (a) Block diagram. (b) Hard-limited power and voltage output/input ratio. Here we use a hard limiter as a first approximation of saturated earth station power amplifiers. (c) Bandlimited and bandlimited–hard limited spectral measurements.

roll-off and that the restored main lobe of the MSK signal extends to $1.5/T_s$. These modulation techniques are studied further in later sections.

Fast spectral roll-off of modulated and hard-limited (nonlinearly amplified) signals is very desirable for power and spectral efficient satellite systems. The objective of intensive research carried out at the University of Ottawa, Hughes Aircraft, Philips Laboratories, JET Propulsion Laboratories, SPAR Aerospace Limited, COMSAT Laboratories, and other institutions is to develop new power-efficient modems. These experimental modems transmit signals through hard-limited, saturated earth station amplifiers. The spectrum of some of these new modulated signals (and the corresponding references) are shown in Fig. 4.42(a). Additional modified PSK and MSK techniques are described in [Atobe et al., 1978, and Yazdani, et al., 1980]. A description of these experimental modulation techniques is beyond the scope of this book. For numerous references, see the *IEEE Transactions on Communications. Solve Problem 4.17.*

4.9.3 Filtering and Limiting Effects on Interphasor Crosstalk

Unfiltered QPSK, offset QPSK, and MSK signals all have constant-amplitude envelope versus time characteristics and therefore are not affected by the ideal limiting amplifier assumed in Fig. 4.41. Thus system performance is the same as if the prelimiter filter and the limiting amplifier were absent. Under this condition there is no intersymbol interference and the E_b/N_o for a P_e of 10^{-4} is 8.4 dB for all three cases.

As the prelimiter filter bandwidth narrows to the stage where significant signal energy is filtered away, the in-phase and quadrature symbol waveforms [$i(t)$ and $q(t)$ in equation (4.131)] become distorted and the signals acquire amplitude modulation (AM). The limiting action modifies the spectral densities, removes the AM acquired during filtering, and in so doing, further distorts the $i(t)$ and $q(t)$ symbol waveforms. As a result of this process, *crosstalk* between the waveforms occurs.

To understand how this crosstalk actually takes place, it is necessary to study the characteristics of the ideal limiter that has been assumed. Let $i_i(t)$ and $q_i(t)$ be the I and Q symbol waveforms into the limiting amplifier, and $i_o(t)$ and $q_o(t)$ be the corresponding equivalent baseband symbol waveforms out of the limiting amplifier.

Since the output of the limiter has a constant amplitude envelope versus time characteristic, we have

$$i_o^2(t) + q_o^2(t) = C \qquad (4.133)$$

where C is a constant dependent only on the output power of the limiting amplifier. Also, as the memoryless infinite bandwidth limiter results in no change of signal phase, then

$$\text{signal phase-out} = \text{signal phase-in}$$

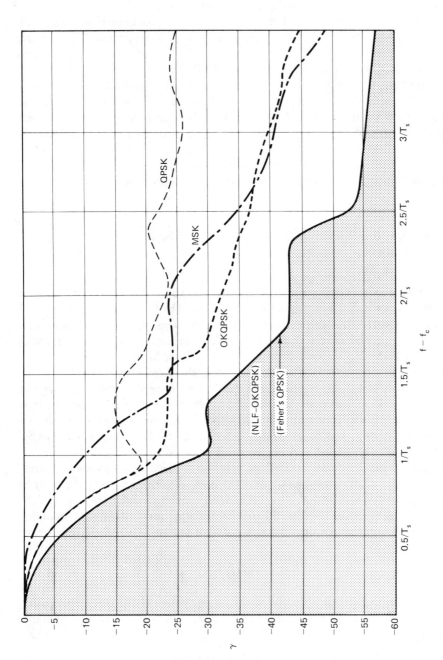

Figure 4.42 (a) Measured, normalized, spectral density of bandlimited and *hard-limited* QPSK, OKQPSK, MSK, and NLF-OKQPSK (Feher's QPSK). Solid line, NLF-OKQPSK (not bandlimited); dash–dotted line, MSK bandlimited by fourth-order Chebyshev $BT_b = 1$; dashed line, OKQPSK bandlimited by fourth-order Butterworth; dotted line, QPSK bandlimited by fourth-order Butterworth.

(b)

Figure 4.42 (*continued*) (b) 14/12-GHz transmit–receive antenna of the University of Ottawa. The performance of new modulation techniques is evaluated under the research direction of Dr. K. Feher. The Canadian ANIK satellites are used in these experiments.

That is,

$$\arctan \frac{q_o(t)}{i_o(t)} = \arctan \frac{q_i(t)}{i_i(t)}$$

Thus

$$\frac{q_o(t)}{i_o(t)} = \frac{q_i(t)}{i_i(t)} \qquad \begin{array}{l}\text{if the principal values of the}\\ \text{arctan are assumed}\end{array} \qquad (4.134)$$

By replacing $q_o(t)$ in equation (4.134) by its equivalence as given in equation (4.133), we obtain

$$i_o(t) = \frac{\sqrt{C}i_i(t)}{[i_i^2(t) + q_i^2(t)]^{1/2}} \qquad (4.135)$$

and by replacing $i_o(t)$ in equation (4.134) by its equivalance as given in equation (4.133), we obtain

$$q_o(t) = \frac{\sqrt{C}q_i(t)}{[i_i^2(t) + q_i^2(t)]^{1/2}} \qquad (4.136)$$

If the **unfiltered** and hence non-amplitude-modulated signal from the QPSK, offset QPSK, or MSK modulator is fed into our ideal limiter, then

$$i_i^2(t) + q_i^2(t) = K^2 \qquad (4.137)$$

where K is a constant. Thus, by equations (4.135) and (4.136), the output symbol waveforms are undistorted, as expected, since

$$i_o(t) = ki_i(t) \qquad \text{and} \qquad q_o(t) = kq_i(t) \qquad (4.138)$$

where $k = C/K = $ constant.

Where the modulator output signal is **filtered,** however, the signal becomes amplitude modulated and

$$i_i^2(t) + q_i^2(t) \neq K^2$$

Thus, by equations (4.135) and (4.136), $i_o(t)$ and $q_o(t)$ become dependent also on $q_i(t)$ and $i_i(t)$, respectively. This interdependent influence of each symbol waveform by the other in quadrature is termed *interphasor crosstalk*. *Solve Problem 4.18.*

4.9.4 Filtering and Limiting Effects on Symbol Waveshapes and Eye Diagrams

Typical QPSK in-phase and quadrature symbol waveforms in their unfiltered, filtered, and filtered then limited forms are shown in Fig. 4.43. The waveforms shown assume only mild filtering. The filtered then limited form follows from the characteristic that the QPSK signal at the limiter output is of a constant amplitude and of identical phase to the limiter input signal. That is, equations (4.133) and (4.134) must be

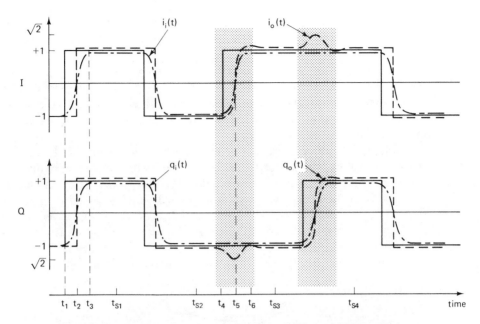

Figure 4.43 Effects of filtering and filtering then limiting on QPSK symbol waveshapes. Solid line, unfiltered; dash–dotted line, filtered; dashed line, filtered then limited. (After [Morais, 1981], with permission. As a further reference see [Morais and Feher, 1980].)

satisfied. Note that this limitation does not imply that the individual $i_o(t)$ and $q_o(t)$ signals must have constant amplitude, only their vector sum. In the figure, the amplitudes of $i_o(t)$ and $q_o(t)$ are both set at value 1 during periods when no transitions are taking place. Thus the combined limiter output signal amplitude, which is the vector sum of the amplitudes of $i_o(t)$ and $q_o(t)$, is of constant value $\sqrt{2}$.

Let us consider what happens during a period such as t_1 to t_3, when both the filtered waveshapes $i_i(t)$ and $q_i(t)$ are changing polarity simultaneously. The limiter output waveshapes, $i_o(t)$ and $q_o(t)$, cannot both decrease simultaneously since, were this to happen, their vector sum would not be constant at $\sqrt{2}$ but would decrease or increase, respectively. The result is therefore that both waveshapes maintain full amplitude except at the polarity transition instant of the input waveshapes, t_2, where both output waveshapes experience instant polarity transition. Thus the output waveshape, $i_o(t)$, is influenced by the polarity changes taking place on the input waveshape, $q_i(t)$, and similarly, $q_o(t)$ is influenced by the changes taking place on $i_i(t)$. This means that waveshape crosstalk occurs.

We now consider what happens during a period such as t_4 to t_6, when only one filtered waveshape changes polarity. During the particular period t_4 to t_6, $i_i(t)$ changes polarity while $q_i(t)$ remains unchanged. As $i_i(t)$ decreases in amplitude, $i_o(t)$ decreases in amplitude and $q_o(t)$ increases in amplitude so that the combined signal output amplitude remains constant and its phase follows that of the input signal. In particular, at time t_5, when the amplitude of $i_o(t)$ is zero, the amplitude of $q_o(t)$ increases to $\sqrt{2}$. Thus, during these periods, in addition to periods such as t_1 to t_3, waveshape crosstalk occurs.

The overall result of the fact that the limiter forces the combined signal at its output to be of constant amplitude and of phase equal to that of its input signal is the creation of filtered then limited waveforms as shown in Fig. 4.43. Note, however, that the effect of limiting on the filtered waveform is minimum at the all-important sampling times t_{sn}, where $n = 1, 2, \ldots$.

Figure 4.44 shows computer-generated eye diagrams for QPSK. Figure 4.44(a) shows the eye diagram associated with the signal when filtered such that $BT_b = 1$, where B is the *double-sided channel filter bandwidth* and T_b is the bit duration. Figure 4.44(b) shows the eye diagram associated with the signal when filtered such that $BT_b = 1$, then limited. The strong crosstalk effects at the waveform transition points and the minimal effects at the sampling points, as previously discussed, are both evident.

Typical offset QPSK I and Q unfiltered, filtered, and filtered then limited symbol

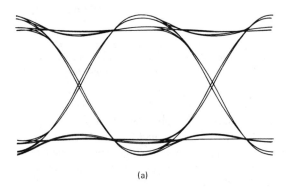

(a)

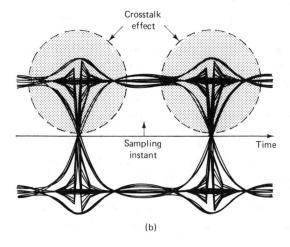

Crosstalk effect

Sampling instant   Time

(b)

Figure 4.44 Simulated eye diagrams for QPSK. (a) Filtered, $BT_b = 1.0$. (b) Filtered, $BT_b = 1.0$, then limited. (After [Morais, 1981], with permission. As a further reference, see [Morais and Feher, 1980].)

waveforms are shown in Figure 4.45(a). We note that the timing of the Q symbol stream is "offset" by one-half of a symbol duration relative to the I stream. As was done for QPSK, we assume only moderate filtering, hence the filtered waveshapes shown. As a result of the symbol streams being offset with respect to each other, only one stream can change polarity over any given transition period. Thus, during a transition such as t_1 to t_2 on the I stream, the limiter output waveshape that is *not* changing polarity increases its amplitude in order that equation (4.133) remain

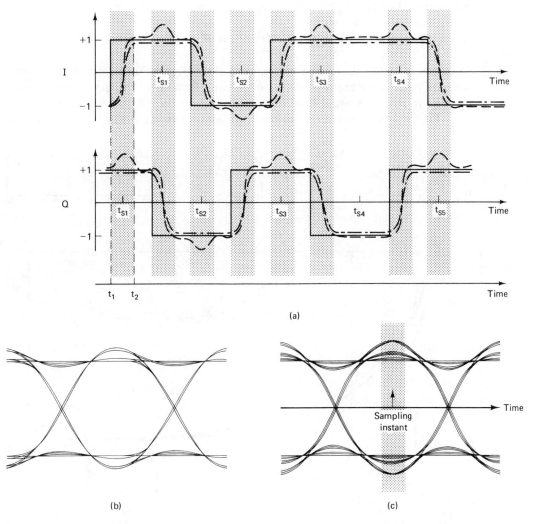

(a)

(b)   (c)

Figure 4.45 (a) Effects of filtering and filtering followed by limiting on offset QPSK symbol waveshapes. Solid line, unfiltered; dash–dotted line, filtered; dashed line, filtered then limited. (b) Simulated eye diagrams for offset QPSK. Filtered, $BT_b = 1.0$. (c) Simulated eye diagrams for offset QPSK. Filtered, $BT_b = 1.0$, then limited. (After [Morais, 1981], with permission. As a further reference, see [Morais and Feher, 1980].)

satisfied. Thus waveform crosstalk occurs. With offset QPSK the center of the polarity transition period for one stream is the sampling instant for the other. Thus a transition on one stream results in maximum crosstalk on the other at the sampling instant. This is the opposite to the crosstalk due to limiting on QPSK, as discussed above.

Figure 4.45(b) and (c) show computer-generated eye diagrams for offset QPSK. They are the equivalent diagrams to Fig. 4.44 for QPSK. Waveform crosstalk in the predicted form of increased amplitude of the affected symbols at the sampling points should be noted.

4.9.5 Filtering and Limiting Effects on P_e Performance

In order to compare the relative performance of QPSK, OKQPSK, and MSK systems, we have determined the degradation both as a function of BT_b and relative to the case where $BT_b = \infty$ for a P_e of 10^{-4}.

Figure 4.46 shows the calculated degradation as a function of frequency for

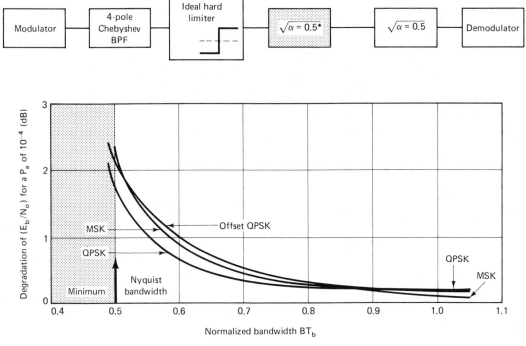

$* \sqrt{\alpha = 0.5} =$ square root of raised-cosine
filter with $\alpha = 0.5$ roll-off
factor x/sin x shaping in
transmitter

Figure 4.46 Degradation, due to filtering then limiting, of $(S/N_b)_{rcvr \cdot input}$ versus normalized prelimiter filter bandwidth. B, double-sideband 3-dB filter bandwidth; T_b, bit duration. (After [Morais, 1981], with permission. As a further reference, see [Morais and Feher, 1980].)

the three modulation systems. The block diagram used for this computer simulation is also shown. However, the conclusion, namely that QPSK has the best performance in severely bandlimited nonlinear channels, applies also to many other practical systems. This is demonstrated in Chapter 9, where regenerative satellite systems are described.

As the transmitter output and receiver input filter combination in our model provides intersymbol-interference-free transmission only for *unfiltered* modulator output signals, the waveform distortions created by the filtering and limiting action result in a degraded P_e versus E_b/N_o performance. Referring to Fig. 4.46, we note that when $BT_b = 0.5$, the degradation is in the order of 2 dB for all three systems. This might seem surprisingly low considering that the prelimiter filter is not phase-equalized and that for this value of BT_b, its double-sided 3-dB bandwidth is equal to the system double-sided Nyquist bandwidth. Note, however, that this filter, being a four-pole Chebyshev type, does not have a very sharp amplitude roll-off outside the passband and exhibits relatively mild in-band phase distortion. Under this condition of mild filtering, we observe that for $BT_b = 0.5$, QPSK is degraded the least, followed by offset QPSK, then MSK. For $BT_b = 1$, the degradation of all three systems is less than 0.2 dB, which is negligible.

4.10 ADJACENT AND COCHANNEL INTERFERENCE EFFECTS ON THE P_e PERFORMANCE OF M-ARY PSK SYSTEMS

4.10.1 Background

The effects of additive white Gaussian noise (AWGN) and intersymbol interference (ISI) on the P_e degradation of binary and four-phase PSK systems were described in previous sections. Degradations caused by imperfect linear filters (e.g., having a linear or parabolic group delay) and by nonlinear saturated amplifiers were investigated. So far, we have assumed that a single digital modulated carrier is transmitted in the system and that the interference caused by other digitally modulated channels is negligible. However, in many systems the interference caused by adjacent channels and co-channels has a predominating effect on the system P_e performance. Interference might be caused by another digital or analog modulated carrier on the same nominal carrier frequency, termed *co-channel interference,* or by *adjacent-channel interference,* where the undesired signal is predominantly outside the passband of the wanted signal. Co-channel interference might be introduced by insufficient vertical and horizontal polarization discrimination, by antenna sidelobe radiation, by satellite earth station-to-terrestrial microwave system caused radiation, or a variety of other factors [CCIR, 1978]. An illustrative example of adjacent channel interference is shown in Fig. 4.47. In this system ideal brick-wall receive channel filtering is assumed. The adjacent-channel interference is introduced predominantly by the spectral density of the two adjacent channels. The transmit filters of these channels are assumed to approximate practical raised-cosine channels having a roll-off factor $\alpha = 0.5$ (note

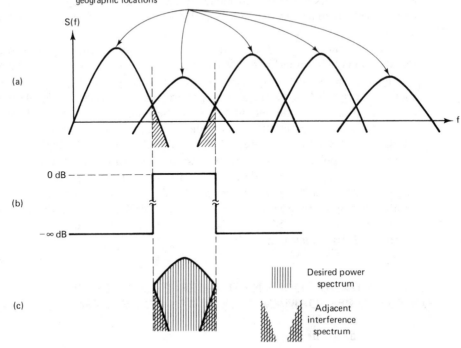

The power of individual modulated signals is different. This variation is due to the different EIRP of the transmit earth stations and the different free-space losses from different geographic locations

Figure 4.47 Adjacent channel interference in a single-channel-per-carrier (SCPC) satellite receive earth station. The cause of the adjacent channel interference is insufficient transmit channel filtering. (a) Power spectral density of a single-channel-per-carrier (SCPC) system displayed at the input of a receive earth station. (b) Ideal brick-wall filter mask of the receive bandpass filter of the desired channel. (c) Power spectral density of the desired and interfering signals.

that the $x/\sin x$ amplitude equalizers are not included). Even though we show an ideal minimum bandwidth filter prior to the demodulator of the required receiver, the adjacent-channel interference falling into the receiver bandwidth is *not* eliminated by this filter.

In this section we study the performance of two-phase, four-phase, eight-phase, . . . (i.e., in general terms, *M-ary*) *PSK systems* in an *adjacent* and *co-channel interference environment*. A vectorial representation of the sinusoidal interference mechanism is followed by an *outline of the major steps* in the derivation of the P_e performance of these systems. Finally, we present the P_e performance curves of differentially coherent *M*-ary DPSK, eight-state PAM-FM, 16-QAM, and quadrature-partial-response systems.

If an in-depth unified analysis is required including the combined effects of AWGN, adjacent, and co-channel interference and intersymbol interference on the

P_e-performance of modulated systems, consult [Fang and Shimbo, 1978, and Glave and Rosenbaum, June 1973]. The performance of PSK systems in intermodulation-caused "impulsive" noise environment is analyzed in [Oshita, 1981]. A comprehensive treatment of the interference environment is presented in [Feher, 1977, Chap. 7].

In the following derivation we assume that the intersymbol interference and other hardware-caused imperfections are negligible.

The vector representation of binary PSK and four-phase QPSK systems operating in an additive white Gaussian noise and interference environment is shown in Fig. 4.48. For both of these systems we assume that the intersymbol-interference-free signal has a zero phase at the sampling instant. For the BPSK signal the error region is the left-half plane, while for the QPSK signal it is three-fourths of the plane (see the shaded regions). Two resultant signal + noise + interference vectors are shown for the QPSK case. The resultant vector Z_1 illustrates error-free reception, and the example Z_2 indicates a received error.

The key step in the P_e performance derivation is the derivation of the probability distribution of the resultant phase. In the BPSK case an error occurs if the phase α of the Z vector is in the range $90° < \alpha < 270°$. Now, after this vectorial presentation we should be ready to proceed with the P_e performance derivation. We follow the method presented in [Rosenbaum and Glave, 1974].

Let the signal of an M-ary PSK system be

$$s(t) = \text{Re} \sum_{k=-\infty}^{\infty} e^{j(\phi_k + \omega_o t)} \, p(t - kT) \qquad (4.139)$$

$$\phi_k \in \left\{ 0, \frac{2\pi}{M}, \cdots, \frac{M-1}{M} 2\pi \right\} \qquad (4.140)$$

Assume that this signal has been degraded by the addition of interference $i(t)$, and Gaussian noise $n(t)$, where the interference is

$$i(t) = \text{Re} \, r(t) e^{j[\xi(t) + \omega_o t]} \qquad (4.141)$$

The *complex envelopes* of these quantities, assuming that ω_o represents the radian carrier frequency, are

$$\bar{S} = \sum_{k=-\infty}^{\infty} p(t - kT) e^{j\phi_k} \qquad (4.142)$$

$$\bar{I} = r(t) e^{j\xi(t)}$$

$$\bar{N} = \{n(t) + j\hat{n}(t)\} e^{-j\omega_o t} \qquad (4.143)$$

Let $p(t)$ be an ideal Nyquist pulse function, without any intersymbol interference in the sampling instants $t_k = kT$, where k is an integer. Thus we have for the signal term in the sampling instants,

$$\bar{S} = p(0) e^{j\phi_k} \qquad t_k = kT \qquad (4.144)$$

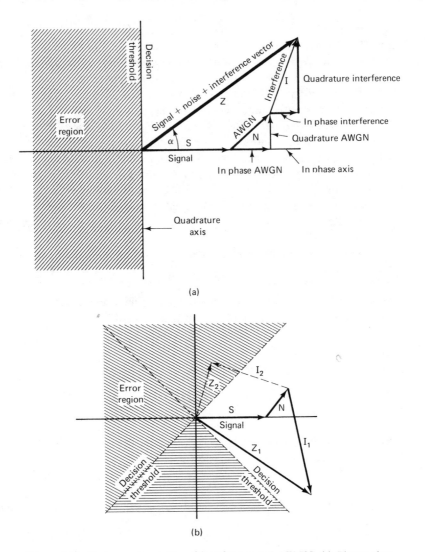

Figure 4.48 Vector representation of interference and AWGN. (a) Binary phase-shift-keyed (BPSK) signal. (b) Four-phase QPSK signal, AWGN, and interference. The vectorial addition of $S + N + I$ gives Z. The illustrated Z_1 vector is in the error-free region, whereas the Z_2 vector is in the error region.

A random process is known as *circular symmetric* (CS) if its complex envelope modulus is independent of the phase, and the phase, modulo-2π, is uniformly distributed. Here we consider *peak limited interference,* with *peak envelope R.* Note that the Gaussian noise is circular symmetric but is not peak limited. Adjacent channel and most co-channel interferences have a random phase relationship with the wanted signal and thus are CS.

The complex envelope input to a coherent phase demodulator in a sampling instant $t = kT$ is

$$\overline{\psi} = p(0)e^{j\phi k} + \overline{N} + re^{j\xi} \tag{4.145}$$

For the *binary* PSK demodulator an error occurs if $\phi_k = 0$ and $\overline{\psi}$ is in the left-half plane [see Fig. 4.48(a)]. That is, an error occurs whenever

$$p(0) + v + r \cos \xi < 0 \tag{4.146}$$

where $v = \text{Re } \overline{N}$ is a zero mean Gaussian random variable with variance σ_n^2. The probability of error is

$$\text{Pr }\{\text{Re }\overline{\psi} < 0\} = \text{Pr }\{p(0) + v + r \cos \xi < 0\} = \frac{1}{2}\text{erfc}\left[\frac{p(0) + r \cos \xi}{\sqrt{2}\,\sigma_n}\right] \tag{4.147}$$

Now we have to obtain the unconditional probability, which is the expected value of (4.147) with respect to r and ξ. To remove the ξ conditioning we expand the erfc ($\cdot$) into its Taylor series about $(p(0)/\sqrt{2}\,\sigma_n) \equiv \sqrt{\gamma}$. We obtain

$$\frac{1}{2}\text{erfc}\left[\sqrt{\gamma} + \sqrt{\gamma}\frac{r}{p(0)}\cos \xi\right]$$

$$= \frac{1}{2}\text{erfc}(\sqrt{\gamma}) + \frac{e^{-\gamma}}{\sqrt{\pi}}\sum_{l=1}^{\infty}\frac{(-1)^l H_{l-1}(\sqrt{\gamma})}{l!}\sqrt{\gamma^l}\left[\frac{r}{p(0)}\right]^l \cos^l \xi \tag{4.148}$$

where H is the Hermite polynomial, related to derivatives of exp $(-x^2)$. Since

$$E\{\cos^l \xi\} = \frac{1}{2\pi}\int_0^{2\pi} \cos^l \xi \, d\xi = \begin{cases} 2^{-l}\binom{l}{\frac{1}{2}l} & l \text{ even} \\ 0 & l \text{ odd} \end{cases} \tag{4.149}$$

and expectation is linear, we then have when $\phi_k = 0$,

$$P_e \,|r = \frac{1}{2}\text{erfc}(\sqrt{\gamma}) + \frac{e^{-\gamma}}{\sqrt{\pi}}\sum_{l=1}^{\infty}\frac{H_{2l-1}(\sqrt{\gamma})\,\gamma^l}{2^{2l}\,l!\,l!}\left[\frac{r}{p(0)}\right]^{2l} \tag{4.150}$$

The unconditioned value of P_e is given by

$$P_e = \int_0^{\infty}(P_e|r)\,dF(r) \tag{4.151}$$

We bound P_e subject to the second-moment constraint,

$$\int_0^{\infty} r^2 \, dF(r) \leq \sigma_r^2 \tag{4.152}$$

and also the peak interference envelope constraint $r \leq R$,

$$\int_{R^+}^{\infty} dF(r) = 0 \tag{4.153}$$

The corresponding upper bound is found to be

$$P_e \leqslant P_e(F_o) = \tfrac{1}{2}\,\text{erfc}\,(\sqrt{\gamma}) + \rho\,\frac{e^{-\gamma}}{\sqrt{\pi}}\sum_{l=1}^{\infty}\frac{H_{2l-1}\,(\sqrt{\gamma})}{2^{2l}l!l!}\cdot[\sqrt{\gamma}\,R/p(0)]^{2l} \tag{4.154}$$

Thus we have obtained *an upper bound for the P_e of binary coherent PSK systems.* However, as shown in [Rosenbaum and Glave, 1974], *it can be easily extended to apply to M-ary coherent PSK systems.* The reason for this follows.

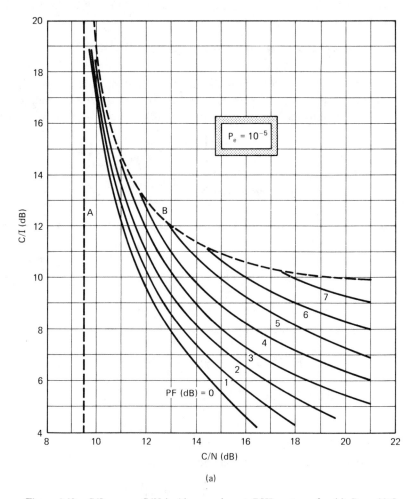

(a)

Figure 4.49 *C/I* versus *C/N* in binary coherent PSK systems for (a) $P_e = 10^{-5}$ and (b) $P_e = 10^{-9}$. To apply to *M*-ary PSK, see equation (4.155) and related text. A, *C/N* without interference; B, interference with characteristics of thermal noise (Gaussian noise); PF, the interference peak factor. The curves are theoretical and take no account of practical system restraints.

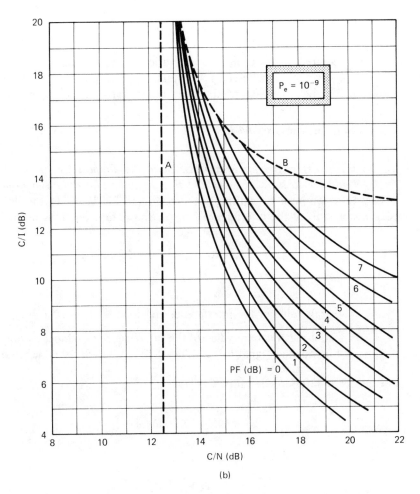

Figure 4.49 (*continued*)

The error region in the complex envelope plane for M-ary CPSK is the union of two symmetric half planes, each located at a perpendicular distance $p(0) \sin \pi/M$ from the signal, $p(0)e^{j\phi_k}$. Because of the CS property of the interference and noise, the probability that the total received complex envelope $\overline{\psi}$ is found in one of those half planes is equal to the binary CPSK error probability, where the signal modulus would instead be $p(0)\pi/M$. In other words, each half plane has occupancy probability equal to the left half-plane probability in the binary case with a foreshortened signal, all other parameters remaining the same. Therefore, the complete M-ary error region has probability that is upper-bounded by twice the binary CPSK P_e, where the values of C/N and C/I each would be reduced by the amount

$$\text{CF} \triangleq -20 \log_{10} \sin \frac{\pi}{M} \qquad (4.155)$$

The required carrier-to-interference ratio C/I, as a function of carrier to AWGN ratio C/N, for binary coherent PSK systems for a $P_e = 10^{-5}$ and $P_e = 10^{-9}$ is shown in Fig. 4.49. This figure illustrates numerical values of equation (4.154).

4.10.2 Application to *M*-ary PSK

To use the binary charts for M-ary PSK, add the amount CF to the abscissa labels; for C/N add CF to the C/I label and then double the ordinate identifications. This is easy to accomplish by shifting the decade and 2X labels to the line below. The *peak factor* (PF) of the interference is used as a parameter. It is defined by

$$\text{PF} \triangleq 20 \log_{10} \frac{R}{\sigma_r} = -10 \log_{10} \rho \qquad (4.156)$$

where

$\qquad R = $ peak value of the interference envelope

$\qquad \sigma_r = $ root-mean-square value of the interference **envelope;** an unfiltered angle-modulated signal has a value of PF $= 0$ under this definition

The carrier-to-interference ratio, C/I, is

$$C/I \triangleq -20 \log_{10} \frac{\sigma_r}{p(0)} \qquad (4.157)$$

and the carrier-to-noise ratio, C/N, is

$$C/N \triangleq 10 \log_{10} \gamma \qquad (4.158)$$

where

$$\sqrt{\gamma} = \frac{p(0)}{\sqrt{2}\,\sigma_n} \qquad (4.159)$$

Example 4.6

Determine the required C/N ratio in a QPSK system if the carrier-to-interference ratio is 13 dB and a $P_e = 10^{-5}$ is required.

(a) Assume that the interference is caused by an unmodulated co-channel interference tone.

(b) Assume that the interference is caused by an adjacent bandlimited QPSK modulated signal having a 5-dB peak factor.

Solution. We refer to Fig. 4.49(a) and equation (4.155).

(a) $(C/I)_{QPSK} = 13$ dB; thus the corresponding $(C/I)_{BPSK}$ is

$$(C/I)_{BPSK} = (C/I)_{QPSK} + 20 \log_{10} \sin \left(\frac{\pi}{M} \right)$$

$$= (C/I)_{QPSK} - 3 \text{ dB} = 10 \text{ dB}$$

For a $(C/I)_{BPSK} = 10$ dB and a PF = 0 dB, we obtain from Fig. 4.49(a) a $(C/N)_{BPSK} = 11.8$ dB requirement. Thus the corresponding $(C/N)_{QPSK} = 11.8$ dB + 3 dB = 14.8 dB. This C/N corresponds to $P_e = 2 \times 10^{-5}$. For 1×10^{-5} a $C/N \approx 15.1$ dB is required.

(b) For a PF = 5 dB, we read from the figure $(C/N)_{BPSK} = 15$ dB; thus for $P_e = 2 \times 10^{-5}$, the $(C/N)_{QPSK} = 18$ dB. Finally, for $P_e = 1 \times 10^{-5}$, we have $(C/N)_{QPSK} = 18.3$ dB.

∎

> *Note. The C/I power ratio is equal in cases (a) and (b). A larger peak factor requires a higher C/N if the same P_e has to be maintained. In other words, interfering signals having a large peak factor degrade the performance of the wanted signal more than low-peak-factor interfering signals do.*

Solve Problems 4.19 and 4.20.

A different P_e performance derivation method to the one presented above is given in [Prabhu, 1969]. Prabhu's derivation applies to *single sinusoidal tone co-channel interference,* for which the results are shown in Fig. 4.50. This presentation provides a good appreciation of the effects of *co-channel* interference into M-ary coherent PSK systems.

The P_e performance derivation of M-ary differential PSK systems (where demodulator does not contain a carrier recovery circuit) is beyond the scope of this text. However, the results are useful for system designers and therefore are shown in Fig. 4.51 [CCIR, 1978 c; Fang and Shimbo, 1973].

Finally, the increase in C/N ratio required to maintain a specified P_e performance in an interference environment for a number of modulation techniques is shown in Fig. 4.52. Note that the susceptibility to interference increases with the number of signaling states, that is, with the spectral efficiency of the system. For further study of the interference effects on digitally modulated signals, including the selective fading environment, see [Feher, 1981].

Effects of earth station and satellite nonlinearities.   When both the required signal and interference are present at the input of a nonlinear device [e.g., earth station high-power amplifier (HPA) or satellite TWT], a nonadditive (multiplicative) degradation is generated. The computation of the interference effect on PSK carriers transmitted over a nonlinear channel requires complex computer simulations, beyond

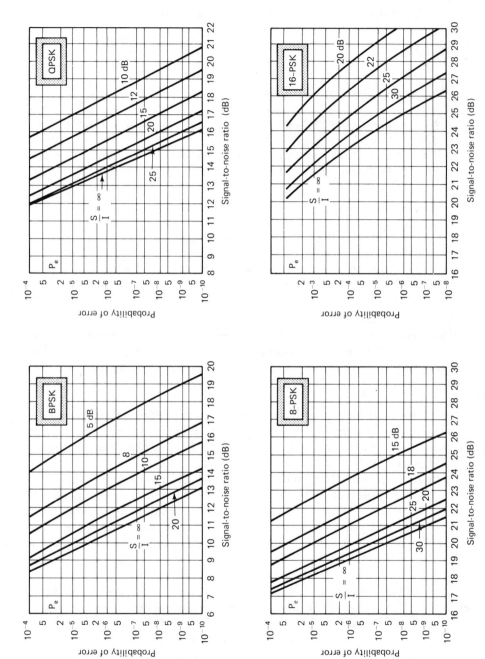

Figure 4.50 $P_e = f(S/N, I)$ in BPSK, QPSK, 8-PSK, and 16-PSK systems where the interference I is a single sinusoidal tone (unmodulated) and S/N represents the carrier-to-noise ratio in the double-sideband Nyquist (symbol rate) bandwidth. (After [Prabhu, 1969], with permission form the *Bell System Technical Journal*, copyright 1969, American Telephone and Telegraph Company.)

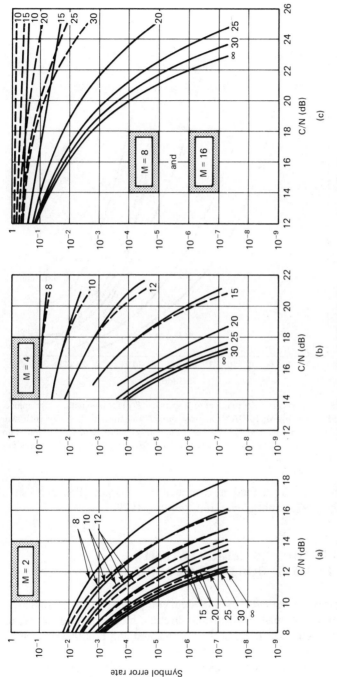

Figure 4.51 P_e performance of *M*-ary differential PSK systems (DPSK) in an AWGN and interference environment. (a) Binary ($M = 2$). Solid line, best θ; dashed line, worst θ. (b) Quarternary ($M = 4$). Solid line, best θ; dashed line, worst θ. (c) Eight-phase ($M = 8$) and 16-phase ($M = 16$). Solid line, $M = 8$; dashed line, $M = 16$. The carrier-to-interference ratio C/I (dB) is shown on each curve. (After [CCIR, 1978c], with permission from the International Telecommunications Union.)

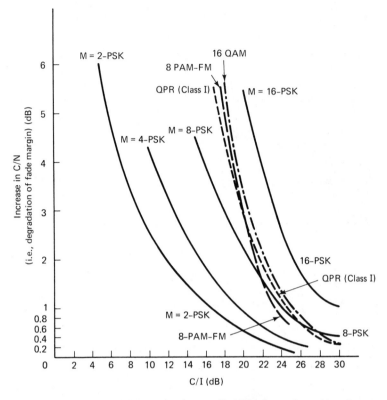

Figure 4.52 Increase In C/N to maintain a specified $P(e)$ for co-channel interference. M-ary PSK; 8 PAM-FM; 16-QAM: increase in C/N to maintain $P_e \leq 10^{-6}$ is shown. Single sinusoidal interference. QPR (Class I-duobinary): increase in C/N to maintain $P_e = 10^{-4}$. Analog FDM-FM interference. (After [Feher, 1981], with permission from Prentice-Hall, Inc.)

the scope of this text. As a further reference, we suggest [Benedetto et al., 1978; Fang and Shimbo, 1973; and Chang and Shimbo, 1979].

4.11 INTEGRATED-CIRCUIT IMPLEMENTATION OF FREQUENTLY USED BUILDING BLOCKS

Schematic diagrams of frequently used modem components (subsystems) are presented in this section. A wide range of electronic circuits, including multipliers, oscillators, phase-locked loops, phase shifters, both conventional and nonlinearly switched filters, and threshold comparator circuits, are described.

Our objective is to present typical circuit configurations in order to give an appreciation of the complex integrated circuits used in modern modulator–demodulator design. The presentation here is limited to brief functional descriptions, with suggested references that contain detailed studies of the electronic subsystems presented. Illustrative circuits are selected to cover both low- and high-bit-rate applications.

Low-bit-rate and low-carrier-frequency modems ($f_b < 100$ kb/s and $f_c < 500$ kHz) are frequently implemented with almost exclusively digital components. For example, instead of a sinusoidal carrier source, a digital carrier is used; instead of analog filters and analog multipliers, digital signal processors are employed. These low-speed modems therefore have an input analog-to-digital converter, a microprocessor to perform the digital signal processing functions, and an output digital-to-analog converter.

An in-depth study of digital signal processing techniques, as required for digital modem implementation, may be found in [Oppenheim and Schafer, 1977]. Recent digital multiprocessor applications in modem design are described in [Murano et al., 1979]. A description of a high-speed (90 Mb/s) modulator design having a fully digital implementation is given in [Wood, 1977].

4.11.1 Multipliers

The multiplier is the most frequently used circuit in coherent modulators and demodulators. In a QPSK modem, such as shown in Fig. 4.17, two multipliers are used for the implementation of the in-phase and quadrature modulators, and two multipliers are also used for demodulation. Additional multipliers are required for the design of the nonlinear processors and phase-locked loops used in the carrier-recovery and symbol timing-recovery subsystems.

A multiplier, as shown in Fig. 4.53, is essentially a gain-controlled amplifier that multiplies the input signal $v_x(t)$ with the external gain controlling signal $v_y(t)$ to produce a resultant output $v_o(t)$. In many applications the gain is externally adjustable.

Also illustrated in Fig. 4.53 are the schematic diagrams of the basic circuits of a two-quadrant multiplier, a four-quadrant multiplier, and a complete monolithic silicon integrated-circuit multiplier. The basic two-quadrant multiplier is functional only in the shaded region. The input signal $v_x(t)$ may have either a positive or negative polarity, whereas the external gain-controlling signal $v_y(t)$ must be positive and have a greater base-to-emitter voltage than the transistor Q_3. The basic four-quadrant multiplier consists of three differential amplifiers. This configuration permits both of the input signals $v_x(t)$ and $v_y(t)$ to have positive and negative polarities.

The multiplier output is given by

$$v_o(t) = k v_x(t) v_y(t) \tag{4.160}$$

where k is a constant.

In the design of coherent modems, four-quadrant multipliers are used. A comprehensive description of analog multipliers is given in [Wait et al., 1975].

Another popular multiplier circuit is the *double-balanced mixer* (DBM) such as that shown in Fig. 4.54. Double-balanced mixers have wide bandwidth and are available over an extremely large frequency range (e.g., in the range 100 kHz to 4

GHz, which is well above the operational frequencies of monolithic integrated-circuit multipliers). As these devices are manufactured in large quantities in integrated packages, their cost is relatively low ($3 to $50).

To achieve almost ideal multiplication, excellent balance and isolation between the local oscillator (LO) radio-frequency (RF) and intermediate-frequency (IF) ports is required. With a careful choice of signal levels (typically specified in manufacturers' data sheets) the high-order intermodulation products and mixer feedthrough (leakage) are minimized. In QPSK modulator designs, the baseband signal is fed to the IF port and the modulated double-sideband-suppressed signal is obtained at the RF port. Multiplication in double-balanced mixers is achieved by changing the *transconductance* of the diodes D_1 to D_4 shown in Fig. 4.54. The output signal $v_{RF}(t)$ is given by

$$v_{RF}(t) = Kv_{LO}(t)v_{IF}(t) \tag{4.161}$$

where K is a constant.

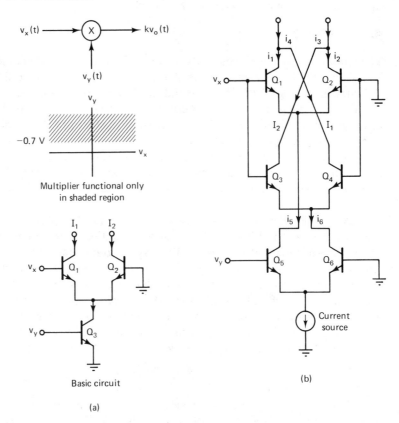

Figure 4.53 Monolithic silicon integrated-circuit multiplier, type CA3091D. (a) Two-quadrant multiplier. (b) Basic four-quadrant multiplier. (c) Schematic diagram of the CA3091D. Resistance values are in ohms. The 3-dB bandwidth of this four-quadrant multiplier is 4.4 MHz. (Courtesy of RCA.)

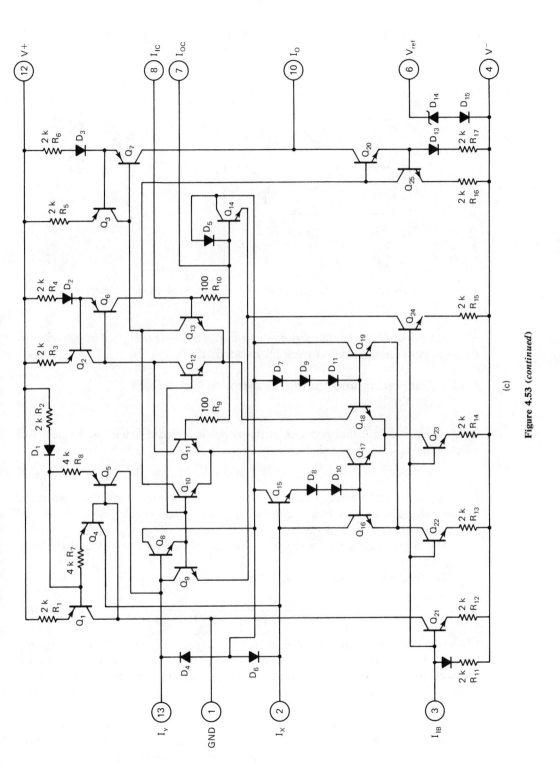

Figure 4.53 *(continued)*

(c)

225

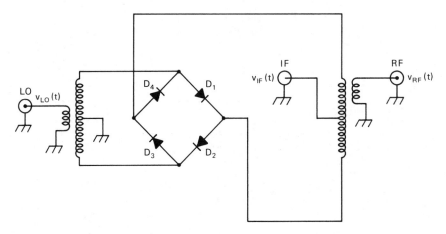

Figure 4.54 Schematic diagram of a typical double-balanced mixer multiplier.

The principles and applications of balanced mixers are described in manufacturers' application notes, such as [Hewlett-Packard, 1967].

4.11.2 Oscillators and Voltage-Controlled Oscillators (VCO)

The VCO is a device that provides at its output a periodic signal the frequency of which is linearly related to the control voltage, v_c, by

$$f_o = f_{fr} + Kv_c \tag{4.162}$$

where f_{fr} is the free-running frequency of the voltage-controlled oscillator when the control voltage is zero. A *local oscillator* (or conventional oscillator) can be viewed as a VCO with zero control voltage.

The circuit diagram of a typical *voltage-controlled oscillator* (VCO) is shown in Fig. 4.55.

The frequency of this VCO is

$$f_o = \frac{I_{AO}}{2C_x(V_{T+} - V_{T-})} + \frac{kv_c}{2C_x(V_{T+} - V_{T-})} \tag{4.163}$$

where the first term is proportional to the dc current of the current source I_A, and the second term is proportional to the control voltage v_c; note that V_{T+}, V_{T-} represent the upper and lower trigger points of the Schmitt trigger. For a derivation and applications of (4.163), see [Grinich and Jackson, 1975; Grebene, 1971; and Wait et al., 1975].

In many demanding applications, high-frequency stability oscillators and voltage-controlled oscillators are required. This is achieved by using voltage-controlled *crystal* oscillators (VCXO). Typical VCXO configurations are shown in Fig. 4.56. In the

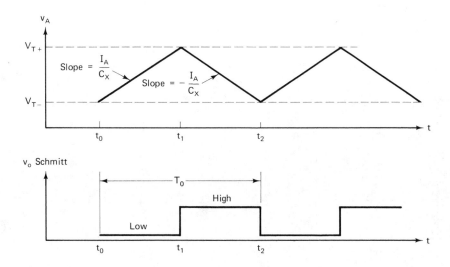

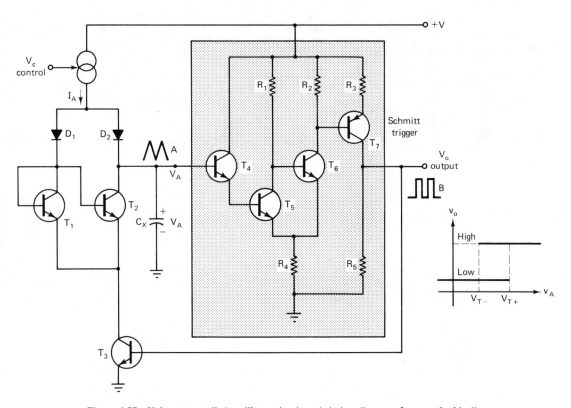

Figure 4.55 Voltage-controlled oscillator circuit and timing diagram; for $v_c = 0$, this diagram represents a conventional oscillator.

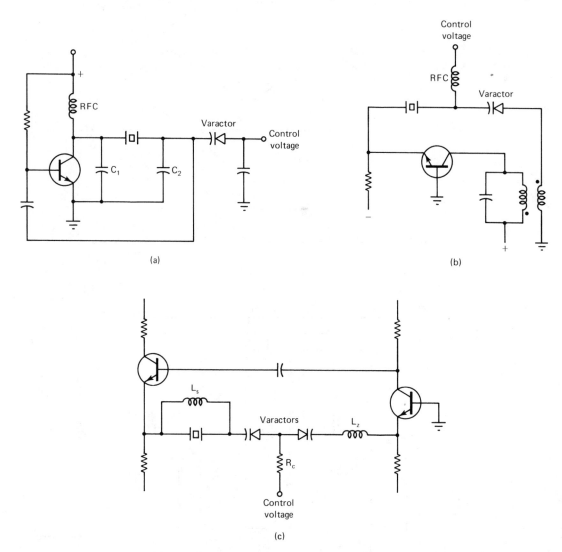

Figure 4.56 Voltage-controlled crystal oscillators (VCXO). (a) Modified Pierce oscillator. (b) Grounded-base oscillator. (c) Emitter-coupled oscillator. (After [Gardner, 1979], with permission from John Wiley & Sons, Inc.)

modified Pierce oscillator the crystal is operated as an inductance. The value of the capacitance of the *varactor* [variable-voltage (i.e., voltage-dependent) capacitance diode] depends on the applied control voltage. This capacitance provides a variation of the total feedback capacitance and thus causes a pulling of the oscillation frequency. The disadvantage of this circuit is that for high-Q crystals, the tuning range is very small. In the circuits shown in Fig. 4.56(b) and (c), a larger tuning range is possible [Gardner, 1979; Gerber and Sykes, 1974; Driscoll, 1973].

4.11.3 Phase-Locked Loops (PLL)

Voltage-controlled oscillators have frequent applications in the design of *phase-locked loops*. The block diagram and schematic diagram of a monolithic integrated-circuit phase-locked loop is shown in Fig. 4.57. The main building blocks of this circuit include a voltage-controlled oscillator, a multiplier, and an externally connected low-pass filter. The multiplier, followed by a low-pass filter, performs phase detection, and thus the term *phase detector* is frequently employed. [Remember that in equations (4.13) and (4.15), we demonstrated the equivalence of multiplier low-pass filters and phase demodulators].

PLL—principle of operation. A rigorous mathematical analysis of phase-locked-loop systems is quite cumbersome. However, from a qualitative point of view, the basic principle of operation can be explained as follows. Assume that there is no signal present at the input of the systems (i.e., $v_s = 0$). With no signal input, the filtered error voltage $v_d(s) = 0$. (Why?) The VCO operates at a set frequency known as the "free-running" VCO frequency. Now, if we apply an input signal, the phase comparator compares the phase and frequency of the input signal V_s and that of the phase-locked-loop output V_o. The resultant signal is an *error signal* which is used as a *control* voltage, $v_d(t)$, to force the VCO frequency to vary in a direction that reduces the error (i.e., the frequency difference between the input signal f_s and the VCO frequency f_o). The input frequency f_s must be sufficiently close to f_o; that is, it must be within the *capture range* in order to achieve lock synchronization. The basic phase-locked-loop equations, assuming that the loop is locked, are

$$\frac{\theta_o(s)}{\theta_i(s)} = H(s) = \frac{K_o K_d F(s)}{s + K_o K_d F(s)} \tag{4.164}$$

$$\frac{\theta_i(s) - \theta_o(s)}{\theta_i(s)} = \frac{\theta_e(s)}{\theta_i(s)} = \frac{s}{s + K_o K_d F(s)} = 1 - H(s) \tag{4.165}$$

$$V_c(s) = \frac{s K_d F(s)\theta_i(s)}{s + K_o K_d F(s)} = \frac{s\theta_i(s)}{K_o} H(s) \tag{4.166}$$

where $H(s)$ is the closed-loop transfer function.

Phase-locked-loop applications include the implementation of narrowband tunable filters and oscillators where the output frequency is an integer or subinteger multiple of the input frequency. Synchronization circuits such as carrier recovery and symbol timing recovery circuits, described in Chapter 7, also use phase-locked loops.

Phase-locked-loop equations, together with principles of operation and applications, are well described in [Gardner, 1979; Grebene, 1971; Spilker, 1977; and Lindsey, 1972]. Application notes published by manufacturers (Motorola RCA, Exar, Signetics, National, etc.) contain many practical circuit designs.

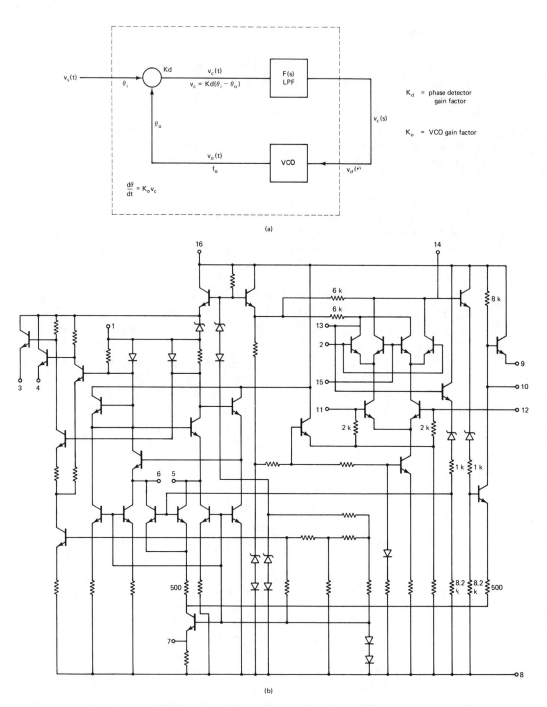

Figure 4.57 (a) Block and (b) schematic diagrams of a popular monolithic integrated circuit. Phase-locked loop. (Circuit diagram courtesy of Signetics, Inc.)

4.11.4 Phase Shifters

To generate quadrature carriers for QPSK modulators and demodulators, 90° phase-shifting circuits are required. A frequently used implementation method is shown in Fig. 4.58. We assume that a sinusoidal local oscillator having an angular frequency ω_c is available. The phase-locked loop has a frequency divider ($\div 2$) circuit included within the loop. The voltage-controlled oscillator output $\sin 2\omega_c t$ is fed into a threshold comparator to convert the sinusoidal signal into a digital square wave having a frequency $2f_c$. The flip-flops are used as frequency dividers. The output timing diagram of the I and Q channels shows the 90° phase shift between the digital carrier waves. A simple output low-pass filter reconverts the digital signals back into sinusoidal carriers. The advantage of this circuit is that it maintains an exact 90° phase shift over a wide range of carrier frequencies. However, one difficulty is that the two LPFs have to be carefully equalized to ensure the exact phase shift.

Quadrature hybrids are also in frequent use for the generation of 90° shifted signals. A quadrature hybrid is a four-port, 3-dB coupler capable of splitting an input signal into two isolated 90° phase-shifted outputs. This device is commonly known as a 3-dB stripline **coupler** or a short-slot waveguide coupler [Liao, 1980]. Commercially available quadrature hybrids provide accurate (i.e., within $\pm 1°$) phase shifts in the frequency range 12 to 300 MHz.

4.11.5 Conventional Filters

Many books and articles on passive LC filter design, active operational amplifier filter implementation, and transversal filter design present a wealth of information on this subject [Wait et al., 1975; Weinberg, 1957; Grinich and Jackson, 1975; Williams, 1975; Irvine, 1981; Christian and Eisenmann, 1966]. Many recent issues of the *IEEE Transactions on Circuits and Systems* contain new design and filter analysis techniques. In Section 3.5.1 a brief discussion of conventional filter design was presented; here a schematic diagram of a simple active filter which approximates the $\alpha = 0.5$ raised-cosine channel characteristics is given. In Fig. 4.59(a) the schematic diagram of an unequalized-phase fourth-order Butterworth low-pass filter having a 3-dB cutoff frequency of 32 kHz is shown. The intersymbol interference caused by this filter, for a 64-kb/s signal, is about 33% [see Fig. 3.9(b)]. With a simple transversal filter preequalizer such as that shown in Fig. 4.59(b), the intersymbol interference can be reduced to about 10%. Note that the setting of the tap coefficients was explained in Section 3.5.2.

4.11.6 Nonlinearly Switched Filter (Feher's Processor)

The principle of operation of the nonlinearly switched filter (which gives **intersymbol-interference- and jitter-free bandlimited output signals**) was explained in Section 3.5.3. The spectral spreading advantages of nonlinearly filtered OKQPSK signals were high-

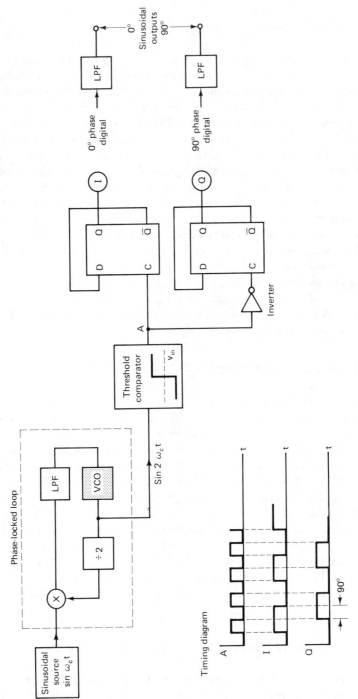

Figure 4.58 Digital 90° phase shifter.

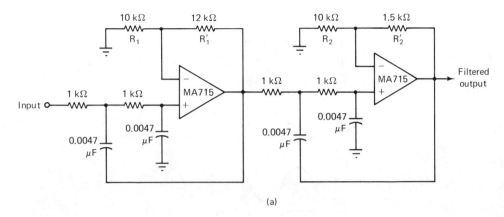

(a)

MA–715 operational
amplifier
SN–7474-flip-flop

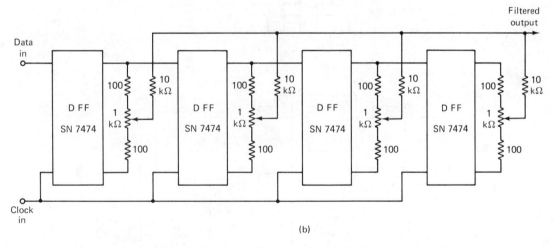

(b)

Figure 4.59   (a) Fourth-order Butterworth low-pass filter and (b) transversal preequalizer circuit. This filter has a 3-dB cutoff frequency of 32 kHz. The cascade of the preequalizer and the active filter approximate an $\alpha = 0.5$ raised-cosine filter (for 64 kbaud data).

lighted in Fig. 4.42. In Fig. 4.60 the schematic and timing diagrams of the nonlinearly switched filter are presented.

In the clock generator section of Fig. 4.60, the sine wave $v(t)$ has a period of $2T$ and is used to generate the CLKA signal at its positive-going edge. This clock and its shifted version CLKB *are then* exclusively OR-ed to produce CLKC and its inversion $\overline{\text{CLKC}}$. The period of these two new timing clocks is T.

At the input of the logic circuit, the data bit $d(kT)$ and its delayed version

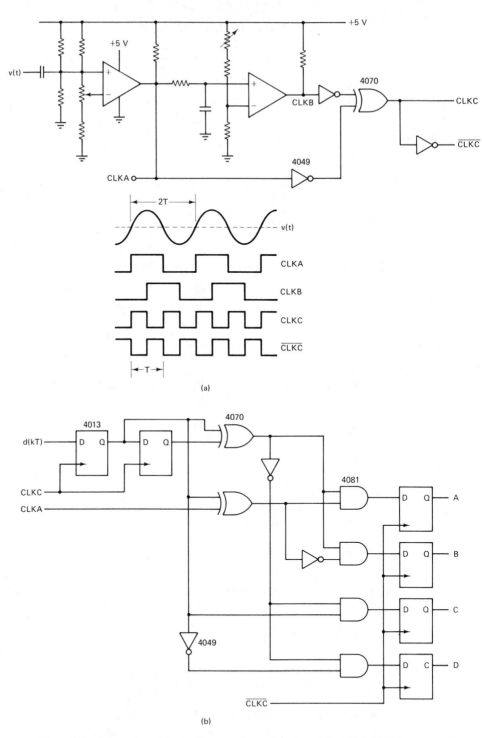

Figure 4.60 Schematic and timing diagram of a nonlinearly switched filter (Feher's processor). (a) Clock generator. (b) Logic circuitry. (c) Function generator and multiplexer.

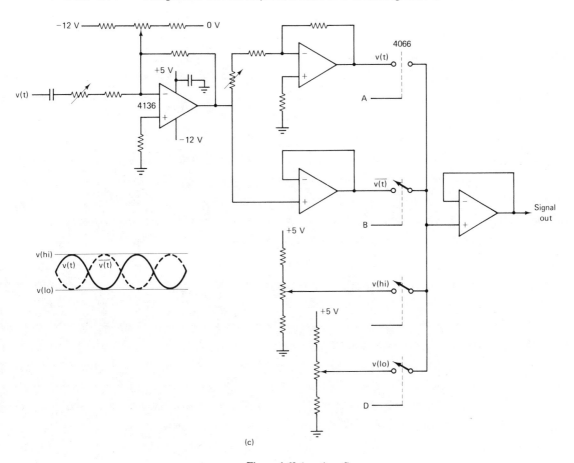

(c)

Figure 4.60 (*continued*)

$d(kT - T)$ are "modulo-2" added. The data bit $d(kT)$ is also modulo-2 added with CLKA. During any given bit interval only one of the four D flip-flops has a high output state. Synchronism is therefore assured by parallel clocking of the flip-flops.

The analog waveform $v(t)$, its inversion $\bar{v}(t)$, and the $\pm$dc signals $v(+a)$ and $v(-a)$ (which are generated at the output of the function generator) are then connected or disconnected by the analog switch (in this case, a CMOS integrated-circuit CD4066).

A detailed description of nonlinearly switched filters and their applications is given in [Feher, 1979, and Huang et al., 1979].

4.11.7 Threshold Comparators

A very high speed threshold comparator circuit diagram is shown in Fig. 4.61. This comparator (Motorola MC1650/MC1651) contains an output latch that provides a sample and hold feature as required in data regenerators. The propagation delay of

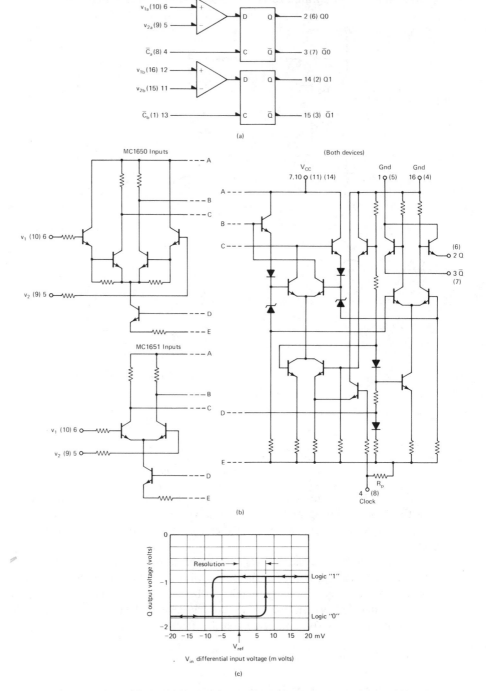

Figure 4.61 High-speed threshold comparator circuit (Motorola MC1650 and MC1651). (a) Positive logic. (b) Circuit schematic (one-half of device shown). (c) Typical transfer curves. (Courtesy of Motorola Semiconductor Products, Inc.)

this comparator is 3.5 ns and the maximal operating rate is about 300 Mb/s. An input voltage differential (or hysteresis) of about ± 7 mV is required to drive the output states into the logic one and logic zero states. Other lower-speed comparators, such as the LF111 and LF311, have a much lower input hysteresis (less than ± 1 mV). Manufacturers' data sheets contain detailed threshold comparator specifications and design information.

PROBLEMS

4.1. Describe the required conditions to satisfy the equivalence between the predetection bandpass filter, BPF_R, and postdetection (postdemodulation) lowpass filter, LPF_R, illustrated in Fig. 4.2. Is a lowpass filter essential in both demodulators? (*Hint:* What would happen if the second-order spectral components were not removed?)

4.2. Sketch the unfiltered and filtered outputs of a BPSK modulator. Your signal time-domain presentation should include waveforms similar to the ones illustrated in Fig. 4.5 for the synchronous case. Here assume an asynchronous modulator.

4.3. Assume that the message sequence, $\{b_k\}$, shown in Fig. 4.7, is differentially encoded and binary-phase modulated. A comparison demodulator (DBPSK in Fig. 4.6) is used. Show that the DBPSK demodulator recovers the $\{b_k\}$ bit sequence correctly. Assume that the channel noise is negligible.

4.4. Sketch the unfiltered power spectrum of a BPSK signal if the unmodulated carrier frequency $f_c = 70$ MHz and the bit rate is $f_b = 6$ Mb/s. Assume that (a) the data source is a periodic 101010 . . . data stream; (b) the data source is an equiprobable random source. Highlight the difference between the discrete and continuous power spectra in cases (a) and (b). Sketch the corresponding transmit power spectra of the Nyquist filtered system. Assume an $\alpha = 0.5$ roll-off factor and include the $(x/\sin x)$-shaped amplitude equalizer in your transmitter.

4.5. An integrate-sample-and-dump (ISD) circuit is a possible implementation of a matched filter receiver. Describe the type of signals for which the ISD receiver is an optimal receiver. Note that this receiver is not an optimal receiver for bandlimited synchronous data transmission. Why? [*Hint:* The ISD circuitry is a possible implementation of $h_o(t) = s(T - t)$ if $s(t)$ is an infinite bandwidth signal.]

4.6. In an $f_b = 10$ Mb/s infinite bandwidth BPSK system the integrator as shown in the correlation receiver of Fig. 4.13 is replaced by a low-pass RC circuit of 3 dB bandwidth f_{3dB}. Calculate the value of f_{3dB} for which the signal-to-rms noise ratio is maximal. Show that for the RC network, the P_e is about 10 times larger than that for the ideal integrator. This corresponds to a 1-dB S/N difference in the $P_e = 10^{-6}$ range.

4.7. Explain in a few sentences the fundamental differences and similarities between the matched filter and the Nyquist receiver. When are these receivers equivalent? Try to answer this question without referring to the book. That is, take the courage to close your book and write down your answer.

4.8. The four phase states of a Gray-coded and binary-coded-decimal, BCD, QPSK system are as follows:

State number	Dibits	Phase state Gray code	Phase state BCD
1	0 0	225°	225°
2	0 1	135°	135°
3	1 1	45°	−45°
4	1 0	−45°	45°

How many bit errors are generated due to adjacent symbol errors? Carefully analyze the Gray-coded and BCD systems. Is the number of bit errors caused by wrong decisions by adjacent symbols independent of the signal state?

4.9. Assume that the spectral shaping of a QPSK modulator is performed with premodulation low-pass filters. The spectral shaping of a second QPSK modulator is performed by postmodulation bandpass filters. Assume that both of these modulators transmit data at the same rate, f_b, and have the same carrier frequency, f_c. If both modulators have the same RF power spectra, which conditions do their respective filters satisfy?

4.10. Explain the difference between QPSK and offset-keyed QPSK systems. Is the differential encoder/decoder of OK-QPSK modems simpler or is it more complex than that of QPSK systems? Which system has larger envelope fluctuations?

4.11. Compare the P_e performance of DQPSK systems with that of DEQPSK systems. Both systems have differential encoders and decoders. What is the physical reason for the significant (about 2 dB) difference in the E_b/N_o requirement?

4.12. What is the C/N ratio requirement of an $f_b = 45$ Mb/s rate DQPSK modem if $P_e = 10^{-4}$ is required? Assume that the receiver noise bandwidth equals 30 MHz.

4.13. The amplitude slope of a QPSK system is 0.4 dB/MHz. The bit rate of the system is $f_b = 45$ Mb/s. Calculate the required E_b/N_o if $P_e = 10^{-4}$ is specified and the only distortion in the system is caused by the amplitude slope.

4.14. The delay slope of an OKQPSK system is 0.6 ns/MHz. How much is the required E_b/N_o if $P_e = 10^{-4}$ is specified and the only system distortion is caused by the delay slope?

4.15. Explain the difference in phase transitions in conventional QPSK, offset QPSK, and MSK. Why are the phase transitions of offset QPSK (OKQPSK) limited to ±90°? What is the reason that there are no abrupt phase transitions in MSK (i.e., why is the MSK a phase continuous system)? [*Hint:* Note the coherence requirement between the frequency deviation and the bit rate, equation (4.125).]

4.16. The main lobe of the MSK System equals $f_b \cdot \frac{3}{4}$, where f_b is the bit rate. Explain why it is possible to achieve a 2-b/s/Hz radio-frequency efficiency. Describe how it is possible to have a minimum-bandwidth radio-frequency filter which is narrower than one-half that of the main lobe. (*Hint:* Revise Nyquist's generalized ISI-free transmission theorem and the equivalence of bandpass and low-pass channels.)

4.17. Compare the bandlimited and hard-limited QPSK, OKQPSK, MSK, and NLF-OKQPSK (Feher's QPSK) spectra with the spectra of unfiltered systems. Determine which modulation scheme has the smallest sideband restorations. How much is the sideband restoration of conventional QPSK at $1.5/T_s$? Assume that the transmit filter is a fourth-order Butterworth filter having 3 dB attenuation at 17.5 KHz and that the bit rate is 64 kb/s. (*Hint:* See Figs. 4.24 and 4.41.)

4.18. A QPSK modulated signal is bandlimited by an ideal Nyquist filter. This filter has a roll-off factor $\sqrt{\alpha} = 0.3$ and includes the $(x/\sin x)$-shaped amplitude equalizer. The filtered signal is hard-limited by an ideal memoryless infinite bandwidth limiter. Assume that an ideal coherent demodulator is used and that the transmission medium is infinite bandwidth, noise free. Is there crosstalk generated in the I and Q channels? If so, explain why.

4.19. Explain the physical reason for larger P_e degradation if the peak factor of the interference is larger.

4.20. A satellite system has to operate at $P_e \leq 10^{-9}$. The available $C/N = 17$ dB. What is the maximum interference power (i.e., minimal C/I ratio) if: (a) the interference consists of a single sinusoidal tone? (b) the interference is the sum of two equal power co-channel sinusoids? Assume that these sinusoids are independent of each other. (*Hint:* Compute the peak factor of two sinusoidal tones.)

5

SPECTRALLY EFFICIENT MODULATION TECHNIQUES FOR SATELLITE SYSTEMS

5.1 INTRODUCTION TO SPECTRAL EFFICIENT (MORE THAN 2 b/s/Hz) MODULATION TECHNIQUES

The principles for design of both binary and four-phase PSK systems were described in Chapter 4. The spectral efficiency of these systems is less than 2 b/s/Hz. Most operational satellite systems are *power limited;* that is, the available ratio of energy per bit to noise density E_b/N_o is insufficient to enable the utilization of modems which have a spectral efficiency of more than 2 b/s/Hz and require a higher E_b/N_o ratio.

The crowded conditions prevailing in many regions of the radio spectrum, combined with the new emphasis on digital satellite transmission, has created a need for improved spectrum utilization techniques. Time-division multiple-access (TDMA) satellite systems, described in Chapters 8 and 9, are among the most powerful spectral and power-efficient system approaches. The intelligent application of *spectral efficient* (more than 2 b/s/Hz) modulation techniques combined with TDMA may provide a further improvement in the spectral efficiency of these satellite systems. With the recent developments in more powerful microwave high-power amplifier designs and the discovery of new nonlinearly amplified modulation techniques, it is probable that some future generations of satellite systems will have a sufficiently high E_b/N_o ratio to enable the use of highly spectral efficient modulation techniques.

240

The terrestrial line-of-sight microwave systems that were developed in the 1950s, 60s, and 70s used predominantly binary and four-phase PSK modulation techniques. In the late 1970s and early 1980s digital terrestrial microwave systems having a spectral efficiency between 3 and 6 b/s/Hz have been developed. A number of satellite systems are expected to follow this more spectrally efficient modulation trend.

In this chapter both the principles and the most important characteristics of M-ary PSK ($M = 8, 16, \ldots$) and *quadrature-amplitude-modulated* (QAM) (also known as *amplitude-phase-keying,* APK) systems are *summarized.* An original method for generating M-ary nonlinearly amplified modulated signals is described which permits saturated output power amplification, hence higher output power, than may be obtained in systems used previously. A concise presentation of quadrature partial-response (QPR) correlative coded systems follows.

For those interested in a more detailed study of spectrally efficient modulation techniques, consult [Feher, K.: *Digital Communications . . .* Prentice-Hall, Inc., 1981, and Feher, K.: *Digital Modulation . . .* Don White, Inc., 1977].

5.2 LINEARLY AND NONLINEARLY AMPLIFIED (SATURATED) *M*-ARY PSK AND QAM (APK) EARTH STATION AND SATELLITE MODEMS

5.2.1 Eight-Phase 8-PSK Systems

The 8-PSK modulation technique can be viewed as an extension of the QPSK system described in Chapter 4. In the classical 8-PSK modulator block diagram, as shown in Fig. 5.1, the f_b rate data are split into three binary parallel streams, each having a transmission rate of $f_b/3$. The two-level to four-level converter provides one of the four possible levels of a polar baseband signal at a and b. If the binary symbol A is a logic one (zero), then the output level A has one of the two possible (positive, negative) signal states. The logic state of the C bit determines whether the higher or lower signal level should be present at A or at B. When $C = 1$, the amplitude of A is greater than that of B; if $C = 0$, the converse is true. The four-level polar baseband signals at a and b are used to double-sideband suppressed-carrier (DSB-SC) amplitude modulate the two quadrature carriers. The modulated 8-PSK signal-state space diagram is also shown in Fig. 5.1. The detailed construction of these signal state-space diagrams is suggested in Problem 5.1. *Solve Problem 5.1.*

A modern approach to the design of an 8-PSK modulator of high-speed (120 Mb/s) transmission, using only digital devices, is illustrated in Fig. 5.2. The f_b rate input binary baseband information is *serial-to-parallel* converted in the data distributor unit. These parallel $f_b/3$ rate data streams switch on and off the logic gates of the high-speed commutative IF multiplexer. Depending on the baseband logic states, one of the eight digital IF vectors is connected to the digital IF output. This digital phase-shifted 8-PSK carrier is filtered by means of a conventional BPF; thus a bandlimited 8-PSK signal is obtained.

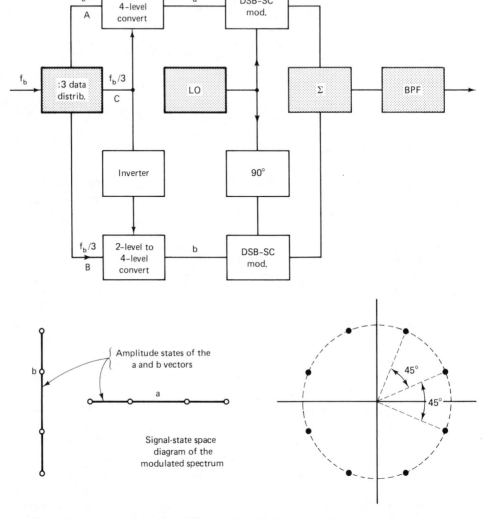

Figure 5.1 Classical eight-phase PSK modulator block and signal-state space diagrams. For construction of the latter diagram, see Problem 5.1. (After [Feher, 1981], with permission from Prentice-Hall, Inc.)

From the signal-state space diagram of the unfiltered 8-PSK signal we conclude that all states have equal amplitudes and are evenly distributed (45° spaced) on a circle. Thus, even if a saturated amplifier is used, which has a significant AM/AM and AM/PM conversion, it will not degrade the modulated unfiltered signal. However, if the nonlinear amplifier is used after the bandpass filter, then due to envelope fluctuations, this amplifier will distort the signal-state space diagram, and a system performance degradation will result.

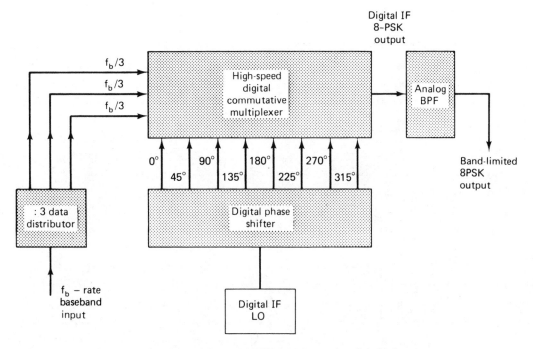

Figure 5.2 High-speed 8-PSK modulator using digital subsystems.

5.2.2 Spectrally Efficient Quadrature-Amplitude-Modulated (QAM) Systems

A block diagram for a QAM suppressed-carrier modulator and its corresponding signal-state space diagram is shown in Fig. 5.3. The information is contained in both the amplitude *and* the phase of the modulated signal; for this reason the term *amplitude phase keying* is also used in the literature.

The f_b rate binary source is commuted into two binary symbol streams, each having a rate of $f_b/2$. The following 2-to-L-level baseband converter converts these $f_b/2$ rate data streams into L-level PAM signals having a symbol rate of

$$f_s = (f_b/2) : (\log_2 L)   \text{symbols/sec}   (5.1)$$

For example, if the source bit rate is $f_b = 10$ Mb/s, then the commuted binary baseband streams have an $f_b/2 = 5$ Mb/s rate. If an $M = 16$-ary QAM modulated signal having a theoretical efficiency of 4 b/s/Hz is desired, these commuted binary streams are converted into $L = 4$-level baseband streams. The resultant 4-level symbol streams of the I and Q channels are 5 Mb/s : $\log_2 4$, or 2.5 M symbols/s. If premodulation LPFs are used, as shown in Fig. 5.3, then the minimum bandwidth (Nyquist bandwidth with $\alpha = 0$) of these filters is 1.25 MHz. The minimum IF bandwidth requirement is identical to the double-sided minimum baseband bandwidth (i.e., 2.5 MHz). This example illustrates that a 10 Mb/s, $M = 16$-ary QAM signal can be

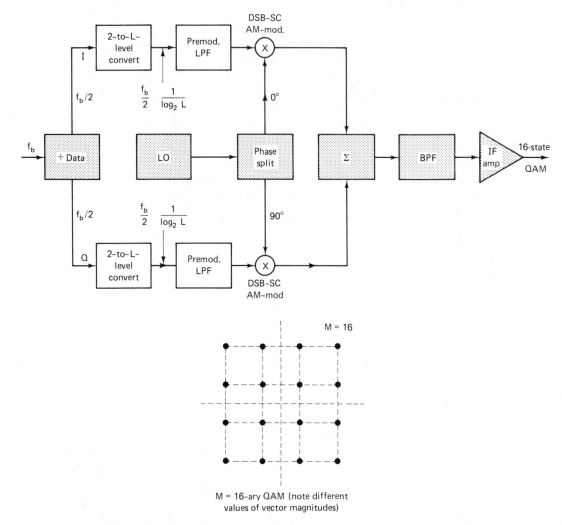

Figure 5.3   *M*-ary QAM modulator block and corresponding signal-state space diagram.

transmitted in a theoretical minimum bandwidth of 2.5 MHz; thus a spectral efficiency of 4 b/s/Hz has been obtained. Practical, high-speed, 400-Mb/s, 16-ary systems have been achieving a bandwidth efficiency of approximately 3.7 b/s/Hz.

5.2.3 Coherent Demodulators for Spectrally Efficient *M*-ary PSK and QAM Signals

The block diagram of a coherent *M*-ary PSK and *M*-ary QAM demodulator is shown in Fig. 5.4. This demodulator incorporates the major components of *M*-ary PSK demodulators. For this reason, with only minor modifications in the signal processing

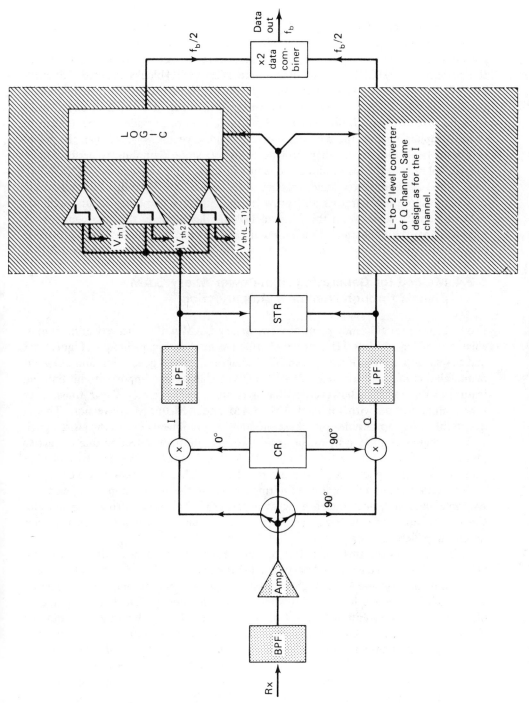

Figure 5.4 *M*-ary QAM or PSK demodulator block diagram. (From [Feher, 1981], with permission of Prentice-Hall, Inc.)

circuity, this demodulator structure could be employed for the demodulation of *M*-ary PSK signals, such as 8-PSK or 16-PSK signals. For optimum performance it is essential to recover the carrier and the symbol rate frequencies. This is achieved in the carrier recovery (CR) and symbol timing recovery (STR) blocks. A detailed description of these blocks is given in Chapter 7. To distinguish between the *L*-demodulated baseband levels, $L - 1$ threshold comparators are required in each of the *L*-to-2-level PAM converters. Individual threshold comparators provide a logic 1 state if, in the sampling instant, the received signal plus noise vector is larger than the preset threshold level. Otherwise, they provide a logic 0 state. Sampling is performed at the symbol rate, at the maximum eye-opening instants. This rate equals $f_b/[2(\log_2 L)]$. The *L*-to-2-level converter logic circuitry accepts the $L - 1$ parallel binary outputs of the threshold comparators and provides the ×2 data combiner inputs with the $f_b/2$ rate binary signal from the *I* channel. A similar input to the data combiner is received from the *Q* channel. Finally, the data combiner, which is a parallel-to-serial converter, provides the desired f_b rate binary signal output. (*Solve Problem 5.2).*

5.2.4 Method for Generating High-Power *M*-ary QAM Signals through Nonlinear Amplification

From the spectral efficiency point of view, *M*-ary QAM with a theoretical maximum efficiency of $\log_2 M$ b/s/Hz is a very attractive modulation technique. Figures 5.3 and 5.4 show a conventional system block diagram for the generation and coherent demodulation of an *M*-ary (e.g., $M = 16$) QAM signal. One important and limiting feature of this method, however, is that it requires an extremely *linear* transmitter power amplification with minimal AM-to-AM and AM-to-PM conversion. This is due to the strong amplitude- and phase-modulated components of the 16-QAM signal.

In digital satellite systems, in order to overcome poor detection due to carrier loss (low E_b/N_o) resulting from large path loss and fading, the generation of high transmitter output power is often preferred. In general, output power is most easily maximized by using an amplitude-limiting device as the power-amplifying element. Examples of such devices are traveling-wave tubes (TWTs) and klystrons in saturation, Class C-operated transistorized amplifiers, and Gunn and Impatt diode injection locked amplifiers (ILAs).

The purpose of this section is to outline a method of generating a 16-state QAM signal and in general an *M*-ary signal that permits nonlinear transmitter output power amplification and hence higher output power than in other methods considered to date. One attractive feature of this modulation method is that despite significant differences with conventional 16-QAM in the method of generating a modulated signal, the same straightforward demodulation techniques apply to both. As this method permits nonlinear amplification, it has been designated *nonlinear amplified M-ary quadrature amplitude modulation,* or NLA-QAM for short [Morais and Feher, 1982]. NLA-QAM can accept either two asynchronous input data streams or, with additional hardware, one single input stream. In the single input stream version, the timing of the data transitions on the quadrature channel can be coincident with respect to the data transitions on the in-phase channel. Alternatively, it can be offset by a unit bit duration, and is then termed *offset-keyed NLA-QAM*.

A block diagram of the NLA-16-ary QAM modulator is shown in Fig. 5.5. Two asynchronous non-return-to-zero (NRZ) data streams $A(t)$ and $B(t)$, of the same nominal bit rate, are fed to the input points 1 and 2. Each stream is converted into two synchronous streams by a serial-to-parallel converter. The outputs from the converter accepting the A stream are labeled I_1 and I_2; the outputs from the second converter are similarly labeled Q_1 and Q_2. I_1 and Q_1 feed "QPSK Modulator 1," and I_2 and Q_2 feed "QPSK Modulator 2," these being standard quaternary phase shift keying (QPSK) modulators. Note, however, that they are both driven by the same carrier source.

As shown in Fig. 5.5, the outputs of the QPSK modulators are amplified by *nonlinear* amplifiers NLA1 and NLA2. Because the QPSK modulated signals are unfiltered, they contain no AM and thus suffer no degradation from the nonlinear amplification. The output voltage level of amplifier NLA1 is arranged to be twice that of the voltage level out of amplifier NLA2. The outputs of the two amplifiers are linearly combined and then filtered through the output bandpass filter to limit the radiated spectrum. We note, therefore that, with NLA-16-QAM, the only transmitter filtering which affects the modulated signal is that beyond the nonlinear amplifiers. This is in contrast with conventional M-ary QAM, where significant premodulation baseband filtering is common.

Principle of operation of NLA-QAM modulators.   The principle of operation is most easily seen by following the path of one input signal only, as the second

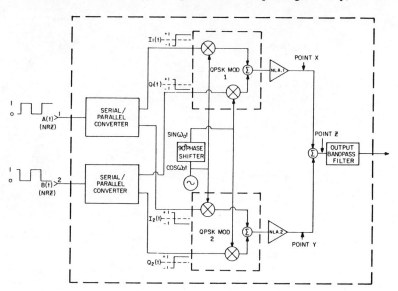

Figure 5.5 NLA-16-QAM modulator. (From D. Morais and K. Feher, ICC-81, Denver, Colo.)

input path is operationally identical. Due to linear combining, this method of analysis permits the application of the superposition principle. Following the path of input signal $A(t)$, we see that at point X of the modulator the signal can be given by

$$S_{XA} = 2GI_1(t) \cos \omega_o t \tag{5.2}$$

where $2G$ is the gain of amplifier NLA1. Similarly, at point Y,

$$S_{YA} = GI_2(t) \cos \omega_o t \tag{5.3}$$

where G is the gain of amplifier NLA2. Thus, at point Z we have

$$S_{ZA} = [2I_1(t) + I_2(t)]G \cos \omega_o t = Z_A(t)G \cos \omega_o t \tag{5.4}$$

where ω_o is the carrier frequency in rad/s. $I_1(t)$ and $I_2(t)$ are synchronous NRZ streams. The possible amplitudes and code levels of $Z_A(t)$ for various combinations of amplitudes or logic states of $I_1(t)$ and $I_2(t)$ are shown in Table 5.1.

Analyzing Table 5.1, we see that as a result of having set the output voltage of NLA1 to be twice that of NLA2, amplitude values of $Z_A(t)$ are equally spaced and symmetrical about zero. Also, when the logic levels of data stream $A(t)$, and hence of $I_1(t)$, are equiprobable, all the levels of $Z_A(t)$ are in turn equiprobable. Thus $S_{ZA}(t)$ is a standard double-sideband suppressed-carrier (DSB-SC) four-level AM signal, with amplitude levels optimum for P_e versus S/N performance.

TABLE 5.1   CODE LEVELS AND AMPLITUDES OF MODULATED SIGNAL $Z_A(t)$ FOR VARIOUS COMBINATIONS OF VALUES OF $I_1(t)$ AND $I_2(t)$

2$I_1(t)$		$I_2(t)$		$Z_A(t)$	
Logic state	Amplitude	Logic state	Amplitude	Code level	Amplitude
0	−2	0	−1	1	−3
0	−2	1	1	2	−1
1	2	0	−1	3	1
1	2	1	1	4	3

A similar analysis of the modulation of input signal $B(t)$ leads to a signal at point Z given by

$$S_{ZB} = [2Q_1(t) + Q_2(t)]G \sin \omega_o t = Z_B(t)G \sin \omega_o t \tag{5.5}$$

where S_{ZB} is also a standard DSB-SC four-level AM signal, but importantly, it is in quadrature with S_{ZA}.

The relationship between the possible amplitude of $S_{ZB}(t)$ and those of $Q_1(t)$ and $Q_2(t)$ is identical to that between $S_{ZA}(t)$ and those of $I_1(t)$ and $I_2(t)$ as given in Table 5.1. Given the above, it is obvious that when input signals $A(t)$ and $B(t)$

are present, we have at point Z two DSB-SC four-level AM signals in quadrature. These signals form a standard 16-state QAM signal, as shown in Fig. 5.3, given by

$$S_Z = G[Z_A(t) \cos \omega_o t + Z_B(t) \sin \omega_o t] \tag{5.6}$$

Solve Problems 5.3 and 5.4.

5.3 QUADRATURE-PARTIAL-RESPONSE (QPR) CORRELATIVE CODED MODULATION TECHNIQUES

The key assumption stipulated in the Nyquist theorem is that the ISI at the sampling instants can be eliminated. In this case the symbols are independent and uncorrelated; thus each symbol can be recovered without resorting to the past sequence. Such systems are often called zero-memory systems. In practice it is very difficult to approach the Nyquist rate of 2 symbols/s/Hz. To achieve this, ideal phase-equalized filters having a roll-off factor $\alpha = 0$, such as described in Chapter 3, would be required.

During the early 1960s *Dr. A. Lender* discovered the *correlative* transmission method. This method, often referred to as *duobinary, modified duobinary,* or *partial response,* found numerous applications in terrestrial microwave, cable, and other data transmission systems. In correlative coded systems a *controlled amount of ISI* is introduced in order to simplify the filter design (particularly the phase-equalization problem) and to enable transmission at the Nyquist rate and even higher rates. An in-depth treatment of correlative techniques as applicable to digital transmission may be found in [Lender, 1981]. Here, in this chapter, a concise *summary* of this technique is presented.

In Fig. 5.6 the cascade of a simple one-tap transversal filter, $H_1(f)$, with a very steep (ideally brick-wall) filter, $H_2(f)$, is shown. The impulse response of $H_1(f)$ is

$$h_1(t) = \delta(t) + \delta(t - T_s) \tag{5.7}$$

and the corresponding transfer function is

$$H_1(f) = F[h_1(t)] = 1 + e^{-j2\pi f T_s} \tag{5.8}$$

thus

$$|H_1(f)| = 2|\cos \pi f T_s| \tag{5.9}$$

The absolute value of $H_1(f)$ assures a continuous spectral roll-off between zero and the Nyquist frequency, $f_N = 1/2T_s$. Close to this frequency the attenuation of this phase-linear filter is very significant. The $H_2(f)$ brick-wall filter can be approximated in practical designs with a steep low-pass filter (e.g., an eighth-order Chebychev filter). In almost all practical filter designs it is very difficult to equalize the phase and the corresponding group delay at the edge of the passband and in the frequency region where the filter has significant attenuation, say more than 3 dB. In the design of a

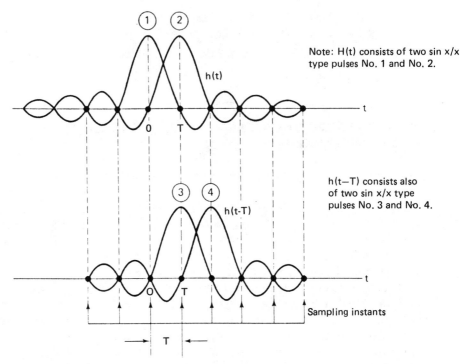

Figure 5.6 Doubinary signal generation. (From [Feher, 1981], with permission of Prentice-Hall, Inc.)

duobinary signal, such as shown in Fig. 5.6, the phase-linear transversal filter $H_1(f)$ has a significant attenuation at the edge of the low-pass filter (i.e., f_N); thus the signal energy around this frequency is negligible. For this reason in the duobinary system it is not required to have a carefully phase-equalized filter.

The transfer function of the cascaded duobinary filter is

$$|H(f)| = |H_1(f)H_2(f)| = \begin{cases} 2T_s \cos \pi f T_s & \text{for } f \leq \dfrac{1}{2T_s} \\ 0 & \text{elsewhere} \end{cases} \qquad (5.10)$$

The corresponding impulse response is obtained from the inverse Fourier transformation of equation (5.10). It is

$$h(t) = \frac{\sin \pi t / T_s}{\pi t / T_s} + \frac{\sin \pi (t - T_s)/T_s}{\pi (t - T_s)/T_s} \qquad (5.11)$$

Thus the response of the $H_1(f)H_2(f)$ filter to *one* input impulse, $\delta(t)$, consists of *two* sin x/x terms. Since the signaling interval is T_s seconds, there will be an overlap of ISI between the symbols. For a synchronous input impulse sequence consisting of positive and negative impulses and having a signaling rate of $f_s = 2f_N = 1/T_s$, the sampled output has three distinct amplitude levels: +2, 0, or −2. In other words, a three-level correlated signal is obtained at the output of the duobinary encoder. This signal is not as sensitive to filter (channel) imperfections as conventional NRZ and PAM, and it is possible to transmit at a rate about 25% higher than the Nyquist rate, without significant degradation. However, as the signal contains three levels, for the same P_e performance the S/N ratio requirement is about 2.5 dB higher than that of binary NRZ [Lender, 1963, 1964, 1981].

Duobinary encoding and other correlative coded signaling is frequently used in digital microwave systems. These systems, known as *quadrature-partial-response* (QPRS) modulated systems, have an implementation complexity comparable to QPSK, yet are less complex than the implementation of eight-phase PSK systems. Correlative QPRS radio systems having duobinary encoders have attained a practical spectral efficiency of 2.25 b/s/Hz [Lender, 1981; Godier, 1977].

A QPRS modulated signal is obtained by quadrature modulation of two three-level duobinary signals. Thus the block diagram of a QPRS modulator is essentially

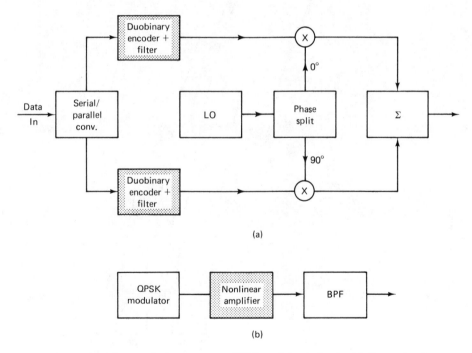

(a)

(b)

Figure 5.7 Two methods to generate QPRS modulated signals. (a) Conventional duobinary QPRS modulator block diagram. (b) QPRS modulation generated by band-limiting a nonlinearly amplified QPSK signal. The cascade of the transmit BPF and the filters in the demodulator approach the duobinary conversion filter described in Fig. 5.6.

the same as that of a QPSK modulator. However, as shown in Fig. 5.7, the QPRS modulator is equipped with duobinary premodulation low-pass filters (encoders) in the I and Q channels. The coherent QPRS demodulator is similar to the one shown in Fig. 5.4.

Nonlinear amplification of QPRS signals is more economical for microwave and satellite systems. This can be achieved by nonlinear (saturated) amplification of an unfiltered QPSK signal followed by duobinary filtering. The postmodulation duobinary filtering may be split between the transmit filter and the demodulation filters [Godier, 1977]. Nonlinearly amplified QPRS and QAM systems use the available

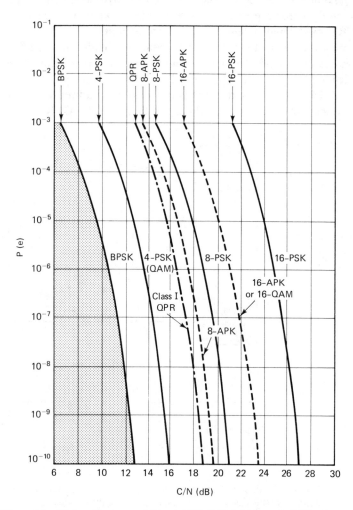

Figure 5.8 $P(e)$ performance of M-ary PSK, QAM, QPR, and APK coherent systems. The rms C/N is specified in the double-sided Nyquist bandwidth. (From [Feher, 1981], with permission of Prentice-Hall, Inc.)

transmit power in an efficient manner. For this reason these modulation techniques may find application in digital satellite communications systems. *Solve Problem 5.5.*

5.4 P_e PERFORMANCE AND SPECTRAL EFFICIENCY OF M-ARY QAM AND QPRS MODULATED SYSTEMS

The probability of error performance, P_e, of M-ary QAM and of QPR systems is derived in [Feher, 1981]. The final results are illustrated in the $P_e = f(C/N)$ curves shown in Fig. 5.8. Note that the mean C/N ratio is specified in the double-sided Nyquist bandwidth, that is, the symbol rate bandwidth. The QPSK system requires a C/N ratio 3 dB higher than that of the BPSK system, in the normalized double-sided Nyquist bandwidth. However, this system has a higher spectral efficiency. The practical spectral efficiency of QPR is 2.25 b/s/Hz, whereas that of 8-PSK and 8-APK systems approaches 3 b/s/Hz. The spectral efficiency in terms of b/s/Hz of radio-frequency bandwidth of various modulated systems is summarized in Fig. 5.9.

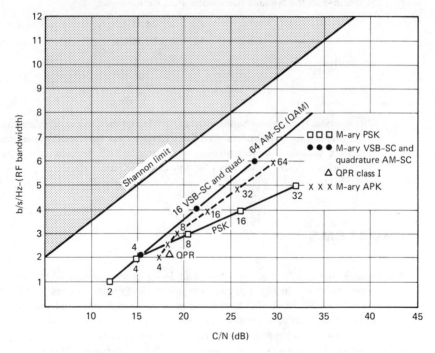

Figure 5.9 Bit rate efficiency (in b/s/Hz) of M-ary coherent PSK, VSB-SC, quadrature AM-SC, APK, and QPR systems as a function of the available C/N at a $P(e) = 10^{-8}$. The average C/N is specified in the double-sided Nyquist bandwidth which equals the symbol rate. Ideal $\alpha = 0$ filtering has been assumed. (From [Feher, 1981], with permission of Prentice-Hall, Inc.)

Example 5.1

Assume that a data rate of 60 Mb/s has to be transmitted through a transponder having a bandwidth of 18 MHz. Which modulation technique would you use for this requirement? How much is the required E_b/N_o for this system if a $P_e = 10^{-8}$ is required?

Solution.   The required spectral efficiency is

$$\frac{60 \text{ Mb/s}}{18 \text{ MHz}} = 3.333 \text{ b/s/Hz}$$

Thus we select a modulation technique having a spectral efficiency of 4 b/s/Hz. From Fig. 5.9 we note that the 16-ary QAM system requires a lower C/N than that of the 16-PSK and that it has a theoretical spectral efficiency of 4 b/s/Hz. Based on these considerations we choose a 16-ary QAM modem. From Fig. 5.8 we note that the theoretical C/N requirement of this modem for $P_e = 10^{-8}$ is 23 dB. The required E_b/N_o is obtained from

$$\frac{E_b}{N_o} = \frac{C}{N_{bw}} \frac{BW}{f_b} \tag{5.12}$$

where BW is the noise bandwidth of the receiver. In our case the required C/N ratio is specified in the Nyquist bandwidth; thus we have

$$\frac{E_b}{N_o} = \frac{C}{N_{15}} \frac{15 \text{ MHz}}{60 \text{ Mb/s}}$$

or, in dB-s,

$$\frac{E_b}{N_{o_{dB}}} = 23 \text{ dB} + 10 \log \frac{15}{60} = 17 \text{ dB}$$

The available transponder bandwidth is 18 MHz and the theoretical minimum bandwidth is 15 MHz. Thus a roll-off factor of 0.2 (20% excess bandwidth) is assumed. Due to imperfect hardware implementation, we assume a 3-dB degradation in the P_e curves. The practical E_b/N_o requirement for this case is 20 dB.   ∎

PROBLEMS

5.1. Assume that the four discrete baseband levels at the outputs of the two-to-four-level converters, shown in Fig. 5.3, can have one of the following values: ±13.07 mV, -13.07 mV, $+5.41$ mV, -5.41 mV. Sketch the resultant modulated signal space diagram for all possible phase states. Determine the resultant phase angles. Would these phase angles be the same if the baseband levels had equidistant spacings?

5.2. Sketch the demodulated eye diagram of a raised-cosine filtered 16-QAM modem, assuming that a roll-off factor $\alpha = 0.3$ is used. How many threshold comparators do we require to detect the I and Q channels? Assume that the received voltages in the sampling instants equal $+100$ mV, $+33$ mV, -33 mV, and -100 mV. Determine the required optimal threshold levels.

5.3. Explain why the performance of a bandlimited QPSK signal is degraded when it is transmitted through a TWT amplifier operating close to saturation. Is the performance of unfiltered QPSK signals degraded when transmitted through nonlinear memoryless output amplifiers?

5.4. Assume that a conventional 16-QAM signal is transmitted through a saturated TWT. What would happen to both the signal state-space diagram and the P_e performance of this system? In nonlinearly amplified 16-QAM systems such as that shown in Fig. 5.5, the system performance is not degraded by the nonlinear (saturated) output amplifier. Explain why this is so.

5.5. A 120-Mb/s QPR system is planned for the uplink of a new digital satellite system. To maximize the uplink power a nonlinear output state is employed. Draw the block diagram of the modulator, the earth station uplink, and the satellite receiver. Assume that a *regenerative* satellite is used, a satellite in which the uplink signal is coherently demodulated. How much is the uplink E_b/N_o requirement if the available uplink bandwidth is 70 MHz? What is the possible advantage of a QPRS system when compared to a QPSK system? What are the disadvantages of QPRS in this hypothetical application?

6

CODING FOR ERROR DETECTION
AND CORRECTION

DR. WILLIAM H. TRANTER

Professor of Electrical Engineering
and Assistant Dean of Engineering for Graduate Studies and Research
University of Missouri, Rolla, Missouri

> *Dr. William H. Tranter is a well-known professor, administrator, and research engineer in the field of digital communications engineering and coding techniques. Professor Tranter presents a clearly written, comprehensive overview of the principles and applications of forward error-correcting coding techniques. I wish to thank him for his valuable chapter.*
>
> *Dr. K. Feher*

6.1 INTRODUCTION

In this chapter a brief overview of coding techniques, applicable to satellite communication systems, is presented. Curves are given that illustrate the system improvement that can be gained through the use of coding techniques. A complete development of algebraic coding theory requires an understanding of modern algebra, a subject that is beyond the scope of this treatment. We can, however, through the use of examples and performance curves, come to appreciate the importance of coding in the design of modern satellite communication systems.

There are two basic alternatives to be explored when coding is considered. These are block coding and convolutional coding. Both of these techniques are important and both have their place in the implementation of modern communication systems.

Although the basic intent of this chapter is to explore the use of error-correcting codes with a noisy communication channel, our understanding and insight will be enhanced if first we take time to examine several important results from the field of information theory. Of particular importance is the concept of channel capacity and the Shannon bound. The Shannon bound provides the ultimate performance limit of a communication system and provides the motivation for coding.

6.2 ENTROPY, MUTUAL INFORMATION, AND CHANNEL CAPACITY

In our brief study of information theory and coding, the assumption will be made that the data are generated by a discrete memoryless source. The output of the source can perhaps best be viewed as a time sequence of source states,

$$A = (a_1, a_2, a_3, a_4, \ldots, a_j, \ldots) \tag{6.1}$$

with each of the states chosen from a finite source alphabet X,

$$X = (x_1, x_2, \ldots, x_N) \tag{6.2}$$

according to some probability rule. As the simplest example, a binary source has a two-element alphabet, defined by

$$X = (0, 1) \tag{6.3}$$

and a typical output could be the sequence

$$A = (1, 1, 0, 0, 1, 0, 1, 1, 1, \ldots) \tag{6.4}$$

The word *memoryless* in the source definition implies that a general source output a_j is independent of all preceding and succeeding source outputs.

6.2.1 Information and Entropy

Shannon, in his landmark paper [Shannon, 1948], defined the *information associated with a source output,* or in general the information associated with any event, to be the negative logarithm of the probability of that event. Thus the *information* associated with source state x_j is

$$I(x_j) = -\log_a p(x_j) \tag{6.5}$$

in which $p(x_j)$ represents the probability of source state x_j. The base of the logarithm, a, is arbitrary and simply determines the units in which information is measured. In today's world, with the extensive use of binary-based digital computers and binary-based transmission systems, the base a is almost universally taken as 2. The corresponding *measure of information* is the bit.

The form of (6.5) is intuitively very pleasing. If the probability of an event or source state, x_j, is very high, then $p(x_j)$ is approximately 1, and the logarithm of 1 is zero. Thus very little information is associated with an almost certain event, source state, or message. However, if the probability of an event is very small, its negative logarithm is large, indicating a great amount of information. As an example, consider the probabilities and corresponding information associated with the following two weather forecasts:

1. The weather tomorrow will be much like it was today.
2. Even though it is early summer, a severe winter storm is on the way and 20 inches of snow is forecast for tomorrow.

Clearly, the second forecast conveys a vast amount of information to anyone planning normal summer activities!

In typical applications, it is not the information associated with source states that is important but rather the *average information* associated with the source. This average information is referred to as *entropy*. The averaging is over all source states. Thus the entropy of a discrete memoryless source, $H(X)$, is

$$H(X) = -\sum_{i=1}^{N} p(x_i) \log p(x_i) \tag{6.6}$$

A simple example will help illustrate the meaning of entropy.

Consider a binary source that produces two symbols, 0 and 1, with probabilities α and $1 - \alpha$, respectively. From equation (6.6), the entropy is a function of α and is

$$H(\alpha) = -\alpha \log_2 \alpha - (1 - \alpha) \log_2 (1 - \alpha) \tag{6.7}$$

The entropy function $H(\alpha)$ is illustrated in Fig. 6.1. Several points are worth noting. The entropy function is symmetric about $\alpha = 0.5$ and achieves the maximum value of one at $\alpha = \frac{1}{2}$. Also, $H(\alpha)$ is equal to zero for $\alpha = 0$ and 1. It is easy to show that in general the entropy possesses a maximum and that the maximum occurs when all source states are equally likely. In equation (6.6), if all source states are equally likely,

$$p(x_i) = \frac{1}{N} \qquad \text{all } i$$

and the entropy is

$$H(X) = -\sum_{i=1}^{N} \frac{1}{N} \log_2 \frac{1}{N}$$

which is

$$H(X) = \log_2 N$$

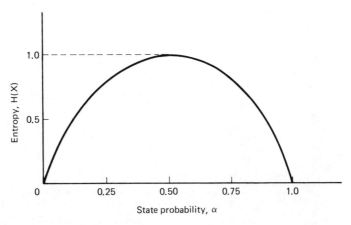

Figure 6.1   Entropy of a discrete binary source.

For any probability distribution on the source states, the source entropy is less than or equal to this value. Note that with $\alpha = 0$ or $\alpha = 1$, the source output is always certain. For $\alpha = \frac{1}{2}$, the source output is maximally uncertain and the entropy is maximum. Therefore, *entropy* is often interpreted as a measure of *uncertainty*.

6.2.2 Mutual Information and Channel Capacity

Consider a channel with N inputs and M outputs. The channel is defined by the set of transition probabilities $p(y_j|x_i)$ for $1 \le i \le N$ and $1 \le j \le M$, where x_i and y_i denote the channel input and output, respectively. The channel is conveniently represented by either the channel transition diagram, shown in Fig. 6.2, or by the channel transition probability matrix

$$[P(Y|X)] = \begin{bmatrix} p(y_1|x_1) & p(y_2|x_1) & \cdots & p(y_M|x_1) \\ p(y_1|x_2) & p(y_2|x_2) & \cdots & p(y_M|x_2) \\ \vdots & \vdots & & \vdots \\ p(y_1|x_N) & p(y_2|x_N) & \cdots & p(y_M|x_N) \end{bmatrix} \qquad (6.8)$$

The channel transition probability matrix, hereafter referred to as the channel matrix, is simply the matrix of the conditional probabilities of the channel outputs given the channel inputs (source states).

By definition, the *mutual information* between channel input and output is

$$I(X;\, Y) = \sum_{i=1}^{N} \sum_{j=1}^{M} p(x_i,\, y_j) \log_2 \frac{p(x_i,\, y_j)}{p(x_i)\, p(y_j)} \qquad (6.9)$$

which can be put in the form

$$I(X;\, Y) = \sum_{i=1}^{N} \sum_{j=1}^{M} p(x_i,\, y_j) \left[\log_2 \frac{1}{p(x_i)} + \log_2 \frac{p(x_i,\, y_j)}{p(y_j)} \right] \qquad (6.10)$$

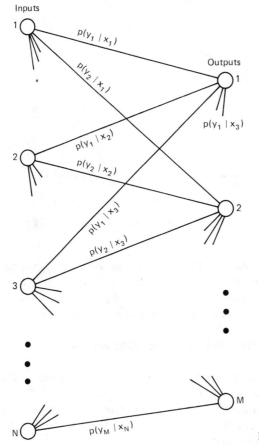

Figure 6.2 Channel transition diagram.

Recognizing that

$$\sum_{j=1}^{M} p(x_i,\, y_j) = p(x_i) \tag{6.11}$$

and

$$\frac{p(x_i,\, y_i)}{p(y_i)} = p(x_i|y_i) \tag{6.12}$$

allows equation (6.10) to be written

$$I(X;\, Y) = \sum_{i=1}^{N} p(x_i) \log_2 \frac{1}{p(x_i)} - \sum_{i=1}^{N}\sum_{j=1}^{M} p(x_i,\, y_j) \log_2 \frac{1}{p(x_i|y_j)} \tag{6.13}$$

The first term in equation (6.13) is simply the *source entropy, H(X).* The second term is the *conditional entropy H(X|Y).* Thus the channel mutual information is

$$I(X;\, Y) = H(X) - H(X|Y) \tag{6.14}$$

A similar development shows that the channel input–output mutual information is given by

$$I(X; Y) = H(Y) - H(Y|X) \qquad (6.15)$$

where

$$H(Y|X) = \sum_{i=1}^{N} \sum_{j=1}^{M} p(x_i, y_j) \log_2 \frac{1}{p(y_j|x_i)} \qquad (6.16)$$

This formulation of mutual information is often easier to work with than equation (6.14) since $p(y_j|x_i)$ are the channel transition probabilities and are given directly by the channel matrix or diagram.

Equation (6.14) has a very simple and enlightning interpretation. The unconditional entropy $H(X)$ is the average uncertainty of the source or source information. The quantity $H(X|Y)$ is the average uncertainty of the source given the channel output Y. The difference between these two quantities is the mutual information and is a measure of the information that has passed through the channel.

The *channel capacity* is the maximum value of the mutual information. Since the channel is fixed, maximization must be with respect to the input (source) probabilities.

Example 6.1

Consider the channel defined by the channel diagram shown in Fig. 6.3(a). This channel is the *binary symmetric channel* (BSC) and plays a central role in the analysis of many digital communication systems. Suppose that the source probabilities are

$$p(x_1) = \alpha$$

and

$$p(x_2) = 1 - \alpha$$

as shown. The capacity will be determined by first computing $I(X; Y)$ using equation (6.15). The value of $H(Y|X)$ is given by

$$H(Y|X) = -\alpha(1 - q) \log_2 (1 - q)$$
$$- \alpha q \log_2 q$$
$$- (1 - \alpha)(1 - q) \log_2 (1 - q)$$
$$- (1 - \alpha) q \log_2 q$$

which is

$$H(Y|X) = -q \log_2 q - (1 - q) \log_2 (1 - q) \qquad (6.17)$$

In the notation of equation (6.7) the mutual information can be written

$$I(X; Y) = H(Y) - H(q) \qquad (6.18)$$

The channel capacity is then found by maximizing this expression with respect to the source probabilities. The entropy $H(Y)$ is maximized when the two outputs occur with equal probability. For this case $H(Y)$ is equal to 1. Since the channel is symmetrical,

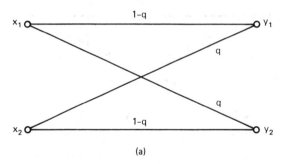

(a)

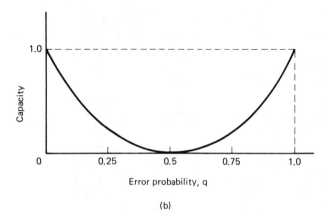

(b)

Figure 6.3   (a) Binary symmetric channel. (b) Capacity.

equally likely outputs imply equally likely inputs. Thus channel capacity is achieved when

$$p(x_1) = p(x_2) = \tfrac{1}{2}$$

and the capacity is

$$C = 1 + q \log_2 q + (1 - q) \log_2 (1 - q) \qquad (6.19)$$

The capacity is illustrated in Fig. 6.3(b).

The unconditional error probability, $P(E)$, for the BSC is the parameter q. The error probability conditioned on input x_i is denoted $p(E|x_i)$ and is $p(y_j|x_i)$ for $i \neq j$. This yields

$$P(E) = p(x_1) p(E|x_1) + p(x_2) p(E|x_2)$$

which is

$$P(E) = \alpha q + (1 - \alpha)q = q \qquad (6.20)$$

When q is 0 or 1, the output is completely determined by the input. If the input carries one bit of information per symbol, the output carries the same information and the channel capacity is one bit per symbol. If q is $\tfrac{1}{2}$, the output of the channel is independent of the channel input and the capacity is zero.

Example 6.2

A satellite relay system consists basically of three channels; the uplink, onboard signal processing, and the downlink. The uplink has an error probability of 0.01. Onboard signal processing takes place with negligible error and the downlink error probability is 0.1. It is often the case that the downlink error probability exceeds the uplink error probability because of the limited satellite transmission power. The three channel capacities are easily computed. The uplink capacity is

$$C_1 = 1 + 0.01 \log_2 0.01 + 0.99 \log_2 0.99$$

which is

$$C_1 = 0.9192 \text{ bit/symbol}$$

In like manner, the capacity of the channel representing the onboard signal processing is

$$C_2 = 1.0 \text{ bit/symbol}$$

and the downlink capacity is

$$C_3 = 0.5310 \text{ bit/symbol}$$

In order to compute the capacity of the overall system it is necessary to first determine the overall error transition probability matrix. This yields

$$[P(Y|X)] = \begin{bmatrix} 0.99 & 0.01 \\ 0.01 & 0.99 \end{bmatrix} \begin{bmatrix} 1.0 & 0 \\ 0 & 1.0 \end{bmatrix} \begin{bmatrix} 0.9 & 0.1 \\ 0.1 & 0.9 \end{bmatrix}$$

which becomes

$$[P(Y|X)] = \begin{bmatrix} 0.892 & 0.108 \\ 0.108 & 0.892 \end{bmatrix}$$

The capacity of the overall system is, therefore,

$$C = 1 + 0.892 \log_2 0.892 + 0.108 \log_2 0.108$$

which is

$$C = 0.5061 \text{ bit/symbol}$$

■

In this example, very high uplink and downlink error probabilities were assumed in order to simplify the computations. Typical satellite systems have uplink and downlink error probabilities which are orders of magnitude smaller.

6.3 SOURCE ENCODING

We saw in Section 6.2 that a discrete binary source has a source entropy of one bit per symbol if both source states are equally likely. If the source states are not equally likely, the average information per source symbol is less than 1 and redundancy exists at the source output. Source encoding allows this redundancy to be systematically removed so that the channel can be used more efficiently. The process is demonstrated by means of a simple example.

Source encoding is accomplished by grouping N of the binary source symbols together to form a new set of 2^N unique source outputs. This set of 2^N symbols is referred to as the Nth-order extension of the binary source. The probability of each output of the extended source is then computed and codewords are assigned according to the probabilities of the extended source outputs. The scheme is to assign the shortest codeword to the most probable output of the extended source and correspondingly longer codewords to less likely outputs of the extended source.

An easily understood, but not optimum, method of assigning codewords is to arrange the outputs of the extended source in order of nonincreasing probability. The set of probabilities is partitioned so that the total probability each side of the partition is as nearly equal as possible. A binary zero code symbol is then assigned to each output lying on one side of the partition and a binary one code symbol is assigned to each output on the other side. This determines the first code symbol of each of the 2^N codewords. The process is continued by subdividing with a new partition each of the subsets generated by the previous partitioning process and assigning a binary zero code symbol to each codeword in the subset lying on one side of the new partition and a binary one to each codeword on the other side of the partition. The process continues until only one symbol of the extended source lies in each partition.

As an example, consider a binary source with outputs A and B having probabilities 0.9 and 0.1, respectively. Figure 6.4 illustrates the binary codewords corresponding to the original source and the first-order and second-order extensions. In Fig. 6.4 $S_i^{(N)}$ represents the symbols of the extended source and $C_i^{(N)}$ represents the corresponding codewords. Also given is the quantity $L^{(N)}/N$, in which $L^{(N)}$ is the average length of the codewords for the Nth-order extension. The average wordlength is given by

$$L^{(N)} = \sum_{i=1}^{2^N} n_i \, Pr(S_i^{(N)}) \tag{6.21}$$

The important observation is that $L^{(N)}/N$ represents the average number of code symbols per source symbol and higher-order extensions yield a decreased symbol rate into to channel. For fixed transmitter power, symbol energy can be increased because of the lower symbol rate and improved system performance results.

Example 6.3

As an example, assume a PSK system with a received power of 10^{-5} W. Further assume that the source rate is 1000 symbols/s. This yields a received symbol energy of 10^{-8} J. With a noise power spectral density N_o of 10^{-9} W/H (-60 dBm/Hz), the uncoded symbol error probability

$$P_E = \tfrac{1}{2} \, \mathrm{erfc} \left(\sqrt{\frac{E_s}{N_o}} \right) \tag{6.22}$$

is

$$P_E = \tfrac{1}{2} \, \mathrm{erfc} \, (\sqrt{10}) = 3.9(10^{-6})$$

i	$S_i^{(1)}$	$Pr(S_i^{(1)})$	$C_i^{(1)}$
1	A	0.9	0
2	B	0.1	1

(a)

i	$S_i^{(2)}$	$Pr(S_i^{(2)})$	$C_i^{(2)}$
1	AA	0.81	0
2	AB	0.09	1 0
3	BA	0.09	1 1 0
4	BB	0.01	1 1 1

(b)

i	$S_i^{(3)}$	$Pr(S_i^{(3)})$	$C_i^{(3)}$
1	AAA	0.729	0
2	AAB	0.081	1 0 0
3	ABA	0.081	1 0 1
4	BAA	0.081	1 1 0
5	ABB	0.009	1 1 1 0 0
6	BAB	0.009	1 1 1 0 1
7	BBA	0.009	1 1 1 1 0
8	BBB	0.001	1 1 1 1 1

(c)

Figure 6.4 Example of source encoding.
(a) Original source $\left(\dfrac{L^{(1)}}{T} = 1.0\right)$.
(b) Second-order extension $\left(\dfrac{L^{(2)}}{2} = 0.645\right)$.
(c) Third-order extension $\left(\dfrac{L^{(3)}}{3} = 0.5333\right)$.

If a second-order extension is taken, the code symbol rate drops to 645 code symbols per second. The result is a symbol energy of $1.55(10^{-8})$ joules. This yields an error probability of

$$P_E = \tfrac{1}{2}\,\text{erfc}\,(\sqrt{15.5}) = 1.3(10^{-8})$$

In a similar manner, the third-order extension results in a symbol error probability of

$$P_E = \tfrac{1}{2}\,\text{erfc}\,(\sqrt{18.8}) = 4.5(10^{-10})$$

Clearly, a significant improvement in system performance is taking place.

■

The quantity $L^{(N)}/N$ approaches a limit as $N \rightarrow \infty$. This result is stated by the *noiseless coding theorem,* which states that for a discrete source, having entropy $H(X)$,

$$\frac{L^{(N)}}{N} = H(X) + e \qquad (6.23)$$

where $e \rightarrow 0$ as $N \rightarrow \infty$.

For the binary source being considered, $H(X) = 0.469$. Thus for very large N the symbol rate at the output of the source encoder is 469 symbols/s. This corresponds to a symbol energy of $2.132(10^{-8})$ J. Thus the limiting value of symbol error probability is

$$P_E = \tfrac{1}{2} \operatorname{erfc} (\sqrt{21.32})$$

or

$$P_E = 3.36 (10^{-11})$$

It should be noted that a third-order extension achieves most of the available system gain.

6.4 CODING FOR RELIABLE COMMUNICATIONS

In Section 6.3 the concept of channel capacity was discussed in some detail. The concept of capacity plays a fundamental role in the performance of communication systems. Shannon showed that if a source produces information at a fixed rate R, and the available channel has capacity C, then a coding and modulation scheme exists such that the decoded error probability can be made arbitrarily small for $R < C$. For $R > C$, the application of coding usually only further degrades system performance.

The capacity computed in the previous section has dimensions of *bits per symbol*. This is easily converted to capacity in *bits per second* by multiplying by the channel symbol rate in symbols per second.

For an additive white Gaussian noise channel, the capacity is given by

$$C = B \log_2\left(1 + \frac{P}{N_o B}\right) \qquad \text{bits/s} \qquad (6.24)$$

where

$$B = \text{bandwidth of the signal}$$
$$P = \text{received signal power}$$
$$N_o = \text{single-sided noise power spectral density}$$

Equation (6.24) is commonly known as the *Shannon–Hartley Law*.

In order to better understand the meaning of the Shannon–Hartley Law it is helpful to place equation (6.24) in a slightly different form. At capacity the bit duration is $T_b = 1/C$ seconds. Thus the energy per bit, E_b, is

$$E_b = PT_b = \frac{P}{C} \qquad (6.25)$$

so that the important quantity E_b/N_o is given by

$$\frac{E_b}{N_o} = \frac{P}{N_o C} \qquad (6.26)$$

This allows equation (6.24) to be written

$$\frac{C}{B} = \log_2\left(1 + \frac{E_b}{N_o}\frac{C}{B}\right) \tag{6.27}$$

Equation (6.27) is plotted in Fig. 6.5.

Figure 6.5(a) illustrates the relationship between E_b/N_o and C/B for $C < B$ for operation at capacity. The important observation is that E_b/N_o approaches a finite limit as $B \rightarrow \infty$. This limit is ln 2, which is approximately -1.6 dB. This value is commonly known as the Shannon bound. The region $C < B$ is referred to as the power-limited region.

For $C > B$, the relationship between E_b/N_o and C/B is illustrated in Fig. 6.5(b). For C/B arbitrarly large, E_b/N_o is arbitrarly large. This region is known as the bandwidth-limited region.

Figure 6.5 shows that operation at capacity is achievable only if the value of E_b/N_o is above -1.6 dB. Thus arbitrarily small error probability can only be achieved for E_b/N_o above -1.6 dB. The performance of this ultimate communication system

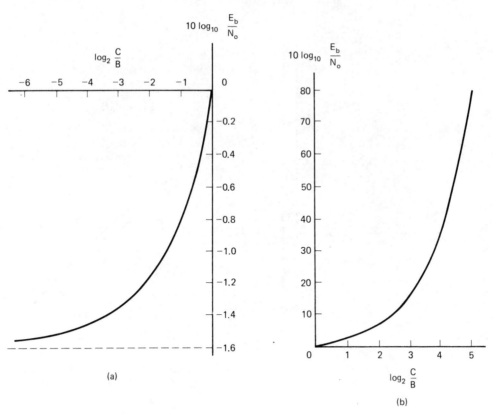

Figure 6.5 Relationship between E_b/N_o and C/B for transmission at capacity. (a) Power-limited case. $C < B$. (b) Bandlimited case. $C > B$.

is shown in Fig. 6.6 together with the previously derived performance of a PSK system. The question of how to approach the Shannon bound has not been addressed. One possible answer lies in the use of coding.

The bound of -1.6 dB implies that all of the information received is used in making a decision on the transmitted symbol. We have shown in previous chapters that typical receiver operation consists of a matched filtering or correlation detection operation. These techniques involve taking a single sample of a waveform produced by performing a linear operation on the received waveform. This sample is then compared to a threshold and a *hard decision* is made on the transmitted symbol (0 or 1). The reliability of that decision is not considered in the process. The cost associated with making hard decisions is approximately 2 dB. The Shannon bound for hard decision receivers is also shown in Fig. 6.6.

Now that the ultimate performance of a digital communication system has been defined, we turn our attention to the coding process.

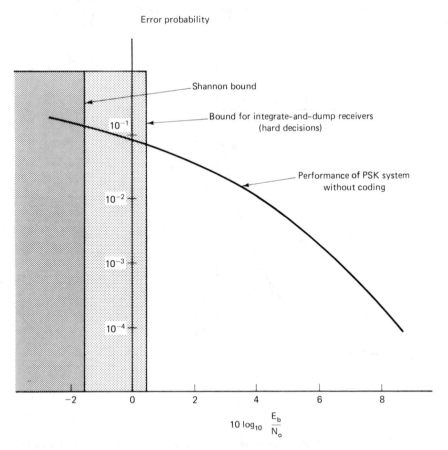

Figure 6.6 Shannon bound.

6.4.1 Introduction to Block Codes

The process of forming a *block code* is illustrated in Fig. 6.7. The binary source is assumed to generate a sequence of symbols at a rate R symbols/s. These symbols are grouped into blocks k symbols long. To each of these k symbol blocks, $n - k$ redundant symbols are added to produce an n-symbol codeword. The $n - k$ redundant symbols are referred to as parity symbols. The result is denoted an (n, k) block code. Since each codeword contains n symbols and conveys k bits of information, the information rate of the encoder output is k/n *bits per symbol.* Thus k/n is referred to as the code rate.

The purpose of the encoder is to map the information sequence

$$\overline{X} = (x_1, x_2, \ldots, x_k) \tag{6.28}$$

onto the codeword

$$\overline{Y} = (y_1, y_2, \ldots, y_n), \tag{6.29}$$

by properly choosing the parity symbols. Decoding is accomplished by determining the most likely transmitted codeword given the received sequence

$$\overline{Z} = (z_1, z_2, \ldots, z_n) \tag{6.30}$$

If all transmitted codewords are equally likely, and the channel is memoryless, this

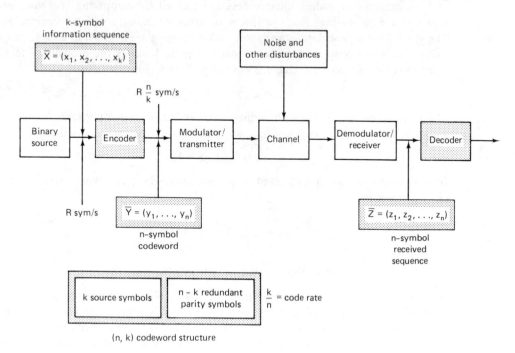

Figure 6.7 Encoding and decoding.

is accomplished by choosing as the most likely transmitted codeword that codeword which is closest in *Hamming distance* to the received codeword. The Hamming distance between the sequence $\bar{Y}$ and $\bar{Z}$ is defined as the weight (the number of binary ones) in the modulo-2 sum of $\bar{Y}$ and $\bar{Z}$. Since $\bar{Y}$ and $\bar{Z}$ are written as vectors, the modulo-2 sum is taken component by component.

In typical applications, coding is used in environments in which the transmitter power is limited. Also, the n-symbol codewords must be transmitted in the same time span in which the k information symbols are generated by the source. If these conditions do not apply, there is little use for coding. Increasing the transmitter power almost always improves system performance. Similarly, if the data stream can be read into a buffer and read out at a reduced rate, system performance is improved because of increased symbol energy.

Let the transmitter power be S watts referred to the receiver and assume that the k symbols are output from the source in T_w seconds. Thus the energy available for each codeword is ST_w joules. The received energy per symbol is ST_w/k without coding. With coding, the energy must be spread over the n-symbol codeword and therefore the received energy per symbol with coding is ST_w/n. Since $n > k$, the symbol energy is reduced with the use of coding; it follows that the symbol error probability with coding is greater than without coding. If the code is properly designed, the redundancy induced with the $n - k$ parity symbols allows sufficient error correction capability for the system to enjoy a *net gain* in performance.

A measure of coding effectiveness is obtained by comparing the word error probability with coding, P_{wec}, to the word error probability without coding, P_{weu}. The symbol error probabilities with and without coding are denoted q_c and q_u, respectively. The *word error probability without coding* is 1 minus the probability that all k information symbols are received correctly. This yields

$$P_{weu} = 1 - (1 - q_u)^k \tag{6.31}$$

The word error probability with coding is somewhat more complicated. Assume that a code has a *minimum distance* d_{min} so that e errors can be corrected, where

$$e = \tfrac{1}{2}(d_{min} - 1) \tag{6.32}$$

If more than e errors in a received codeword always lead to a word error, the word error probability is

$$P_{wec} = \sum_{i=e+1}^{n} \binom{n}{i} q_c^i (1 - q_c)^{n-i} \tag{6.33}$$

in which $\binom{n}{i}$ represents the number of all possible patterns of i errors in an n-symbol codeword and is equal to $n!/i!(n - 1)!$ and q_c denotes the symbol error probability with coding. Codes for which (6.33) holds *exactly* are known as perfect codes. The only *perfect codes* are the single-error-correcting Hamming codes and the Golay code, both of which will be explored in some detail in following sections. For other codes there are particular codewords for which correct decoding can occur with more than e errors. For these codes equation (6.33) provides a useful upper bound.

Example 6.4

A particularly useful code is the (23, 12) code known as the Golay code. This code corrects all patterns of three errors. It follows from the preceding discussion that with coherent PSK transmission, the uncoded symbol error probability is

$$q_u = \frac{1}{2} \operatorname{erfc}\left(\sqrt{\frac{ST_w}{12N_o}}\right)$$

and that with coding the symbol error probability is

$$q_c = \frac{1}{2} \operatorname{erfc}\left(\sqrt{\frac{ST_w}{23N_o}}\right)$$

The word error probability without coding is

$$P_{\text{weu}} = 1 - (1 - q_u)^{12}$$

and the word error probability with coding is

$$P_{\text{wec}} = \sum_{i=4}^{23} \binom{23}{i} q_c^i (1 - q_c)^{23-i}$$

These results are plotted in Fig. 6.8. The increase in symbol error probability with coding is clear, as is the decrease in word error probability. More will be said about the Golay code throughout this chapter.

It should be noted that the q_u and q_c curves differ by 2.8 dB, which is of course $10 \log_{10}(n/k)$ for $n = 23$ and $k = 12$.

∎

6.4.2 Repetition Codes

A very simple block code, but one that points out very well the concept of error correction and code performance, is the $(n, 1)$ block code known as the repetition code. In the repetition code each of the $n - 1$ parity symbols is equal to the information symbol. The code rate is $1/n$, which is very low for large n. The relationship between the information symbols and the codewords is shown in Fig. 6.9. The minimum distance of the code is n, and therefore for large n, repetition codes have very powerful error-correction capabilities. Since the minimum distance of the code is n, a total of $e = \frac{1}{2}(n - 1)$ errors in a received codeword can be corrected.

Example 6.5

Suppose that a data symbol is transmitted without coding over a BSC. The symbol error probability is denoted q_u. Now assume that the same data sequence is transmitted over the same BSC using a rate $1/n$ repetition code. Since the repetition code can correct $e = \frac{1}{2}(n - 1)$ errors per n-symbol codeword, the word error probability for the coded system is, from the preceding section,

$$P_{\text{wec}} = \sum_{i=e+1}^{n} \binom{n}{i} q_c^i (1 - q_c)^{n-i} \tag{6.34}$$

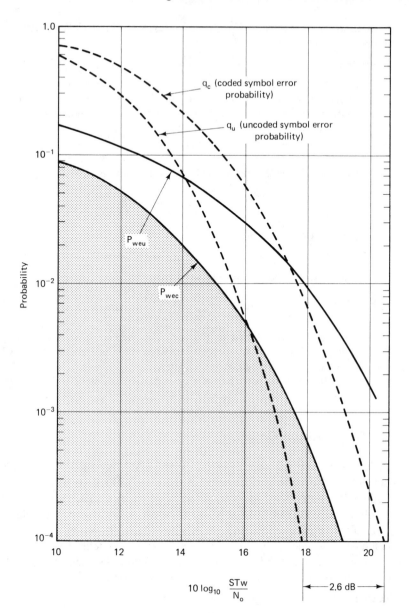

Figure 6.8   Symbol and word error probabilities of (23, 12) Golay code.

where q_c is the *symbol error probability with coding*. As an example, for $n = 5$,

$$P_{wec} = 10q_c^3(1 - q_c)^2 + 5q_c^4(1 - q_c) + q_c^5 \qquad (6.35)$$

For small values of q_c we have

$$P_{wec} \approx 10q_c^3 \qquad (6.36)$$

Rate	Distance	Errors corrected per codeword	Codewords
$\dfrac{1}{3}$	3	1	0 0 0 1 1 1
$\dfrac{1}{5}$	5	2	0 0 0 0 0 1 1 1 1 1
. . .	. . .	. . .	. . .
$\dfrac{1}{n}$	n	$\dfrac{1}{2}(n-1)$	0 0 $\cdots$ 0 1 1 $\cdots$ 1

Figure 6.9 Repetition code.

Initially, it would appear that a considerable improvement has taken place. As a matter of fact, in the typical environments of fixed transmitter power and fixed information rate, *system performance is always made worse with the use of repetition codes.*

To illustrate this phenomenon, consider repetition coding used with a PSK system. For the uncoded system, the received transmitted power multiplied by the word time, ST_w, is the received bit energy, E_b, since the word time is the transmission time for one symbol (or bit). Thus

$$q_u = \frac{1}{2}\,\text{erfc}\left(\sqrt{\frac{E_b}{N_o}}\right) \tag{6.37}$$

With repetition coding, the energy E_b is spread over the n symbols in a codeword. Thus

$$q_c = \frac{1}{2}\,\text{erfc}\left(\sqrt{\frac{E_b}{nN_o}}\right) \tag{6.38}$$

Substituting q_c into equation (6.34) yields the word error probability with repetition coding. It should be noted that since repetition codewords each convey one bit of information, the word error probability is equal to the bit error. The code performance is illustrated in Fig. 6.10 for several values of n. Clearly, increasing n only serves to degrade system performance.

■

The repetition code has served to illustrate that error-correcting capability and performance do not always go hand in hand. The reason for the poor performance of the repetition code is the low rate. We now look at a simple encoding procedure that yields error-correcting capability with reasonable rates.

6.4.3 Linear Block Codes

In the preceding section we saw that an encoder maps a k-symbol information sequence onto an n-symbol codeword:

$$\overline{Y} = (y_1, y_2, \ldots, y_e, y_n)$$

A **linear code** is defined as one in which the *lth* component of the codeword $\overline{Y}$ can

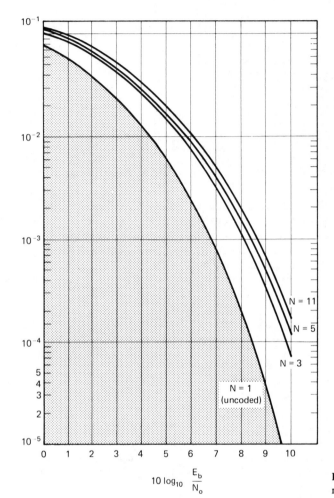

$$10 \log_{10} \frac{E_b}{N_o}$$

Figure 6.10 Error probability for repetition codes.

be written as a linear combination of the k information symbols. This is best expressed in matrix form as

$$\overline{Y} = \overline{X}\,\overline{G}$$

in which the matrix $\overline{G}$ is known as the *generator matrix* of the code, and $\overline{X}$ is the information sequence. It is clear that in order for the matrix multiplication to be defined $\overline{G}$ must have k rows and n columns. Thus the *generator matrix* has the general form

$$\overline{G} = \begin{bmatrix} g_{11} & g_{12} & \cdots & g_{1k} & \cdots & g_{1n} \\ g_{21} & g_{22} & \cdots & g_{2k} & \cdots & g_{2n} \\ \vdots & \vdots & & \vdots & & \vdots \\ g_{k1} & g_{k2} & \cdots & g_{kk} & \cdots & g_{kn} \end{bmatrix} \qquad (6.39)$$

It follows from the definition of $\overline{G}$ that the lth component of the codeword $\overline{Y}$ is given by

$$y_l = \sum_{i=1}^{k} x_i g_{il} \qquad 1 \le l \le n \qquad (6.40)$$

in which the symbol $\textstyle\sum$ denotes *modulo-2 addition*. The process of code design is the process of determining the elements g_{il} of the generator matrix.

Without loss of generality we can consider the code to be systematic. A **systematic code** is one in which the first k symbols of the codeword $\overline{Y}$ constitute the information sequence $\overline{X}$. Thus $y_j = x_j$ for $j \le k$, which implies that $g_{il} = 1$ for $i = l$ and $g_{il} = 0$ for $i \ne l$. Therefore, a *systematic code* has a generator matrix of the form

$$\overline{G} = \begin{bmatrix} 1 & 0 & \cdots & 0 & g_{1,k+1} & \cdots & g_{1,n} \\ 0 & 1 & \cdots & 0 & g_{2,k+1} & \cdots & g_{2,n} \\ \vdots & \vdots & & \vdots & \vdots & & \vdots \\ 0 & 0 & \cdots & 1 & g_{k,k+1} & \cdots & g_{k,n} \end{bmatrix} \qquad (6.41)$$

or

$$\overline{G} = [\,\overline{I}_k \mid \overline{P}\,] \qquad (6.42)$$

in which $\overline{I}_k$ is the $k \times k$ identity matrix and $\overline{P}$ represents the last $(n-k)$ columns of the generator matrix.

Closely related to the generator matrix is the parity check matrix. The *parity check matrix*, $\overline{H}$, is constructed from the generator matrix by writing

$$\overline{H} = \begin{bmatrix} \overline{P} \\ \hline \overline{I}_{n-k} \end{bmatrix} \qquad (6.43)$$

so that the parity check matrix has the form

$$\overline{H} = \begin{bmatrix} g_{1,k+1} & \cdots & g_{1,n} \\ g_{2,k+1} & \cdots & g_{2,n} \\ \vdots & & \vdots \\ g_{k,k+1} & \cdots & g_{k,n} \\ 1 & & 0 \\ \vdots & & \vdots \\ 0 & \cdots & 1 \end{bmatrix} = \begin{bmatrix} \overline{P} \\ \hline \overline{I}_{n-k} \end{bmatrix} \qquad (6.44)$$

Note that for systematic codes the parity check matrix can be written by inspection from the generator matrix.

Decoding of a linear code takes place by first multiplying the received sequence at the receiver output, $\overline{Z}$, by the parity check matrix. This multiplication yields a vector $\overline{S}$,

$$\overline{S} = \overline{Z}\overline{H}$$

known as the *syndrome*. Since binary codes are being considered, the *received sequence* $\bar{Z}$ is the modulo-2 sum of the transmitted codeword $\bar{Y}$ and the *n-symbol error sequence* $\bar{E}$. In other words, the syndrome can be expressed

$$\bar{S} = (\bar{Y} \oplus \bar{E})\bar{H} = \bar{Y}\bar{H} \oplus \bar{E}\bar{H} \tag{6.45}$$

We now examine the syndrome in more detail.

In order to better understand the meaning of the syndrome, and its use in the error correction process, consider the first term on the right-hand side of equation (6.45), $\bar{Y}\bar{H}$. We will denote $\bar{Y}\bar{H}$ by $\bar{S}^{(1)}$. It follows from equation (6.44) that the jth component of $\bar{S}^{(1)}$ can be written

$$S_j^{(1)} = \left(\sum_{i=1}^{k} y_i g_{i,k+j}\right) \oplus y_{k+j} \qquad j = 1, 2, \ldots, n-k \tag{6.46}$$

Since the only systematic codes are being considered, $y_i = x_i$ for $i \leq k$. Therefore,

$$\bar{S}_j^{(1)} = \left(\sum_{i=1}^{k} x_i g_{i,k+j}\right) \oplus y_{k+j} \qquad j = 1, 2, \ldots, n-k \tag{6.47}$$

From equation (6.40) it follows that

$$\left(\sum_{i=1}^{k} x_i g_{i,k+j}\right) = y_{k+j}$$

Therefore,

$$\bar{S}_j^{(1)} = y_{k+j} \oplus y_{k+j} = 0 \qquad j = 1, 2, \ldots, n-k \tag{6.48}$$

Since all $(n-k)$ components of $\bar{S}^{(1)}$ are zero,

$$\bar{S}^{(1)} = [0] \tag{6.49}$$

so that the syndrome is given by

$$\bar{S} = \bar{E}\bar{H} \tag{6.50}$$

Thus an all-zero syndrome denotes that the received sequence is a member of the set of codewords. This means that either no errors were made in transmission of the n-symbol codeword *or* that an error pattern occurred which mapped the transmitted codeword onto a different codeword. If the code has minimum distance d_{min}, at least d_{min} errors are required to map the transmitted codeword onto a different codeword.

Decoding is accomplished by relating to each syndrome the minimum weight error pattern that satisfies $\bar{S} = \bar{E}\bar{H}$. This error pattern is then modulo-2 added to the received sequence in order to obtain the most likely transmitted codeword.

6.4.4 Single-Error-Correcting Hamming Codes

To illustrate a single-error-correcting code, consider the parity check matrix in the following form

$$\bar{H} = \begin{bmatrix} \bar{h}_1 \\ \bar{h}_2 \\ \vdots \\ \bar{h}_j \\ \vdots \\ \bar{h}_n \end{bmatrix} \tag{6.51}$$

in which $\bar{h}_j$ is an $(n - k)$ symbol row vector. With this representation equation (6.50) becomes

$$\bar{S} = [e_1 \quad e_2 \quad \cdots \quad e_j \quad \cdots \quad e_n] \begin{bmatrix} \bar{h}_1 \\ \bar{h}_2 \\ \vdots \\ \bar{h}_j \\ \vdots \\ \bar{h}_n \end{bmatrix} \tag{6.52}$$

It follows that if the error sequence is all zero except for a 1 in the jth position, denoting a single error in the jth symbol of the n-symbol codeword, then

$$\bar{S} = \bar{h}_j \tag{6.53}$$

Therefore, a single error in the jth position of the received codeword yields the jth row vector of the parity check matrix for a syndrome. If all single errors are to be detected, it follows that all row vectors making up the parity check matrix must be distinct. The all-zero vector must be excluded since an all-zero syndrome denotes an *assumed* correct transmission. Since there are $2^{(n-k)}$ different binary sequences of length $n - k$, it follows that the parity check matrix has $2^{(n-k)} - 1$ rows. Thus n and k satisfy the relationship

$$n = 2^{(n-k)} - 1 \tag{6.54}$$

where, of course, $n - k$ is the number of parity check symbols.

Example 6.6. The (7, 4) Hamming Code

As an example, consider a code with $n - k = 3$ parity check symbols. There are $2^3 - 1 = 7$ possible sequences of length 3, excluding the all-zero sequence. These seven sequences of length 3 form the rows of the parity check matrix. It follows that $n = 7$ and $k = 4$. For a (7, 4) systematic code, the first four rows of the parity check matrix are the four sequences having more than one binary 1. The order of these four rows is arbitrary and does not affect code performance. Thus one possible parity check matrix for a single-error-correcting code is

$$\bar{H} = \begin{bmatrix} 0 & 1 & 1 \\ 1 & 0 & 1 \\ 1 & 1 & 0 \\ 1 & 1 & 1 \\ 1 & 0 & 0 \\ 0 & 1 & 0 \\ 0 & 0 & 1 \end{bmatrix} \tag{6.55}$$

For this choice of parity check matrix, the generator matrix is given by

$$\bar{G} = \begin{bmatrix} 1 & 0 & 0 & 0 & 0 & 1 & 1 \\ 0 & 1 & 0 & 0 & 1 & 0 & 1 \\ 0 & 0 & 1 & 0 & 1 & 1 & 0 \\ 0 & 0 & 0 & 1 & 1 & 1 & 1 \end{bmatrix} \qquad (6.56)$$

Assume an information sequence $\bar{X} = [1\ 0\ 1\ 1]$. The corresponding codeword is

$$\bar{Y} = \bar{X}\bar{G} = [1\ \ 0\ \ 1\ \ 1\ \ 0\ \ 1\ \ 0]$$

If a transmission error occurs in the fifth position, the received sequence is

$$\bar{Z} = \bar{Y} \oplus \bar{E} = [1\ \ 0\ \ 1\ \ 1\ \ 1\ \ 1\ \ 0]$$

and the syndrome is

$$\bar{S} = \bar{Z}\bar{H} = [1\ \ 0\ \ 1\ \ 1\ \ 1\ \ 1\ \ 0] \begin{bmatrix} 0 & 1 & 1 \\ 1 & 0 & 1 \\ 1 & 1 & 0 \\ 1 & 1 & 1 \\ 1 & 0 & 0 \\ 0 & 1 & 0 \\ 0 & 0 & 1 \end{bmatrix} = [1\ \ 0\ \ 0]$$

Since the syndrome is the fifth row of the parity check matrix, an error is indicated in the fifth position.

Now assume an error pattern consisting of errors in the second and fifth positions so that the error vector is given by

$$\bar{E} = [0\ \ 1\ \ 0\ \ 0\ \ 1\ \ 0\ \ 0]$$

The received sequence is

$$\bar{Z} = [1\ \ 1\ \ 1\ \ 1\ \ 1\ \ 1\ \ 0]$$

and the corresponding syndrome is

$$\bar{S} = \bar{Z}\bar{H} = [0\ \ 0\ \ 1]$$

indicating (incorrectly) an error in the seventh position. It should be noted that the syndrome is the modulo-2 sum of the second and fifth rows of the parity check matrix. The reason for the failure of the code, of course, is that it takes a distance 5 code to correct double errors and the code generated by (6.55) is a distance 3 code. Note that an incorrect transmission was detected, however.

■

6.4.5 Cyclic Codes

Of all the linear block codes, the most useful and popular are the cyclic codes. A cyclic code is one for which an end-around shift of a codeword yields another codeword. In other words, if

$$(y_1, y_2, y_3, \ldots, y_n)$$

is a codeword, it follows that for a cyclic code,

$$(y_n, y_1, y_2, \ldots, y_{n-1})$$

is also a codeword. Note that only $n - 1$ codewords can be generated by cyclic shifts of a single codeword. Therefore, it takes several different codewords to generate a complete set of codewords by cyclic shifts.

The principal reason that cyclic codes are of major importance is that encoding and decoding can be implemented using simple shift register circuits and a small amount of additional logic. Encoding is accomplished by using either a k-stage shift register or an $(n - k)$-stage shift register. Syndrome calculation for decoding can also be accomplished using a k-stage or $(n - k)$-stage shift register.

A number of excellent treatments of cyclic codes are contained in the literature [Berlekamp, 1968; Gallager, 1968; Lin, 1970; Peterson, 1961; Clark and Cain, 1981]. Each of these textbooks provides the necessary mathematical background for implementing both the encoder and the decoder. Thus we shall simply illustrate the technique by means of a simple example.

Example 6.7

An encoder for a (7, 4) cyclic code using a three-stage shift register is illustrated in Fig. 6.11(a). The operation of the encoder is summarized in Fig. 6.11(b) for the input

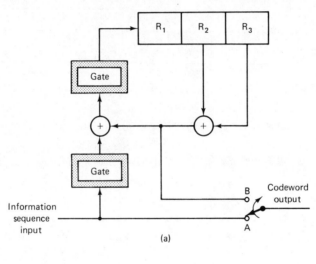

(a)

Shift	Input	Gate	Switch	R_1	R_2	R_3	Output
1	1	C	A	1	0	0	1
2	0	C	A	0	1	0	0
3	1	C	A	0	0	1	1
4	0	C	A	1	0	0	0
5	−	0	B	0	1	0	0
6	−	0	B	0	0	1	1
7	−	0	B	0	0	0	1

(b)

Figure 6.11 Encoding a (7, 4) cyclic code. (a) Example encoder for (7, 4) cyclic code. (b) Table of register contents for encoder.

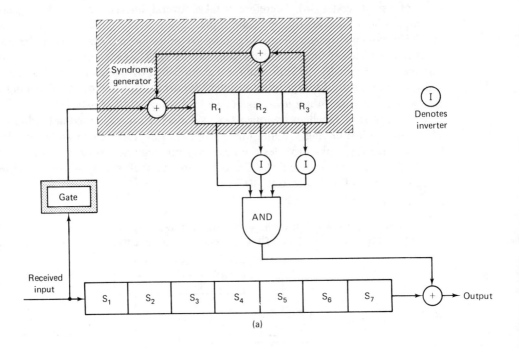

Shift	Input	Gate	R_1	R_2	R_3	AND	S_1	S_2	S_3	S_4	S_5	S_6	S_7	Output
1	1	C	1	0	0	–	1	–	–	–	–	–	–	–
2	0	C	0	1	0	–	0	1	–	–	–	–	–	–
3	1	C	0	0	1	–	1	0	1	–	–	–	–	–
4	1	C	0	0	0	–	1	1	0	1	–	–	–	–
5	0	C	0	0	0	–	0	1	1	0	1	–	–	–
6	1	C	1	0	0	–	1	0	1	1	0	1	–	–
7	1	C	1	1	0	–	1	1	0	1	1	0	1	–
8	–	0	1	1	1	0	–	1	1	0	1	1	0	1
9	–	0	0	1	1	0	–	–	1	1	0	1	1	0
10	–	0	0	0	1	0	–	–	–	1	1	0	1	1
11	–	0	1	0	0	1	–	–	–	–	1	1	0	0
12	–	0	0	1	0	0	–	–	–	–	–	1	1	0
13	–	0	1	0	1	0	–	–	–	–	–	–	1	1
14	–	0	1	1	0	0	–	–	–	–	–	–	–	1

(b)

Figure 6.12 Decoding a (7, 4) cyclic code. (a) Example decoder for (7, 4) cyclic code. (b) Table of register contents for decoding.

word 1 0 1 0. As indicated in Fig. 6.11(b), the gates are closed (denoted by C) and the switch is in position A during the first four shifts. During this period the information sequence is shifted into the modulator and channel and into the register. The gates are then opened (denoted by 0) and the switch is placed in position B. Three parity symbols are then output to complete the codeword. For the information sequence 1 0 1 0, the parity sequence 0 1 1 is generated. Thus the complete codeword is 1 0 1 0 0 1 1.

The operation of the decoder is investigated by assuming an error in the fourth position so that the sequence 1 0 1 1 0 1 1 is received. The decoder is illustrated in Fig. 6.12(a). Fourteen shifts are necessary for processing the complete received word. The state of the gate the AND gate, and all registers are given in Fig. 6.12(b) for all 14 shifts. The AND gate gives a binary one output only when all three inputs are binary ones. Thus the contents of the R register must be 1 0 0 in order for a one to be generated at the output of the AND gate. This condition exists on the eleventh shift and, as can be seen, this inverts the fourth symbol as the received sequence is being shifted out of the decoder.

If the generator and parity matrices are constructed for the given encoder, it is clear that an error in the fourth position generates a syndrome of 0 1 1. Thus the upper part of the decoder is actually a syndrome generator.

■

We now briefly turn our attention to a very important and flexible class of cyclic codes, the BCH codes.

6.4.6 BCH Codes

The *BCH* (*Bose–Chaudhuri–Hocquenghem*) *code* is perhaps the most important code in the class of cyclic codes. The design of these codes is straightforward and for a given block length, n, codes can be designed with a wide range of rates and error-correcting ability. Specifically, if e is the number of correctable errors per codeword and m is an arbitrary integer, the number of symbols per codeword is

$$n = 2^m - 1 \qquad m > 2 \tag{6.57}$$

and the number of parity symbols per codeword is defined by the bound [Lin, 1970]

$$n - k \leq me \tag{6.58}$$

Since e errors per codeword can be corrected, it follows that the minimum distance is given by

$$d > 2e + 1 \tag{6.59}$$

The relationship among n, k, and e is illustrated in Table 6.1 for a number of BCH codes. The value of k for given values of n and e is not easily determined. However, for small e equality holds in (6.58). Observation of Table 6.1 illustrates that for $n = 63$, equality in (6.58) holds for $e \leq 4$. It should be noted that for $e = 1$, the values of n and k are those values of n and k that define the Hamming code. Indeed, the Hamming code is a single-error-correcting BCH code.

TABLE 6.1   PARAMETERS n, k, AND e FOR BCH CODES

n	k	e	n	k	e	n	k	e	n	k	e
7	4	1		16	11		15	27		123	19
15	11	1		10	13		8	31		115	21
	7	2		7	15	255	247	1		107	22
	5	3	127	120	1		239	2		99	23
31	26	1		113	2		231	3		91	25
	21	2		106	3		223	4		87	26
	16	3		99	4		215	5		79	27
	11	5		92	5		207	6		71	29
	6	7		85	6		199	7		63	30
63	57	1		78	7		191	8		55	31
	51	2		71	9		187	9		47	42
	45	3		64	10		179	10		45	43
	39	4		57	11		171	11		37	45
	36	5		50	13		163	12		29	47
	30	6		43	14		155	13		21	55
	24	7		36	15		147	14		13	59
	18	10		29	21		139	15		9	63
				22	23		131	18			

Source: W. W. Peterson, *Error-Correcting Codes,* MIT Press, Cambridge, Mass., 1961, used with permission.

Since the BCH codes are cyclic codes, encoding and decoding is accomplished using simple shift register circuits. The performance of several BCH codes is an additive Gaussian noise channel is illustrated in a subsequent section.

6.4.7 The Golay Code

The *Golay code* is important since it is the only multiple-error-correcting code ($e > 1$) that is also a **perfect code.** The Golay code is a (23, 12) cyclic code that corrects all patterns of three or fewer errors. Closely related to the perfect (23, 12) Golay code is the (24, 12) Golay code, which is obtained from the (23, 12) Golay code by adding an overall parity check symbol. The (23, 12) Golay code has a minimum distance of 7 and the (24, 12) Golay code has a minimum distance of 8. Thus, in addition to correcting all patterns of three errors, the (24, 12) Golay code detects all patterns of four errors with a trivial reduction in code rate. The (24, 12) code is therefore popular for many applications.

The performance of the Golay code is examined in a subsequent section.

6.4.8 Word Error and Bit Error Probabilities

In the preceding section, the improvement in system performance obtained through the use of coding was determined by comparing the word error probabilities with coding and without coding. This was done for a given code. When two different codes are to be compared, the evaluation process is more complicated than simply comparing word error probabilities, since the two codes generally have different rates and a different number of information symbols (bits) per codeword.

The exact relationship between word error probability and bit error probability is dependent on the generator matrix of the code. However, simple bounds on the bit error probability are easily derived. Consider 1 s of transmission. The number of codewords transmitted during this interval is $1/T_w$. Since each codeword contains k information symbols, the total number of information symbols transmitted is k/T_w. The number of word errors is P_{wec}/T_w. If Γ denotes the number of information symbol errors per word error, the bit error probability is

$$P_{\text{bec}} = \frac{\Gamma P_{\text{wec}}/T_w}{k/T_w} = \frac{\Gamma P_{\text{wec}}}{k} \tag{6.60}$$

which is simply the ratio of the number of information symbols in error to the total number of information symbols transmitted. The problem, of course, is to determine the appropriate value for the parameter Γ.

As a worst case, assume that each word error results in k information symbol errors. This yields the upper bound

$$P_{\text{bec}} < P_{\text{wec}} \tag{6.61}$$

The lower bound is obtained by considering the most favorable situation in which each word error results in only one information symbol error. For this case Γ is 1 and we have

$$P_{\text{bec}} > \frac{1}{k} P_{\text{wec}} \tag{6.62}$$

For small values of k, the bounds are rather tight and the bit error probability is closely approximated by the word error probability.

There is a very simple approximation which is quite useful for high E_b/N_o environments. In such an environment, the symbol error probability is quite small and word errors are therefore probably due to $e + 1$ symbol errors. Of these $e + 1$ symbol errors, $(e + 1)(k/n)$ are, on the average, information symbol errors. Thus

$$\Gamma = (e + 1)\frac{k}{n}$$

and the approximation

$$P_{\text{bec}} \simeq \frac{e + 1}{n} P_{\text{wec}} \tag{6.63}$$

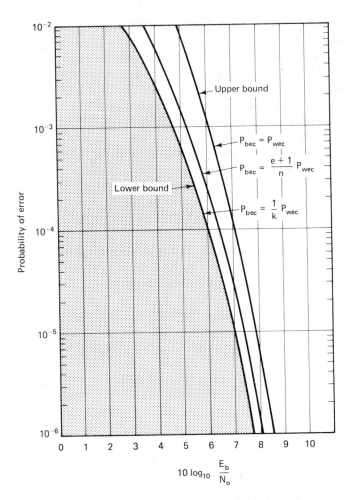

Figure 6.13 Bit error probability for (23, 12) Golay code.

results. The approximations of expressions (6.61), (6.62), and (6.63) are illustrated in Fig. 6.13 for the (23, 12) Golay code.

6.4.9 Comparative Performance of Block Codes

The performance of a number of block codes is illustrated in Fig. 6.14. The assumption that a word error results in errors in all information symbols was used so that the illustrated performance represents worst-case bounds. The approximation of (6.33) was also used.

It can be seen from Fig. 6.14 that the (7, 4) and (15, 11) codes, which are single-error-correcting Hamming codes, offer moderate improvement for values of E_b/N_o greater than approximately 8 dB. The (127, 113) code is a double-error-correct-

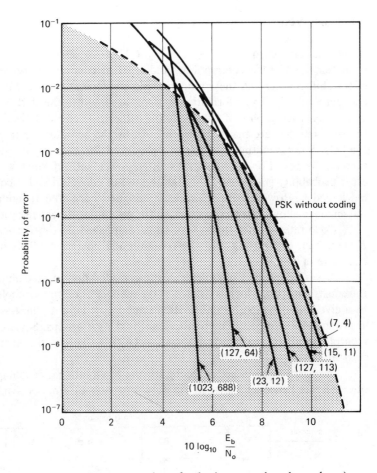

Figure 6.14   Comparison of codes (worst-case bounds are shown).

ing BCH code and improves performance by over an order of magnitude for a fixed value of E_b/N_o in excess of 8 dB. A (112:127 $\cong$ 7/8) code has been specified for use in the 120-Mb/s rate INTELSAT-V TDMA system [Muratani et al., 1978]. The (23, 12) code is the triple-error-correcting Golay code. The (127, 64) and (1023, 688) codes are BCH codes capable of correcting 10 and 36 errors per codeword, respectively.

6.5 CONVOLUTIONAL CODES

We now consider a different type of encoder, in which the information symbols are not grouped together in blocks for encoding. This is the *convolutional encoder* and has great potential for satellite applications.

6.5.1 Convolutional Encoding

A convolutional encoder is shown in Fig. 6.15. It consists of a K-stage shift register, v modulo-2 adders, a commutator, and a set of connections between the K stages of the shift register and the v modulo-2 adders. Operation of the basic convolutional encoder is simple. The information symbols are input to the shift register one symbol at a time. The outputs of the modulo-2 adders, determined by the connections to the shift register, are then sampled in turn by the commutator to produce v output symbols. Since v output symbols are produced for each input symbol, the rate of the code is $1/v$. For constant transmitter power and information rate, the symbol error probability is increased through the use of convolutional encoding just as with block codes. However, for a properly designed code, the redundancy induced by the encoder allows error correction and a net improvement in system performance.

Code rates greater than $\frac{1}{2}$ can be accomplished by shifting k symbols into the K-stage shift register between the commutation operations. This obviously yields a code rate of k/v.

A very important parameter in the consideration of convolutional encoding is the constraint span, which is defined as the number of output symbols that are affected by a given input symbol. If information symbols are input to the K-stage shift register in groups of k symbols. The register can hold K/k groups. Since each group yields v output symbols, the constraint span is $(K/k)v$. This is the memory time of the encoder.

In order to gain a better feeling for the convolutional encoding operation, consider the rate $\frac{1}{3}$ encoder shown in Fig. 6.16. For each information symbol, the sequence

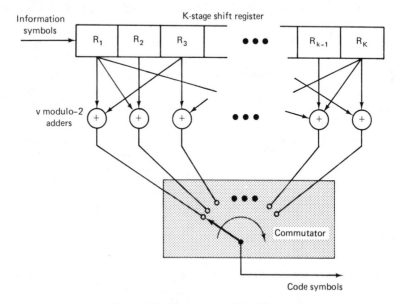

Figure 6.15   General convolutional encoder.

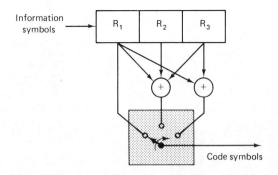

Figure 6.16   Example systematic rate $\frac{1}{3}$ convolutional encoder.

(v_1, v_2, v_3) is generated. It follows from Fig. 6.16 that v_1, v_2, and v_3 are given by

$$v_1 = R_1$$

$$v_2 = R_1 \oplus R_2 \oplus R_3$$

$$v_3 = R_1 \oplus R_3$$

in which R_i denotes the contents of the ith register. Since the first symbol in the output sequence is the information symbol, this particular convolutional code is *systematic*. Thus v_2 and v_3 can be viewed as parity symbols.

The output sequence for an arbitrary input sequence is often determined with the aid of a code tree. The code tree for the encoder of Fig. 6.16 is shown in Fig. 6.17. The branches correspond to input symbols and a branch upward corresponds to an input zero and a branch downward corresponds to an input one. The three symbols on a given branch denote the output sequence corresponding to that branch. Thus the input sequence 1 0 1 1 generates the output sequence 1 1 1 0 1 0 1 0 0 1 0 1. Note that after nine output symbols the code tree is symmetric about the dashed line. This results since the constraint length is 9.

Conceptually, decoding is accomplished by taking the received sequence and finding the path through the code tree that lies closest in Hamming distance to the received sequence. This is, of course, not practical for long sequences since decoding a sequence of r symbols requires that 2^r branches of the code tree be searched for the minimum Hamming distance. Many algorithms have been developed to facilitate the decoding process. These are mentioned briefly later.

6.5.2 Threshold Decoding—An Example

We now consider a simple example for which a reliable performance estimate can be simply derived. The coder is shown in Fig. 6.18(a). The register contents are assumed to be

$$R_1 = x_n$$

$$R_2 = x_{n-1}$$

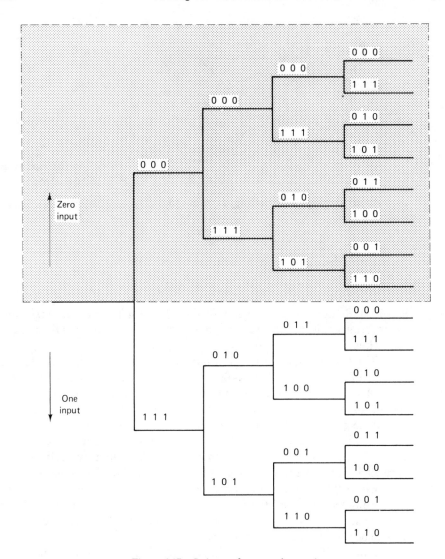

Figure 6.17 Code tree for example encoder.

so that the output is the two-symbol sequence

$$(x_n, \; x_n \oplus x_{n-1})$$

The decoder is a threshold decoder first developed by [Massey, 1963]. The input to the decoder is the two-symbol sequence

$$(x_n \oplus e_n^1, \; x_n \oplus x_{n-1} \oplus e_n^2)$$

in which e_n^1 is the error induced by the channel during transmission of the first symbol and e_n^2 is the error induced by the channel during transmission of the second

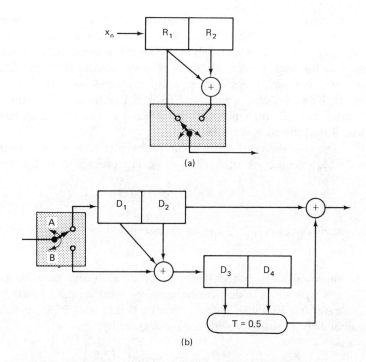

Figure 6.18   Threshold encoding and decoding example. (a) Encoder. (b) Decoder.

symbol. The switch on the decoder is in position A when the first symbol is input to the decoder and in position B when the second symbol is input to the decoder. Thus

$$D_1 = x_n \oplus e_n^1$$

and it follows that

$$D_2 = x_{n-1} \oplus e_{n-1}^1$$

Since the contents of D_3 is given by

$$D_3 = x_n \oplus x_{n-1} \oplus e_n^2 \oplus D_1 \oplus D_2$$

it follows that

$$D_3 = e_n^1 \oplus e_{n-1}^1 \oplus e_n^2$$

and

$$D_4 = e_{n-1}^1 \oplus e_{n-2}^1 \oplus e_{n-1}^2$$

In reasonable signal-to-noise ratio environments, D_3 and D_4 provide sufficient information for making a reliable decision.

If D_3 and D_4 are *both* equal to 1, there are two possibilities. The first possibility is that the e^1_{n-1} is in error. The second possibility is that e^1_n or e^2_n is in error *and* e^1_{n-2} or e^2_{n-1} is in error. For small channel error probability, q_c, the probability that e^1_{n-1} is 1 is approximately q_c. The other event requires that two errors are made in the sequence $e^1_n \; e^2_n \; e^1_{n-1} \; e^2_{n-1} \; e^1_{n-2}$ and the probability of this event is q^2_c. Thus, with high probability, e^1_{n-1} is 1 if D_1 and D_2 are both 1. This determination is made by establishing a threshold level of $\frac{1}{2}$ as shown. If the threshold is exceeded, an error is (with high probability) detected in the preceding information symbol. (Recall that the code is systematic.)

The probability of error is the probability that more than one error is made in the sequence $e^1_n \; e^2_n \; e^1_{n-1} \; e^2_{n-1} \; e^1_{n-2}$. The probability of this event is

$$P_E = \sum_{i=2}^{5} \binom{5}{i} q^i_c (1 - q_c)^{5-i}$$

For small values of q_c, P_E can be approximated by

$$P_E \simeq 10q^2_c \qquad (6.64)$$

The improvement in system performance is significant for small values of q_c.

As with block codes, the value of q_c must be determined by considering the code rate. For example, if the code rate is $1/v$ and PSK modulation is used, the symbol error probability with coding is given by

$$q_c = \tfrac{1}{2} \operatorname{erfc} \left[\sqrt{\frac{E_b}{vN_o}} \right] \qquad (6.65)$$

The detailed diagram, the principal of operation, and the performance of the rate $\frac{3}{4}$ convolutional encoder with threshold decoding used in the INTELSAT satellite is described in Chapter 10.

6.5.3 Performance of Convolutional Codes

At the present time, the most popular decoder for convolutional codes is the Viterbi algorithm. The **Viterbi algorithm** is a *maximum likelihood technique* and was first published by Viterbi in 1967 [Viterbi and Omura, 1979]. It is most practical for convolutional codes having a relatively short constraint length. A description of the algorithm is beyond the scope of this chapter but excellent treatments are contained in the textbooks by [Viterbi and Omura, 1979, and Clark and Cain, 1981]. In addition, both of these textbooks contain descriptions of several other convolutional coding methods.

Computer simulation is a popular method for studying the performance of systems utilizing convolutional codes, and a large number of results have been published. Of particular note are the results published by Heller and Jacobs, which were obtained by considering a power-limited satellite system with PSK modulation and Viterbi decoding [Heller and Jacobs, 1971]. They considered both hard-decision and soft-decision decoders and their results agree with the 2-dB performance penalty associated with hard-decision receivers discussed in Section 6.3.

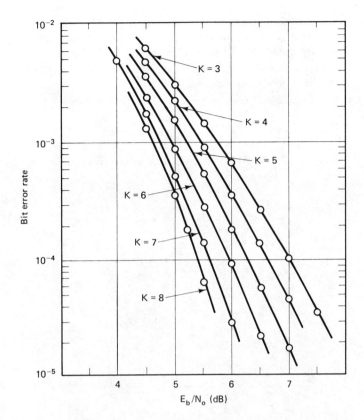

Figure 6.19 Performance of rate $\frac{1}{2}$ convolutional codes. (After J. A. Heller and I. M. Jacobs, "Viterbi Decoding for Satellite and Space Communication," *IEEE Trans. Commun. Technol.*, © October 1971, pp. 835–848, with permission.)

An example of performance results obtained by Heller and Jacobs are illustrated in Fig. 6.19. In these performance curves the parameter K *represents the number of stages* in the shift register and the code rate is $\frac{1}{2}$.

PROBLEMS

6.1. A discrete source has outputs x_i, $1 \leq i \leq N$, with corresponding probabilities $p(x_i)$. Prove, by direct differentiation, that the source entropy

$$H(X) = -\sum_{i=1}^{N} p(x_i) \log_2 p(x_i)$$

has a maximum value of $\log_2 N$ and that the maximum occurs when all source outputs are equally likely.

6.2. A friend draws a card from a well-shuffled deck. How much information does he receive? How many questions, answerable yes or no, must you ask your friend in order to determine the card completely? Completely formulate the questions.

6.3. A binary source has outputs A and B with probabilities 0.75 and 0.25, respectively. Determine the source entropy.

6.4. A signal source defined by the probability density function

$$p(x) = \frac{1}{\sqrt{2\pi}} \exp{(0.5x^2)}$$

is input to a 3-bit A/D converter. The A/D converter is defined by the following table:

Output	Input
0 0 0	$x < -2$
0 0 1	$-2 < x < -1$
0 1 0	$-1 < x < -0.5$
0 1 1	$-0.5 < x < 0$
1 0 0	$0 < x < 0.5$
1 0 1	$0.5 < x < 1$
1 1 0	$1 < x < 2$
1 1 1	$2 < x$

Determine the source entropy.

6.5. The source in Problem 6.3 is the input to a channel defined by the matrix

$$[P(Y|X)] = \begin{bmatrix} 0.9 & 0.1 \\ 0.2 & 0.8 \end{bmatrix}$$

Determine the channel input–output mutual information.

6.6. The binary erasure channel is defined by

$$[P(Y|X)] = \begin{bmatrix} 1-q & q & 0 \\ 0 & q & 1-q \end{bmatrix}$$

Determine the channel capacity using

$$I(X;\ Y) = H(X) - H(X|Y)$$

6.7. Repeat Problem 6.6 using

$$I(X;\ Y) = H(Y) - H(Y|X)$$

6.8. An N-input noiseless channel is defined by the channel matrix

$$[P(Y|X)] = [I_N]$$

in which $[I_N]$ is the $N \times N$ identity matrix. Derive the capacity of the channel.

6.9. Prove that if a channel has statistically independent inputs and outputs, the capacity is zero.

6.10. A satellite system is to be designed so that the overall capacity is 0.7 bit per symbol. Transmission is to be PSK. The downlink error probability is 0.02 and on-board signal processing takes place with negligible error. Determine the minimum acceptable value of E_b/N_o on the uplink.

6.11. Construct Fig. 6.4 for $N = 4$.

6.12. A binary source has outputs A and B with probabilities 0.75 and 0.25, respectively. Determine a source code for the third-order extension of the source. Repeat for the fourth-order extension.

6.13. Repeat Problem 6.12 assuming that the source probabilities are 0.6 and 0.4.

6.14. Prove that if $\overline{Y}_1$ and $\overline{Y}_2$ are codewords,

$$\overline{Y}_3 = \overline{Y} \oplus \overline{Y}_2$$

is also a codeword. This is known as the group property of codewords.

6.15. Derive all 16 codewords generated by equation (6.56). Prove that (6.56) generates a distance 3 code.

6.16. Derive a generator matrix and parity check matrix for a (15, 11) Hamming code.

6.17. Describe the generator matrix for a rate $1/n$ repetition code.

6.18. Show that the rate of a Hamming code approaches 1 as n grows without bound.

6.19. Determine the generator matrix for the encoder illustrated in Fig. 6.11.

6.20 By finding all 16 codewords generated by the encoder illustrated in Fig. 6.11, show that the encoder generates a cyclic code.

6.21. Design a rate $\frac{1}{2}$ convolutional encoder with a three-stage shift register. Choose the connections so that the output with a binary one input is the complement with a binary zero input.

7

SYNCHRONIZATION SUBSYSTEMS: ANALYSIS AND DESIGN

DR. LEWIS E. FRANKS

Professor of Electrical and Computer Engineering
University of Massachusetts, Amherst, Massachusetts

> *It is a pleasure to have Dr. Lewis E. Franks as the author of this chapter. Professor Franks has had numerous original contributions in the fields of signal theory and synchronization systems. He presents an in-depth study of carrier recovery and symbol timing synchronization techniques.*
>
> *Dr. K. Feher*

7.1 INTRODUCTION

In digital data communication, there is a hierarchy of synchronization problems to be considered. First, assuming that a carrier-type system is involved, there is the problem of *carrier synchronization,* which concerns the generation of a reference carrier with a phase closely matching that of the data signal. This reference carrier is used at the data receiver to perform a coherent demodulation operation, creating a baseband data signal. Next comes the problem of synchronizing a receiver clock with the baseband data-symbol sequence. This is commonly called *bit synchronization,* even when the symbol alphabet happens not to be binary.

Depending on the type of system under consideration, problems of *word, frame,* and *packet synchronization* will be encountered further down the hierarchy. A feature

294

that distinguishes the latter problems from those of carrier and bit synchronization is that they are usually solved by means of special design of the message format, involving the repetitive insertion of bits or words into the data sequence solely for synchronization purposes. On the other hand, it is normally regarded as desirable that carrier and bit synchronization be effected without multiplexing special timing signals onto the data signal, which would use up a portion of the available channel capacity. Only timing-recovery problems of this type are discussed in this chapter. This excludes those systems wherein the transmitted signal contains an unmodulated component of sinusoidal carrier (such as with "on–off" keying).

For modulation formats that exhibit a high bandwidth efficiency (i.e., have a large "bits per cycle" figure of merit), we find the accuracy requirements on carrier and bit synchronization increasingly severe. Unfortunately, it is also in these high-efficiency systems that we find it most difficult to extract accurate carrier phase and symbol timing information by means of simple operations performed on the received signal. The pressure to develop higher-efficiency data transmission has led to a dramatically increased interest in timing recovery problems and, in particular, in the ultimate performance that can be achieved with optimal recovery schemes.

We shall examine problems associated with determining the phase of a suppressed carrier in the one-dimensional modulation format where the sinusoidal carrier is simply multiplied by a zero-mean message signal, and in the two-dimensional format where the in-phase (I) and quadrature (Q) components are modulated by two separate message signals. When the message signal is a binary process, the one- and two-dimensional formats correspond to **BPSK** and **QPSK**, respectively. We regard the message signal, in general, as a synchronous, baseband **PAM** signal where the symbol rate is not constrained to have any particular relationship with the carrier frequency. The timing recovery problem associated with the baseband signal is to determine the time positions of individual data pulses relative to the ticks of a clock operating at the symbol rate.

Taken together, the carrier and bit synchronization problem is to determine, based only on operations performed on the signal received,

$$z(t) = \text{Re} \left[\beta(t - \tau)e^{j\theta}e^{j2\pi f_o t} \right] + \text{noise} \qquad (7.1)$$

good estimates of the parameters θ and τ, which we consider here to remain constant over relatively long intervals of time. In equation (7.1), $\beta(t)$ is the complex envelope representation of the modulated carrier signal. Analysis of phase and timing recovery schemes is greatly simplified using complex envelope notation and Appendix 7A presents a brief review of the basic aspects; in particular, the effects of bandpass filtering operations and bandpass multiplication operations in terms of complex envelope notation. This appendix also discusses the statistical properties of complex envelope representations for random signal and noise processes.

Errors in estimation of θ and τ degrade the overall performance of the system. Ultimately, we are concerned with the effect of carrier phase and symbol timing errors on the individual bit or symbol error probability performance of the communication system. There are several ways in which the system designer can display the degradation in error probability resulting from inexact synchronization. However, for initial design considerations, the format of Figs. 7.1 and 7.2 [Stiffler, 1971] seems most appropriate. Figure 7.1 shows bit error probability as a function of signal-to-noise ratio, R, for the one-dimensional, binary modulation format of BPSK. The parameter distinguishing various performance curves in Fig. 7.1 is the rms value of *phase jitter* (so called because the estimated phase $\hat{\theta}$ fluctuates with time). Thus $\sigma_\phi^2 = E(\theta - \hat{\theta})^2$. We see that there is a rather strong "threshold effect" at the higher values of signal-to-noise ratio. For an rms phase jitter between 0.3 and 0.5 rad, there is a great variation of error probability. Some comments about how these performance graphs are obtained are necessary here. It is a relatively simple matter to calculate the error probability conditional on a fixed value, ϕ, of carrier phase error. The average error probability is then evaluated (using numerical integration techniques), assuming that ϕ is a random variable with a Gaussian distribution. This Gaussian assumption appears to be quite reasonable, as far as average error probability calculations are concerned, for most mechanisms producing phase error if the amount of phase error is not too large.

Performance graphs in the format of Fig. 7.1 are useful for pinpointing design limits on allowable jitter variance. However, for displaying overall performance they may be misleading, because it might be expected that rms phase jitter would decrease uniformly with increase of signal-to-noise ratio. In some circumstances this is true, but in others, the phase jitter may be partly due to effects which are independent of the additive channel noise. With different modulation and phase recovery schemes and various kinds of signal shaping (Fig. 7.1 assumes instantaneous phase changes), we get a variety of P_e versus signal-to-noise ratio graphs and a universal set of design graphs seems out of the question. In practice, a combination of analysis, simulation, and experimental measurement is used to come up with an allocation of permissible rms phase jitter.

The situation for rms error in symbol synchronization (i.e., *timing jitter*) is quite similar. Figure 7.2 shows the degradation in P_e performance for PBSK with increasing rms values of the relative error, $\xi = (\tau - \hat{\tau})/T$, in the receiver symbol clock. We see that an rms timing jitter greater than about $40/\pi$ percent of a symbol period produces a substantial degradation in error probability at the higher signal-to-noise ratios. These graphs were also obtained assuming that ξ is a slowly fluctuating (essentially constant over a T-second interval) random variable with a Gaussian density function.

When we consider two-dimensional modulation formats (e.g., QPSK rather than BPSK), the allowable phase jitter must be reduced for the same P_e performance, because the receiver must choose from four phase states rather than two. Also, with a phase error in QPSK we introduce crosstalk between the I and Q channels of the system, that is, an error which is independent of the additive noise and which places tighter requirements on the allocation of permissible phase jitter. The procedure for P_e calculations for other PSK formats is presented in various communication theory texts [Lindsey and Simon, 1973; Spilker, 1977]. For a more general QAM case, such as four-level PAM on both I and Q channels (a popular format for terrestrial

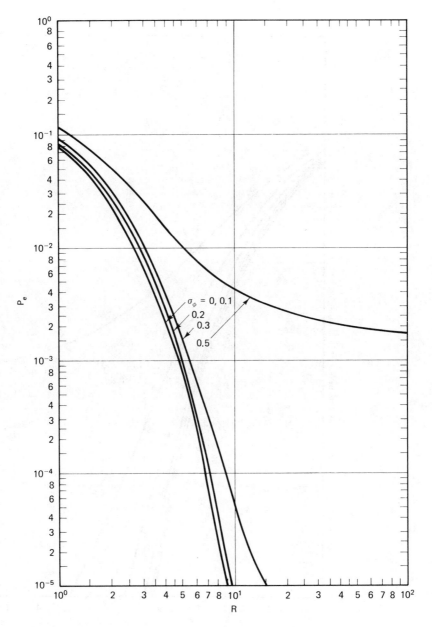

Figure 7.1 Binary PSK symbol error probability with imperfect *carrier* synchronization. *R*, signal-to-noise ratio. (After [Stiffler, 1971], with permission from Prentice-Hall, Inc.)

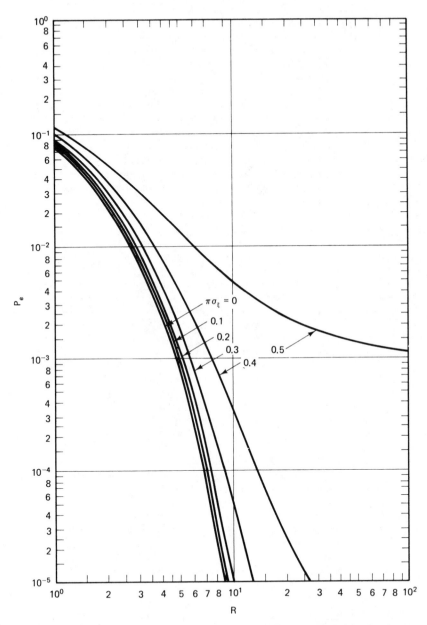

Figure 7.2 Binary PSK symbol error probability with imperfect *symbol* synchronization. *R*, signal-to-noise ratio. (After [Stiffler, 1971], with permission from Prentice-Hall, Inc.)

digital radio systems), the specifications can become quite severe (e.g., less than $1°$ rms jitter for a single communication link).

In order to proceed with an investigation of the performance of various phase- and timing-recovery synchronization schemes, we shall assume that the system de- signer has a target value of rms jitter and that he is seeking a relationship between rms jitter and various system parameters. The small-jitter situation is of greatest practical interest and we employ linearization techniques which result in manageable formulas for rms phase jitter and timing jitter. These formulas reveal the jitter depen- dence on signal-to-noise ratio, bandwidth of the recovery circuits, and on the band- width and other properties of the shape of the data pulse employed. We can also compare various recovery strategies for jitter performance; in particular, the improve- ment that results from adopting a *data-aided* strategy in considered in some detail.

Many systems recover the θ and τ parameters in a two-step process; first, a reference carrier is recovered and the signal is demodulated to baseband and second, a baseband timing-recovery circuit is used to extract the timing information. In some situations, however, the improved performance obtainable by joint recovery of the two parameters may be very useful. The last section of the chapter includes an analysis of joint carrier phase and symbol timing recovery and some configurations of tracking loops based on these ideas. It would perhaps be helpful if the reader had some back- ground on the theory and design of the phase-locked loop (PLL) and the text [Gardner, 1979] is a good starting point. Some of the details in deriving the jitter formulas presented here are omitted; however, the problems included at the end of the chapter should aid the reader to develop the missing steps.

7.2. CARRIER PHASE RECOVERY

7.2.1 Carrier Recovery for Unmodulated Signals

First, we consider the case of an *unmodulated* carrier in a background of white Gaussian noise having a (double-sided) power spectral density of N_o volts²/hertz. We do this primarily to establish a reference point for the performance of more practical phase-recovery schemes. Certainly, the unmodulated-carrier case presents the least challenge for accurate phase recovery. All that is required is a bandpass filter (BPF) or a PLL, which rejects the noise components everywhere except in the immediate vicinity of the carrier frequency. For this case, shown in Fig. 7.3, we let the received signal plus noise at the input of the BPF have the **complex envelope,**

$$\alpha(t)e^{j\theta} = [1 + u_I(t) + j\,u_Q(t)]e^{j\theta} \qquad (7.2)$$

where the in-phase and quadrature Gaussian noise terms are independent, zero-mean, and $E[u_I(t)u_I(s)] = E[u_Q(t)u_Q(s)] = 2N_o\,\delta(t-s)$. See equation (7A.15). It is merely a matter of convenience to reference the I and Q components of the noise to the signal phase, θ; their joint statistical properties are not affected by a rotation of

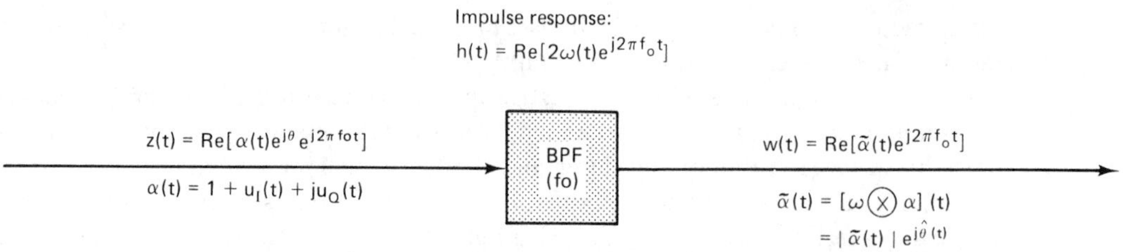

Impulse response:

$h(t) = Re[2\omega(t)e^{j2\pi f_o t}]$

$z(t) = Re[\alpha(t)e^{j\theta}e^{j2\pi f_o t}]$

$\alpha(t) = 1 + u_I(t) + ju_Q(t)$

BPF
(fo)

$w(t) = Re[\tilde{\alpha}(t)e^{j2\pi f_o t}]$

$\tilde{\alpha}(t) = [\omega \bigotimes \alpha]\,(t)$

$= |\tilde{\alpha}(t)|\,e^{j\hat{\theta}(t)}$

Figure 7.3 Bandpass filtering of sinusoidal signal plus noise.

coordinates. We denote the phase angle of the recovered carrier, $w(t)$, by $\hat{\theta}$. Then the phase error is given by*

$$\phi(t) = \hat{\theta}(t) - \theta = \tan^{-1}\left\{\frac{Im\,[\omega \bigotimes \alpha]}{Re\,[\omega \bigotimes \alpha]}\right\} \tag{7.3}$$

Assuming that the BPF transfer function is symmetric about f_o, then the lowpass equivalent impulse response, $\omega(t)$, is real and the numerator in the argument of the arctangent function is a "noise-only" term. In the denominator, the signal term [corresponding to the "1" in equation (7.2)] predominates at relatively high signal-to-noise ratios. Thus, assuming that phase error is relatively small, we replace the denominator by its mean value and then use the $\tan^{-1} x \approx x$ approximation, giving

$$\phi(t) = \frac{[\omega \bigotimes u_Q]}{[\omega \bigotimes 1]} \tag{7.4}$$

We shall always assume that the BPF has unity gain in the center of its passband [i.e., $\Omega(0) = 1$]. Then $\omega \bigotimes 1 = 1$ in the denominator of equation (7.4). Now it is a simple matter to calculate the variance of the phase jitter, and we get

$$\text{Var } \phi(t_o) = \iint_{-\infty}^{\infty} \omega(t_o - t)\omega^*(t_o - s)2N_0\,\delta(t-s)\,dt\,ds$$

$$= 2N_0 \int_{-\infty}^{\infty} |\omega(t_o - t)|^2\,dt = 2N_0 \int_{-\infty}^{\infty} |\Omega(f)|^2\,df \tag{7.5}$$

$$\overset{\Delta}{=} 2N_0\,B$$

which we note is independent of t_o [i.e., $\phi(t)$ can be regarded as a stationary process].

In (7.5), we have defined a bandwidth parameter, B, for the BPF. This is often called the *noise bandwidth* of a filter because the rms value of the filtered white noise is the same as would result from an idealized rectangular passband with gain,

---

* We let $\omega \bigotimes \alpha$ represent the convolution of the complex impulse response of the low-pass equivalent for the BPF with the complex envelope, $\alpha(t)$, of the signal received, $z(t)$. The reader should keep in mind that $\omega \bigotimes \alpha$ is, in general, a time-varying quantity even though the t-dependence is suppressed in this abbreviated notation. An exception to this is the case of convolution with a constant value [e.g., $\omega \bigotimes 1 = 1$ in equation (7.4)].

$\Omega(0)$, and width B.* If the BPF is replaced by a PLL, then B is called the *loop bandwidth,* and its value depends on the loop gain of the PLL.† Clearly the steady-state rms phase jitter can be made as small as desired by reducing the loop bandwidth. On the other hand, too small a bandwidth will result in problems such as:

1. Inability to track instabilities in the transmitted carrier (i.e., θ is not actually a constant value in practice).
2. Mistuning of the BPF; hence distorting the symmetry condition and introducing additional terms into the numerator and denominator of (7.4).
3. Sustained transient phase error (i.e., prolonged acquisition time for phase recovery).

For these reasons, the phase-recovery problem cannot be solved by making the bandwidth of the recovery circuits very small. There is always a compromise between steady-state jitter performance and the amount of bandwidth required to satisfy the requirements of agility of acquisition and tracking mentioned above. The viewpoint that we adopt in this chapter is that the system designer is presented with a specified minimum recovery bandwidth as well as a maximum rms jitter and that other system parameters and synchonization strategies must be adjusted to meet the requirements. In some cases, we can get analytical solutions for system parameters which minimize rms jitter for a specified recovery bandwidth.

To express equation (7.5) in terms of input signal-to-noise ratio, we encounter a difficulty because of the infinite power of the white noise. Therefore, we consider noise power only in the band to be occupied by the signal. The applications here assume digital data transmission at a rate of $1/T$ symbols per second. The minimum (Nyquist) bandwidth for a signal of this nature, assuming double-sideband modulation of the carrier, is $1/T$ hertz; therefore, the input noise power is taken as $2N_0/T$ volts². As the signal power in this unmodulated carrier case is $\frac{1}{2}$ volts², we can express the phase jitter in equation (7.5) in terms of signal-to-noise power ratio, R, as

$$\text{Var } \phi = (BT)\left(\frac{1}{2R}\right) \qquad R = \frac{T}{4N_0} \qquad (7.6)$$

and this becomes the reference performance for carrier phase recovery.

7.2.2 Carrier Modulated by a Zero-Mean PAM Signal

Now we consider the case of a zero-mean baseband PAM signal modulating the carrier. Then the received signal plus noise can be expressed as

$$\alpha(t)e^{j\theta} = [a(t) + u_I(t) + ju_Q(t)]e^{j\theta}$$

$$a(t) = \sum_{k=-\infty}^{\infty} a_k g(t - kT) \qquad (7.7)$$

---

* A single-resonator BPF having a bandwidth B' between -3-dB points has a noise bandwidth of $B = (\pi/2)B'$.

† Sometimes, as a matter of definition, the loop bandwidth is taken as one-half of this value.

where the $\{a_k\}$ is the data sequence and $g(t)$ is the baseband data pulse. This bandpass signal does not possess a carrier component. Unlike the previous case, its mean value vanishes. Of course, its mean-square value does not vanish, and in fact this quantity has a periodic fluctuation (at a frequency of $2f_o$) which, if measured, will indicate the phase of the suppressed carrier. Considering only the signal components,

$$E[z^2(t)] = \tfrac{1}{2} E[a^2(t)] + \tfrac{1}{2} E[a^2(t)] \operatorname{Re} e^{j2\theta} e^{j4\pi f_o t} \tag{7.8}$$

The real random process, $z(t)$, belongs to the class known as *cyclostationary processes* [Franks, 1969], which have statistical moments that vary periodically with time. It is the presence of these periodic moments that allows phase recovery by means which resemble moment estimation procedures. Assuming $E[a^2(t)]$ constant in (7.8), which is true only in special circumstances, the mean-square signal has a periodicity at $2f_o$ and with a phase just double that of the desired recovered carrier. In the PAM signal case, $a(t)$ is also a cyclostationary process with period T. Hence the mean-square signal (7.8) has components at $2f_o$ plus and minus integer multiples of $1/T$.

7.2.3 Squarer Bandpass-Filter Carrier Recovery Circuit

These facts suggest the popular carrier-phase recovery circuit shown in Fig. 7.4. The BPF centered at $2f_o$ retains the mean value of the squarer output and reduces its variance to acceptable values. The input BPF at f_o is required to limit the noise variance at the squarer output to a finite value. We assume that this BPF does not alter the signal; then the phase error in $w(t)$ is given by $2\phi = 2\hat{\theta} - 2\theta$, where, using the same type of approximations leading to equation (7.4), we get

$$\phi(t) = \frac{[\omega \otimes (a + u_I) \, u_Q]}{\omega \otimes E[a^2]} \tag{7.9}$$

where $u_I(t)$ and $u_Q(t)$ are the I and Q components, relative to θ, of the white noise passed by the input BPF. These components are independent, and we let $E[u_I(t)u_I(s)] = E[u_Q(t)u_Q(s)] = k(t - s)$. Recalling that $\Omega(0) = 1$, we have

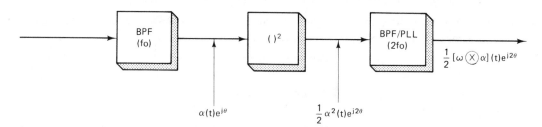

Figure 7.4   Squarer/BPF carrier recovery.

$$\text{Var } \phi(t_o) = [E(a^2)]^{-2} \int\!\!\int \omega(t_o - t)\omega^*(t_o - s)$$
$$\times \{E[a(t)a(s)]k(t - s) + k^2(t - s)\} \, dt \, ds \tag{7.10}$$

For a specific illustration of this result, we consider independent, ± 1 binary data (i.e., a BPSK signal). The extention to more general PAM formats is quite straightforward. The autocorrelation of the PAM signal is calculated as follows (see Problem 7.5):

$$E[a(t)a(s)] = E \sum_k \sum_m a_k a_m g(t - kT)g(s - mT)$$

$$= \sigma_a^2 \sum_k g(t - kT)g(s - kT) \tag{7.11}$$

$$= \frac{\sigma_a^2}{T} \sum_{\ell} \int_{-\infty}^{\infty} G(f)G\left(\frac{\ell}{T} - f\right) e^{j2\pi f(t-s)} \, df e^{j2\pi\ell s/T}$$

where the summation indices in (7.11) run from $-\infty$ to $+\infty$ and $\sigma_a^2 \triangleq E[a_k^2]$. Now let us consider the minimum-bandwidth PAM signal where $g(t) = \text{sinc }(t/T)$ and only the $\ell = 0$ term in (7.11) is nonzero. With an input BPF having a $1/T$ bandwidth, we get

$$E[a(t)a(s)] = g(t - s)$$
$$k(t - s) = \frac{2N_o}{T} g(t - s) \tag{7.12}$$

Furthermore, we assume that the output BPF has a bandwidth much smaller than $1/T$, so that only the low-frequency power density ($= 4N_o^2/T$ volts²/hertz) of $k^2(\tau)$ needs to be considered in an approximate evaluation of equation (7.10). The result is (see Problem 7.4)

$$\text{Var } \phi(t_o) = T\left[\frac{2N_o}{T} + \left(\frac{2N_o}{T}\right)^2\right]\int_{-\infty}^{\infty} |\Omega(f)|^2 \, df \tag{7.13}$$

The input signal power is again $\frac{1}{2}$ volts² in this case, so we let $R = T/4N_o$ and using the noise bandwidth definition, the phase-jitter performance of the squarer circuit is

$$\text{Var } \phi = BT\left[\frac{1}{2R} + \frac{1}{4R^2}\right] \tag{7.14}$$

Comparison with equation (7.6) shows that, at high signal-to-noise ratio, the performance is equivalent to that for the unmodulated carrier case. The second term in equation (7.14) introduces an additional phase jitter referred to as the *squaring loss* [Lindsey and Simon, 1973] in this type of carrier recovery circuit. The rms phase jitter for the two cases is compared in Fig. 7.5 for a specific value of the relative bandwidth of the output BPF.

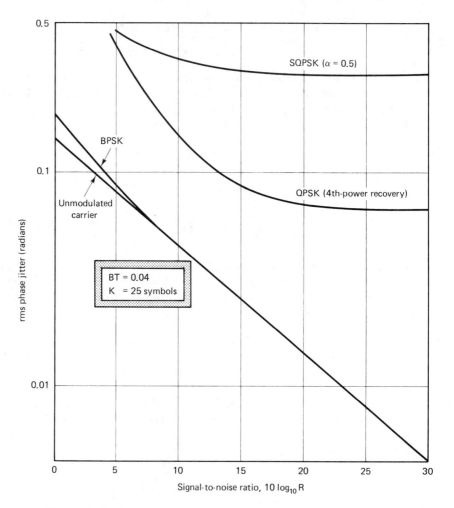

Figure 7.5 Carrier-phase-recovery performance for a fixed recovery bandwidth.

7.2.4 Costas Phase Tracking Loop for Suppressed Carrier Signals

A tracking loop implementation of this recovery scheme is obtained by using the product of the I and Q components of the signal to drive the VCO as shown in Fig. 7.6. This configuration is called a *Costas loop* [Stiffler, 1971; Lindsey and Simon, 1973] and its performance is identical to that of a squarer followed by an ordinary PLL in place of the output BPF in Fig. 7.4. The LPFs in the I and Q arms of the Costas loop can take the place of the input BPF, although as a practical matter it may be advisable to limit the input spectrum to some degree to avoid overloading the demodulators. When the loop filter is omitted, $F(f) = 1$, the Costas loop performs like a single-resonator output BPF.

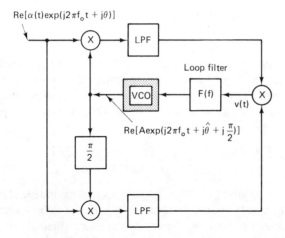

Re[$\alpha(t)$exp(j2$\pi f_o t + j\theta$)]

LPF

Loop filter

VCO   F(f)   X   v(t)

Re[Aexp(j2$\pi f_o t + j\hat{\theta} + j\frac{\pi}{2}$)]

$\frac{\pi}{2}$

LPF

Figure 7.6 Costas phase-tracking loop for suppressed-carrier signals.

Next we consider the two-dimensional modulation scheme of quadrature amplitude modulation (QAM), where both I and Q components of the carrier are modulated with a baseband signal. Assuming that the baseband signals are independent PAM signals, we have

$$\alpha(t) = a(t) + u_I(t) + jb(t) + ju_Q(t)$$

$$a(t) = \sum_k a_k g_I(t - kT)$$

$$b(t) = \sum_k b_k g_Q(t - kT)$$

(7.15)

Normally, we have $g_I(t) = g_Q(t) = g(t)$; however, the case of staggered QAM (SQAM), where $g_Q(t) = g_I(t - T/2)$, is of interest and will be considered later. A popular format for data transmission by terrestrial microwave radio is four-level data in each channel (e.g., a_k, b_k = ± 1, ± 3), giving a data rate of $4/T$ bits per second. Our interest here centers on satellite communication, where QPSK signaling, a_k, b_k = ± 1, is employed. However, the jitter expressions are just as easily derived for the more general case, so we present them in that manner here. Specific results for the QPSK case assume that the four phase states are equiprobable and independent.

The square-law recovery circuit fails in the QPSK case because the mean-square signal corresponding to (7.8) becomes

$$E[z^2(t)] = \tfrac{1}{2} E[a^2(t) + b^2(t)]$$

$$+ \tfrac{1}{2} \text{Re} \left[E\{a^2(t) - b^2(t) + j2a(t)b(t)\}e^{j2\theta} e^{j4\pi f_o t} \right]$$

(7.16)

and the periodic term vanishes because of the balanced statistical nature of the I and Q signal components. One can intentionally unbalance the QAM signal to make carrier recovery possible, but this leads to inefficient signal formats. In packet data transmission, a part of the *preamble sequence* can be devoted to carrier recovery by making $b(t) = 0$, thus unbalancing the signal.

7.2.5 Fourth-Law Carrier Recovery for QPSK and QAM Signals

As an alternative, we consider continuous carrier-phase tracking of the QPSK signal by changing the nonlinearity in Fig. 7.2 to a fourth-power device and tuning the output BPF to $4f_o$, as shown in Fig. 7.7. Thus only the filtered components of $\alpha^4(t)e^{j4\theta}$ appear at the output of the BPF. Letting $4\phi = 4\hat{\theta} - 4\theta$ and using the previous small-jitter approximations, we get

$$\phi(t) = \frac{\omega \otimes [(a + u_I)^3(b + u_Q) - (a + u_I)(b + u_Q)^3]}{\omega \otimes E[a^4 + b^4 - 6a^2b^2]} \qquad (7.17)$$

Notice that the numerator of equation (7.17) now contains a noise-independent term given by the filtered version of $a^3b - ab^3 = ab(a^2 - b^2)$. In the infinite bandwidth case where $g(t)$ is a rectangular pulse of T-seconds duration, then $a^2 = b^2 = 1$ and this term vanishes, but of course the noise terms would blow up in this case. With bandlimiting, the noise-independent term does not vanish and it gives rise to a jitter component called *pattern jitter* (because phase error depends on the data sequence) or *self-noise* of the synchronizer [Gardner, 1980]. Calculation of the variance of the phase error is a very tedious process because of the many higher-order moments involved. We present here the results for the minimum-bandwidth case, where $g(t)$ = sinc (t/T). Signal power = 1 volt^2 in this case, so we let $R = T/2N_o$ and the variance of phase error is given by

$$\text{Var } \phi = BT \left[0.1125 + 1.4625\frac{1}{R} + 24.469\frac{1}{R^2} + 21.094\frac{1}{R^3} + 2.531\frac{1}{R^4} \right] \qquad (7.18)$$

In addition to the pattern-jitter term, this expression shows the *quadrupling loss* inherent to the fourth-power recovery circuit. A graph of this expression is shown for comparison in Fig. 7.5. An interesting variation on the fourth-power recovery circuit, which exhibits improved performance by reducing the occurrences of cycle skipping, has recently been reported [Kurihara et al., 1981].

7.2.6 Extended Costas Loop

An extended version of the Costas tracking loop can be implemented by forming the numerator term in (7.17) with baseband multipliers and squarers to drive the VCO as shown in Fig. 7.8 [Simon, 1978]. In the carrier-recovery schemes considered, there is a phase ambiguity of π radians in the BPSK case and $\pi/2$ radians in the QPSK case, resulting in a polarity ambiguity in the demodulated I and Q signals.

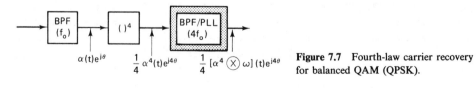

Figure 7.7 Fourth-law carrier recovery for balanced QAM (QPSK).

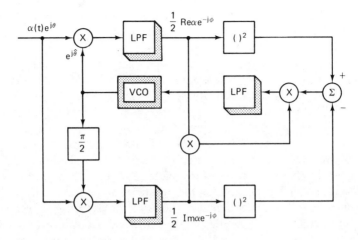

Figure 7.8 Extended Costas loop for QPSK carrier recovery.

This polarity ambiguity may be resolved if the data sequence contains some predetermined segments. Another approach to resolving the ambiguity is to encode the data as phase changes from symbol to symbol rather than as absolute values of phase (i.e., by differential encoding, described in Chapter 4).

7.3 BASEBAND TIMING RECOVERY

7.3.1 Problem Statement

The problem of symbol timing recovery in a baseband PAM signal is to determine the proper sampling instants to extract the data values from

$$x(t) = \sum_m a_m g(t - mT - \tau) \qquad (7.19)$$

with a minimum of intersymbol interference. If $g(t)$ is an even Nyquist pulse, $g(kT) = 0$ for $k \neq 0$, then sampling at $kT + \tau$ produces no intersymbol interference, since $x(kT + \tau) = a_k g(0)$. If the data have a zero mean value, $E[a_k] = 0$, then $E[x(t)]$ has no periodic components. However, $x(t)$ is a cyclostationary process and, assuming independent data symbols, $E[a_k a_j] = 0$ for $k \neq j$, we can use (7.11) to obtain

$$E[x^2(t)] = \frac{\sigma_a^2}{T} \sum_l A_l e^{j(2\pi l/T)(t-\tau)}$$

$$\text{where } A_l = \int_{-\infty}^{\infty} G(f) G\left(\frac{l}{T} - f\right) df, \quad \sigma_a^2 = E[a_k^2] \qquad (7.20)$$

Now if the data pulse has a bandwidth in excess of the minimum (Nyquist) bandwidth of $1/2T$, then $E[x^2(t)]$ has periodically varying components due to the $\ell \neq 0$ terms in (7.20). If the excess bandwidth is less than 100% [i.e., $G(f) = 0$ for $|f| > 1/T$], then $E[x^2(t)]$ contains a sinusoidal component at a frequency of $1/T$ with a phase of $(-2\pi\tau/T)$. An effective timing-recovery circuit can therefore be implemented by the previously discussed squarer/BPF arrangement, where the center frequency of the BPF is now tuned to $1/T$. The zero crossings of the filtered timing wave, $w(t)$, can be used as indicators of the proper sampling instants, as shown in the block diagram of Fig. 7.9. As indicated in this diagram, the data pulse entering the timing-recovery circuit may be either partially or fully equalized and further pulse shaping may be introduced by the prefilter $E(f)$ preceding the square-law nonlinearity. We assume that $g(t)$ represents the data pulse at the prefilter output. For a real, symmetric BPF transfer function there is an offset of $T/4$ in the zero crossings of $w(t)$ from the desired sampling instants, which must be incorporated into the clocking circuitry. If the BPF is replaced by a PLL, the required offset results because the VCO output locks in quadrature with the input. An oscillographic display of the timing wave, $w(t)$, for a random data sequence is shown in Fig. 7.10. For this example, $H(f)$ is a single-resonator BPF with a Q of approximately 50 and a prefilter was employed to reduce the variations in the zero crossings. Selection of the proper prefilter characteristic will be discussed shortly.

7.3.2 Timing Jitter Analysis

Because of the random nature of $x(t)$ and because of additive noise at the input to the timing recovery circuit, the zero crossings of $w(t)$ do not coincide exactly with the regularly spaced zero crossings of $E[w(t)]$, thereby producing a *timing jitter*. Let t_0 denote a zero crossing of the mean timing wave, $E[w(t_0)] = 0$; then assuming that the jitter is relatively small, the actual zero crossing $\hat{\tau}$ can be approximated by

$$\hat{\tau} = t_0 - \frac{w(t_0)}{E[\dot{w}(t_0)]} \qquad \text{where } w(t) = [h \otimes (x + u)^2], \quad \dot{w}(t) = \frac{d}{dt}w(t) \qquad (7.21)$$

The denominator term in the jitter expression can be determined from equation (7.20), letting $H(1/T) = 1$, with the result that

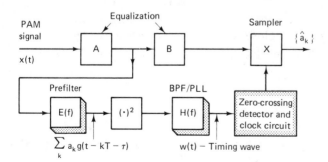

PAM
signal

$x(t)$

Equalization

A

B

Sampler

X $\{\hat{a}_k\}$

Prefilter

$E(f)$ $(\cdot)^2$

$\sum_k a_k g(t - kT - \tau)$

BPF/PLL

$H(f)$

Zero-crossing
detector and
clock circuit

$w(t)$ − Timing wave

Figure 7.9   Squarer/BPF symbol timing recovery for baseband PAM signal.

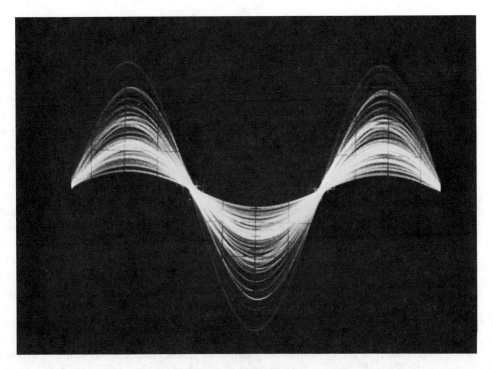

Figure 7.10   Oscillographic display of the timing wave process.

$$E[\dot{w}(t_0)] = \frac{4\pi\sigma_a^2 A_1}{T^2} \tag{7.22}$$

This equation is the primary indicator of the effectiveness of timing recovery by the squarer/filter technique. The size of A_1 in equation (7.20) depends directly on the amount of excess bandwidth of the $g(t)$ pulse since the integrand in (7.20) is equal to the product of $G(f)$ and $G(1/T - f)$. With very small excess bandwidth, A_1 will be small and we can expect poor timing-recovery performance (see Problem 7.7). This phenomenon will be illustrated quantitatively in the later discussion of a similar revovery scheme.

Calculation of the rms timing jitter involves calculation of the variance of the $w(t_0)$ term in (7.21), and this depends jointly on the functions $G(f)$ and $H(f)$. Assuming these functions both to be real, the variance expression takes the form [Franks, 1974]

$$\text{Var } w(t) = V_o + V_2 \cos\left[\frac{4\pi(t - \tau)}{T}\right] \tag{7.23}$$

This expression takes on its minimum value at $t = t_0$ (i.e., at the mean sampling instant). Results of rms jitter calculations for various $G(f)$ and $H(f)$ functions, including the effects of mistuning, have been published in the reference cited above.

It is not possible to give a compact jitter formula in terms of the noise bandwidth of $H(f)$, as in the phase-recovery case, because the spectral components of the pattern jitter terms vanish at $f = 1/T$ [Franks, 1980]. Because of this, the actual rolloff shape of the $H(f)$ characteristic has a major significance in the numerical results. A result of particular interest is that it has been found that proper selection of $G(f)$ and $H(f)$ can increase the V_2 term in (7.23) until it is equal to the V_o term in the absence of noise. This means that it is possible to eliminate the pattern jitter entirely. The conditions for *eliminating the pattern jitter* are that $G(f)$ be a bandpass character- istic symmetric about $1/2T$ with a bandwidth not exceeding $1/2T$ and that $H(f)$ be symmetric about $1/T$. These symmetry conditions are illustrated in Fig. 7.11. Shaping of the $G(f)$ function can be accomplished to the degree desired by the prefiltering function $E(f)$ in Fig. 7.9. Even making $E(f)$ a differentiator can result in substantial improvements in pattern jitter. A $G(f)$ function that is perfectly antisym- metric about $1/2T$ will also eliminate pattern jitter and this condition is sometimes approached with a prefilter characteristic exhibiting a transmission zero at $1/2T$.

At this point we feel it appropriate to interject a warning about possible ap- proaches to analysis of jitter variance. It is often stated that the effect of the squarer is to create a discrete component (line) in the power spectrum at $1/T$. Following up on this, several investigators have proceeded to calculate the continuous part of the spectrum of the squarer output in the vicinity of the line in order to arrive at the rms fluctuations at the output of $H(f)$. The error inherent in this approach is the neglect of the cyclostationary character of the timing wave. From equation (7.23) with $V_2 \simeq V_o$, we see that there is a considerable time variation in the variance of $w(t)$. Using the average value of this function, rather than its minimum value, for jitter calculations leads to erroneous results. Another helpful viewpoint is to express the squarer output in terms of I and Q components relative to a "carrier" frequency of $1/T$ [Gardner, 1980; Franks, 1980]. Then we find that jitter depends primarily on the Q component. Using a power spectral density approach, the distinction between the I and Q components becomes obliterated and the incorrect results are obtained.

7.3.3 Data-aided Timing Recovery

A considerable improvement in timing-recovery performance can be obtained by em- ploying a *data-aided* technique. This can be regarded as a "bootstrap" approach

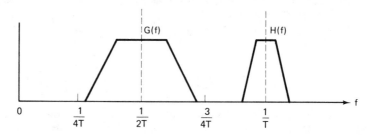

Figure 7.11   Symmetry conditions on filtering functions to eliminate pattern jitter.

because it assumes that the data sequence $\{a_k\}$ is already known prior to timing recovery. However, there are two practical considerations supporting this approach. First, timing recovery may be accomplished during periods when known data sequences are transmitted, such as during part of a preamble of a data packet. Second, continuous tracking with random data can employ this principle if the outputs of the data detectors at the receiver are used in place of the true data values. The rationale here is that the data detected are sufficiently accurate to aid the timing-recovery process substantially. This is the same viewpoint as that used in decision-directed equalization of data channels and, in fact, it has been proposed that equalization and timing recovery could be accomplished jointly [Kobayashi, 1971; Ungerboeck, 1974]. Data-aided recovery of carrier phase has also received considerable attention [Lindsey and Simon, 1971; Matyas and McLane, 1974; Simon and Smith, 1974; Mengali, 1976].

The motivation for the data-aided approach is based on the maximum-likelihood strategy for estimation of the parameter τ in (7.19) based on observations on the signal received, $z(t) = x(t) + u(t)$, where $u(t)$ is white Gaussian noise with a (double-sized) power spectral density of N_o volts²/hertz. Appendix 7.B presents some of the details of the maximum-likelihood estimation strategy. The proposed timing-recovery circuit is based on choosing the value of $\hat{\tau}$ to maximize the quantity

$$\Lambda(\hat{\tau}) = \sum_{k=0}^{K-1} a_k q_k(\hat{\tau}) \qquad \text{where } q_k(\hat{\tau}) = \int_{-\infty}^{\infty} z(t)g(t - kT - \hat{\tau}) \, dt \qquad (7.24)$$

We see that the $q_k(\hat{\tau})$ quantities can be interpreted as the samples (at $t = kT + \tau$) of the response of the matched filter [impulse response $= g(-t)$] to the input PAM signal plus noise. The parameter K specifies how many data symbols are to be used in the estimation process and its value will be related directly to the loop bandwidth of the recovery scheme.

A small-jitter, linearized version of this estimator can be examined by maximizing a three-term Taylor series expansion of $\Lambda(\hat{\tau})$ about an already close estimate, t_o, of the value of τ. If we further approximate $\ddot{\Lambda}(t_o)$ by its mean, noise-free value, the estimate $\hat{\tau}$ is expressed as

$$\hat{\tau} = t_o - \frac{\dot{\Lambda}(t_o)}{E[\ddot{\Lambda}(t_o)]} \qquad (7.25)$$

which should be compared with equation (7.21).

Calculation of rms timing jitter is done by evaluating the variance of $\hat{\tau}$ in equation (7.25) with $t_o = \tau$. This calculation is quite straightforward and the result is (see Problem 7.10)

$$\text{Var} \, (\hat{\tau} - \tau) = \frac{N_o}{K\sigma_a^2[-\ddot{r}(0)]} + \frac{F}{K^2[\ddot{r}(0)]^2} \qquad (7.26)$$

where

$$r(t) \overset{\Delta}{=} \int_{-\infty}^{\infty} g(t + s)g(s) \, ds \qquad (7.27)$$

$$F \overset{\Delta}{=} \sum_{k=0}^{K} \sum_{m \in K'} \ddot{r}^2(mT - kT) \qquad (7.28)$$

The pulse $r(t)$ in (7.27) is the time-ambiguity function [Franks, 1969] of the data pulse $g(t)$. Note that $R(f) = |G(f)|^2$ is the energy spectrum of the data pulse. Alternatively, $r(t)$ can be viewed as the response of the matched filter to a single data pulse. We also see that $\ddot{r}(0) = -\int_{-\infty}^{\infty} \dot{g}^2(t)\, dt$, so that the denominator terms in equation (7.26) are given by the energy in the time derivative of the data pulse. It is not surprising that a data pulse with large slope values should provide good timing recovery.

The second term in equation (7.26) represents the pattern jitter component and it is directly attributable to the implementation approximation described in Appendix 7.B. In a true maximum-likelihood estimator, this term would not be present. The term F depends only on the properties of $\ddot{r}(t)$, and it tends to increase very slowly for large K, so that the overall behavior of the second term in (7.26) is approximately as $1/K^2$. The notation $m \in K'$ means that the sum is taken over all indices for which $k < 0$ and $k \geq K$ (i.e., just the terms not used in the other sum).

A tracking-loop implementation of the data-aided estimator results by using the values of $\dot{\Lambda}(\tau_m)$, omitting the K-term summation, to drive a voltage-controlled clock (VCC) as shown in Fig. 7.12. The summation is provided, in effect, by the integrating action of the loop. The VCC phase is updated every T seconds by an amount proportional to $a_m \dot{q}_m(\tau_m)$. A linearized, small-jitter approximation leads to a first-order difference equation for time jitter.

$$\tau_{m+1} - \tau_m = \mu A \tau_m + u_m \qquad (7.29)$$

where $A = -E[a_m \ddot{q}_m]$ and μ is a step-size parameter which determines the loop bandwidth. We can regard the u_m as essentially independent random noise variables driving the loop. Solution of equation (7.29) gives

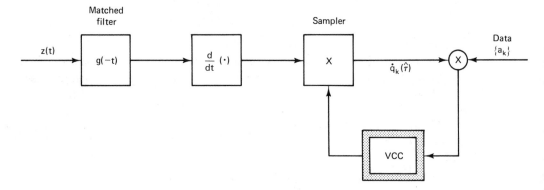

Figure 7.12   Data-aided baseband timing recovery loop.

$$\tau_m = \sum_{k=0}^{\infty} \beta^k u_{m-k} \tag{7.30}$$

with $\beta = 1 - \mu A \simeq 1$. Using standard discrete-time filtering theory, the noise bandwidth of this operation is

$$B = \frac{1}{T}\left(\frac{1-\beta}{1+\beta}\right) \simeq \frac{\mu A}{2T} = \frac{-\mu \sigma_a^2 \ddot{r}(0)}{2T} \tag{7.31}$$

On the other hand, the jitter calculation (7.26) used a filtering operation where the "exponential" weighting terms in equation (7.30) are merely replaced by constant weights which are unequal to zero for only K terms (see Problem 7.2). The noise bandwidth of a discrete-time filter of this nature is given by $B = 1/KT$. Using this correspondence, equation (7.26) also characterizes the performance of a tracking loop. Notice that the noise-dependent part of the timing jitter is proportional to loop bandwidth, whereas the pattern jitter increases roughly as the square of the loop bandwidth. The latter fact can be related to the earlier mentioned property that the spectral components of the pattern jitter vanish at $1/T$.

7.3.4 Maximum-likelihood Estimation

Maximum-likelihood estimation also provides us another viewpoint on non-data-aided estimation. If we replace a_k in equation (7.24) by $q_k(\hat{\tau})$, which would be a good approximation under ideal conditions, we can maximize the quantity

$$\Lambda(\hat{\tau}) = \frac{1}{2}\sum_{k=0}^{K-1} q_k^2(\hat{\tau}) \tag{7.32}$$

and proceed with the same approach as that used in the analysis of the data-aided case. The calculations are more complex for this case, but some simplifications result if we assume that $r(t)$ is a Nyquist pulse [i.e., $r(kT) = 0$ for $k \neq 0$]. From equation (7.20), the average power in the PAM signal is equal to $\sigma_a^2 r(0)/T$, so we let the signal-to-noise ratio parameter be $R = \sigma_a^2 r(0)/T$, where $r(0)$ is the energy of a single data pulse. Again assuming an excess pulse bandwidth not exceeding 100%, the calculations give

$$\text{Var}\, \frac{\hat{\tau} - \tau}{T} = \frac{4(1/2\pi)^4 [r(0)/T]^2 F[(1+1/R)^2]}{K^2 [C_1/T^3]^2} \tag{7.33}$$

$$+ \frac{2(1/2\pi)^2 [r(0)/T]^2 [1/R + 2/R^2]}{K[C_1/T^3]}$$

where F is the same term defined in equation (7.28). The terms in (7.33) are grouped so that each of the bracketed terms is dimensionless, so that the formula can be easily applied for an arbitrary data rate. The C_1 term corresponds to the A_1 term (7.20), but for the $r(t)$ pulse; that is,

$$C_1 \overset{\Delta}{=} \int_{-\infty}^{\infty} R(f) \, R\left(\frac{1}{T} - f\right) df \qquad (7.34)$$

Thus, as in the previously discussed non-data-aided approach using the squarer/filter circuit, the performance is strongly dependent on excess bandwidth and deteriorates rapidly as excess bandwidth approaches zero.

7.3.5 Performance Comparison of Synchronization Circuits

For a specific illustration of these results and a comparison with the data-aided case, we chose

$$r(t) = T \cos \frac{\pi \alpha t}{T} \, \mathrm{sinc} \, \frac{t}{T} \qquad (7.35)$$

The corresponding $R(f)$, the energy spectrum of $g(t)$, is shown in Fig. 7.13. This shape is chosen because it clearly maximizes the value of C_1 for any Nyquist shape with a monotonically decreasing roll-off and a relative excess bandwidth factor of α. For this pulse shape, the data-aided jitter performance (7.26) becomes

$$\mathrm{Var} \frac{\hat{\tau} - \tau}{T} = \frac{3}{\pi^2 K (1 + 3\alpha^2)} \left[\frac{1}{R}\right] + \frac{9 \, F}{\pi^4 K^2 (1 + 3\alpha^2)^2} \qquad (7.36)$$

and for the non-data-aided case, equation (7.33) becomes

$$\mathrm{Var} \frac{\hat{\tau} - \tau}{T} = \frac{2}{\pi^2 K \alpha} \left[\frac{1}{R} + \frac{2}{R^2}\right] + \frac{4 \, F}{\pi^4 K^2 \alpha^2} \qquad (7.37)$$

Comparing equations (7.36) and (7.37), a squaring-loss effect is apparent; however, the primary difference is due to the $1/\alpha$ behavior of the noise-dependent jitter and the $1/\alpha^2$ behavior of the pattern jitter in the non-data-aided case. The dependence of rms jitter on signal-to-noise ratio of these timing-recovery schemes is illustrated in Fig. 7.14. The dependence upon the excess bandwidth factor is illustrated in Fig.

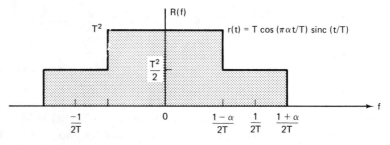

Figure 7.13 Energy spectrum of data pulses with good timing recovery properties. α is the excess-bandwidth factor.

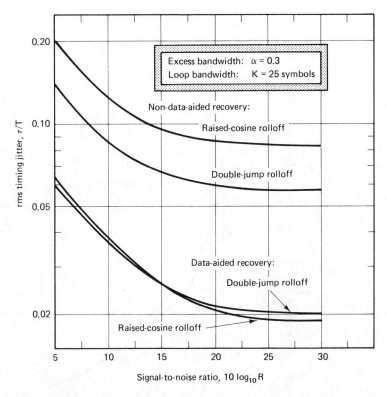

Figure 7.14   Baseband symbol timing-recovery performance for fixed signal and recovery bandwidths.

7.15. The effects of changes in data pulse shape are also shown in Fig. 7.15, where rms jitter resulting from the popular "raised-cosine" frequency roll-off function is presented. The $r(t)$ pulse for this case is given by

$$r(t) = T \frac{\cos(\pi \alpha t/T)}{1 - (2\alpha t/T)^2} \text{ sinc } \frac{t}{T} \qquad (7.38)$$

The jitter performance of the pulse having the "double-jump" spectral roll-off of Fig. 7.13 is substantially better, due primarily to a C_1 value being larger by a factor of 2 in comparison with the raised-cosine roll-off case. The double-jump spectral roll-off pulse has another advantage in that it is close to the Nyquist pulse which exhibits the maximum tolerance to small offsets from the correct sampling instants [Franks, 1968].

A further comparison of the effect of pulse shape is included in Fig. 7.15 by showing results for non-data-aided baseband timing jitter for a pulse having a trapezoidal energy spectrum. In this case,

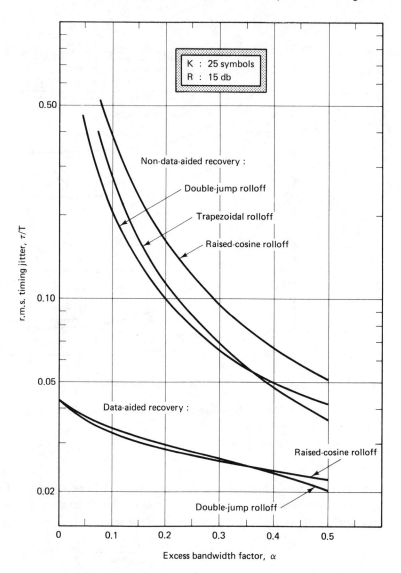

Figure 7.15 Baseband symbol timing-recovery performance for fixed signal-to-noise ratio and recovery bandwidth.

$$r(t) = T \operatorname{sinc} \frac{\alpha t}{T} \operatorname{sinc} \frac{t}{T} \qquad (7.39)$$

As might be expected from the nature of the spectral roll-off shape, the jitter results in the trapezoidal case are intermediate to those of the double-jump and raised-cosine cases.

7.3.6 Implementation of Non-data-aided Tracking Loops

Implementation of the non-data-aided tracking loop is developed along the same lines as discussed for the data-aided case, but using a $\Lambda(\hat{\tau})$ function given by equation (7.32) instead of (7.24). This results in the double-arm structure shown in Fig. 7.16, with a remarkable resemblance to the Costas loop. As with the Costas loop, there is an equivalent configuration involving a squarer and an ordinary PLL, and this equivalent non-data-aided tracking scheme is shown in Fig. 7.17. For binary ± 1 data, the non-data-aided log-likelihood function is more accurately expressed as (see Appendix 7.B)

$$\Lambda(\hat{\tau}) = \sum_{k=0}^{K-1} \ln \cosh \frac{q_k(\hat{\tau})}{N_o} \tag{7.40}$$

and since the quantity used to update the VCC phase is proportional to a single term in $\dot{\Lambda}(\hat{\tau})$, the VCC should be driven by the sequence, $[\dot{q}_k(\hat{\tau})/N_o] \times [\tanh(q_k(\hat{\tau})/N_o)]$. This means that the upper arm in the configuration shown in Fig. 7.16 should contain a tanh $(\cdot)$ nonlinearity [Gitlin and Salz, 1971]. Often the tanh $(\cdot)$ nonlinearity is replaced by a sgn $(\cdot)$ nonlinearity for ease of implementation. Not only is the sgn $(\cdot)$ nonlinearity easy to realize, but with it the multiplier is also much simpler, being merely a polarity-reversal gate.

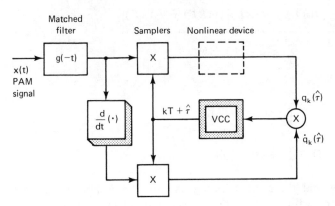

Matched filter   Samplers   Nonlinear device

Figure 7.16 Non-data-aided timing loop. The nonlinear device may be inserted in the upper arm to give a closer approximation to ML estimation, or to simplify implementation.

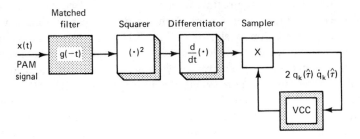

Figure 7.17 Alternative implementation of non-data-aided timing-recovery loop.

7.4 JOINT RECOVERY OF CARRIER PHASE AND SYMBOL TIMING

7.4.1 Problem Statement and Analysis

Instead of using a two-step process where θ is estimated first, then τ is extracted from the demodulated baseband signal, we can use a simultaneous estimation process for θ and τ. The joint estimates certainly will not be worse than individual estimates, and we consider first a particular case which illustrates the joint estimation can be very much superior. For this, we take the example of QAM (QPSK) phase estimation using the square-law nonlinearity, where we assume additionally that the exact symbol timing is known. From equation (7.16) we saw that this type of estimation fails because the denominator vanishes in the balanced QAM case. It also vanishes, for the regular QAM format, if it is observed only at particular time instants. However, if one of the baseband PAM signals, I or Q, is offset in time, then summing up samples taken every T seconds gives a nonzero result, and a phase measurement can be made. The optimal amount of time offset is one-half a symbol period, and the resulting signal is called *staggered* QAM (SQAM). The quadriphase versions are called SQPSK or OQPSK (for *offset* QPSK).

Suppose that we take as our phase estimate

$$\phi = \frac{\sum_{k=0}^{K-1} [a(kT) + u_I(kT)] [b(kT) + u_Q(kT)]}{\sum_{k=0}^{K-1} E[a^2(kT) - b^2(kT)]} \tag{7.41}$$

where

$$a(t) = \sum_m a_m g(t - mT)$$

$$b(t) = \sum_m b_m g\left(t - mT - \frac{T}{2}\right) \tag{7.42}$$

In (7.42) we are assuming that the symbol timing parameter τ is known, and we have arbitrarily assigned the value $\tau = 0$. The phase error (7.41) corresponds to (7.9) with the continuous-time filtering operation of convolution with $\omega(t)$, replaced by the discrete-time operation of summing over the past K sample values. The tracking loop implementation is shown in Fig. 7.18. Of course, the samplers in each arm could be replaced by a single sampler at the output of the baseband multiplier. Some of the references on phase recovery show the LPFs in the arms as "integrate-and-dump" filters which approximate the action of a matched filter for the infinite band-width case of a T-second rectangular $g(t)$ pulse. These are sometimes referred to as (although not by circuit theorists) as "active" arm filters. In any case, the effect of this kind of arm operation is equivalent to the appropriate time-invariant LPF transfer function followed by a sampler.

318

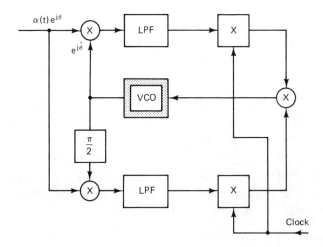

Figure 7.18 SQPSK carrier recovery using Costas loop with synchronized arm samplers.

We evaluate the performance of the phase estimator of equation (7.41) by first using (7.11) to evaluate the denominator term. We get

$$\sum_{k=0}^{K-1} E[a^2(kT) - b^2(kT)] = \frac{4K\sigma_a^2}{T} A_1$$

$$\text{where } A_1 = \int_{-\infty}^{\infty} G(f)G\left(\frac{1}{T} - f\right) df \qquad (7.43)$$

Then calculation of the variance of the numerator term in equation (7.41) leads directly to the variance expression for phase jitter. We present the results here for the particular case of a $G(f)$ having the same shape as $R(f)$ shown in Fig. 7.13. To calculate the noise terms, we assume that the LPFs are flat with a bandwidth of $(1 + \alpha)/2T$, but we use the same signal-to-noise ratio parameter $R = T/2N_o$ for comparison with the other QPSK results. The jitter variance for SQPSK ($\sigma_a^2 = \sigma_b^2 = 1$) is

$$\text{Var } \phi = \frac{1}{K\alpha^2}\left[(1 - \alpha) + 2\left(\frac{1}{R}\right) + (1 + \alpha)\left(\frac{1}{R}\right)^2\right]$$

$$\text{with } g(t) = \cos\left(\pi \alpha t/T\right) \text{ sinc } \frac{t}{T} \qquad (7.44)$$

This is compared with the other phase estimation results in Fig. 7.5. We notice that excess bandwidth is required for this scheme to work and that the situation deteriorates very rapidly as the excess bandwidth factor α approaches zero. For a raised-cosine spectral roll-off shape, the variance values would be approximately doubled because of the reduced size of the A_1 term.

Now we consider joint estimation of θ and τ based on a maximum-likelihood strategy. From equation (7B.7) for the data-aided QPSK case, we want to find $\hat{\theta}$ and $\hat{\tau}$ which simultaneously maximize

$$\Lambda(\hat{\theta}, \hat{\tau}) = \sum_{k=0}^{K-1} a_k q_k(\hat{\theta}, \hat{\tau}) + b_k p_k(\hat{\theta}, \hat{\tau}) \tag{7.45}$$

where

$$q_k(\hat{\theta}, \hat{\tau}) = \operatorname{Re}\left[e^{-j\hat{\theta}} \int_{-\infty}^{\infty} \alpha(t) g(t - kT - \hat{\tau}) \, dt \right]$$

$$p_k(\hat{\theta}, \hat{\tau}) = \operatorname{Im}\left[e^{-j\hat{\theta}} \int_{-\infty}^{\infty} \alpha(t) g(t - kT - \hat{\tau}) \, dt \right] \tag{7.46}$$

We can assume, without loss of generality, that the true values are $\theta = 0$ and $\tau = 0$. Then maximizing a Taylor series expansion, as in equation (7.25), our estimates are the simultaneous solutions of

$$\Lambda_{\theta\theta}(0, 0)\, \hat{\theta} + \Lambda_{\theta\tau}(0, 0)\, \hat{\tau} = -\Lambda_{\theta}(0, 0)$$

$$\Lambda_{\theta\tau}(0, 0)\, \hat{\theta} + \Lambda_{\tau\tau}(0, 0)\, \hat{\tau} = -\Lambda_{\tau}(0, 0) \tag{7.47}$$

where the subscripts on Λ denote partial derivatives with respect to $\hat{\theta}$ and $\hat{\tau}$. Now replacing the left-hand-side coefficients in (7.47) by their mean values, we get

$$\hat{\theta} = \frac{-\Lambda_{\theta}(0, 0)}{E\Lambda_{\theta\theta}(0, 0)} \qquad \hat{\tau} = \frac{-\Lambda_{\tau}(0, 0)}{E\Lambda_{\tau\tau}(0, 0)} \tag{7.48}$$

because $E\Lambda_{\theta\tau}(0, 0) = 0$ for the QAM format.* Calculating the other expectations, assuming that $\sigma_a^2 = \sigma_b^2 = 1$,

$$E\Lambda_{\theta\theta}(0, 0) = -2Kr(0)$$

$$E\Lambda_{\tau\tau}(0, 0) = +2K\ddot{r}(0) \tag{7.49}$$

where $r(t)$ is the time-ambiguity function of the data pulse as defined in equation (7.27). Then (7.48) becomes

$$\hat{\theta} = \frac{1}{2Kr(0)} \sum_{k=0}^{K-1} a_k \operatorname{Re}\left[-j \int_{-\infty}^{\infty} \alpha(t) g(t - kT) \, dt \right]$$

$$+ b_k \operatorname{Re}\left[-\int_{-\infty}^{\infty} \alpha(t) g(t - kT) \, dt \right]$$

$$\hat{\tau} = \frac{1}{2K\ddot{r}(0)} \sum_{k=0}^{K-1} a_k \operatorname{Re}\left[\int_{-\infty}^{\infty} \alpha(t) \dot{g}(t - kT) \, dt \right] \tag{7.50}$$

$$+ b_k \operatorname{Re}\left[-j \int_{-\infty}^{\infty} \alpha(t) \dot{g}(t - kT) \, dt \right]$$

---

* For an SSB/PAM carrier signal, this coupling term in estimation of θ and τ does not vanish and for rapid acquisition of these parameters, it is important that the tracking loops are designed to incorporate the coupling effect [Meyers and Franks, 1980; Mengali, 1977; Mancianti et al., 1979].

7.4.2 Performance of Joint Recovery QPSK Synchronizers

For the QPSK case, we assume independent, binary PAM signals for the I and Q components of $\alpha(t)$, and a signal-to-noise ratio parameter $R = r(0)/2N_o$, then the rms jitter performance for the joint recovery, data-aided QPSK case is calculated to be

$$\mathrm{Var}\ \phi = \frac{1}{K}\left(\frac{1}{2R}\right) \tag{7.51}$$

$$\mathrm{Var}\ \frac{\tau}{T} = \frac{r(0)}{K[-T^2\ddot{r}(0)]}\left(\frac{1}{2R}\right) + \frac{F}{2K^2[T\ddot{r}(0)]^2} \tag{7.52}$$

Notice that the phase-jitter term has no pattern-jitter component or squaring-loss term. The timing jitter term is practically identical to the data-aided baseband timing recovery jitter. For the particular data pulse with an excess bandwidth factor of α and an energy spectrum as shown in Fig. 7.13, the timing jitter expression (7.52) becomes

$$\mathrm{Var}\ \frac{\tau}{T} = \frac{3/2\pi^2}{K(1+3\alpha^2)}\left(\frac{1}{2R}\right) + \frac{(3/\pi^2)^2 F}{K^2(1+3\alpha^2)^2} \tag{7.53}$$

The jitter performance versus signal-to-noise ratio is illustrated in Fig. 7.19. The effect of excess bandwidth on the timing jitter can be inferred from the results for data-aided baseband recovery shown in Fig. 7.15.

A tracking loop implementation of the joint recovery scheme follows directly from an examination of a single term in $\Lambda_\theta(\hat{\theta}, \hat{\tau})$ and $\Lambda_\tau(\hat{\theta}, \hat{\tau})$. Thus the VCO phase is updated every T seconds by an amount proportional to

$$v_1(kT) = a_k \frac{\partial}{\partial\hat{\theta}}\ q_k(\hat{\theta}, \hat{\tau}) + b_k \frac{\partial}{\partial\hat{\theta}}\ p_k(\hat{\theta}, \hat{\tau})$$

$$= a_k\ \mathrm{Re}\left[-je^{-j\hat{\theta}}\int_{-\infty}^{\infty}\alpha(t)g(t - kT - \hat{\tau})\ dt\right] \tag{7.54}$$

$$+ b_k\ \mathrm{Re}\left[-e^{-j\hat{\theta}}\int_{-\infty}^{\infty}\alpha(t)g(t - kT - \hat{\tau})\ dt\right]$$

The quantity $\mathrm{Re}\left[-je^{-j\hat{\theta}}\int_{-\infty}^{\infty}\alpha(t)g(t - kT - \hat{\tau})\ dt\right]$ is interpreted as the response of the bandpass matched filter for the data pulse $g(t)$ demodulated against a carrier phase of $\hat{\theta} + \pi/2$ and then sampled at $kT + \hat{\tau}$. This is the sample that appears at point 1 in Fig. 7.20 since the VCO locks in phase quadrature to the input carrier. This gets multiplied by a_k to form the first term in $v_1(kT)$ driving the VCO. The second term is the product of b_k and the demodulated and sampled signal at point 2. The VCC phase is updated by an amount proportional to

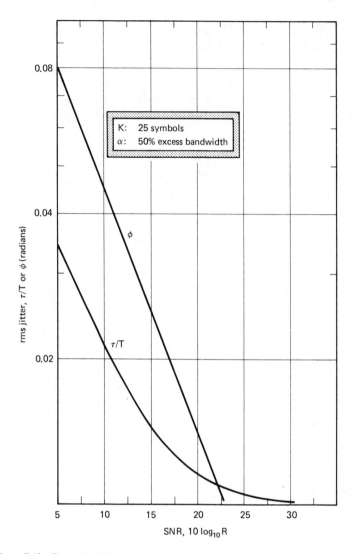

Figure 7.19   Data-aided joint recovery of carrier phase and symbol timing for QPSK.

$$v_2(kT) = a_k \frac{\partial}{\partial \hat{\tau}} q_k(\hat{\theta}, \hat{\tau}) + b_k \frac{\partial}{\partial \hat{\tau}} p_k(\hat{\theta}, \hat{\tau})$$

$$= a_k \; \mathrm{Re} \left[-\epsilon^{-j\hat{\theta}} \int_{-\infty}^{\infty} \alpha(t)\dot{g}(t - kT - \hat{\tau}) \, dt \right] \qquad (7.55)$$

$$+ b_k \; \mathrm{Re} \left[j\epsilon^{-j\hat{\theta}} \int_{-\infty}^{\infty} \alpha(t)\dot{g}(t - kT - \hat{\tau}) \, dt \right]$$

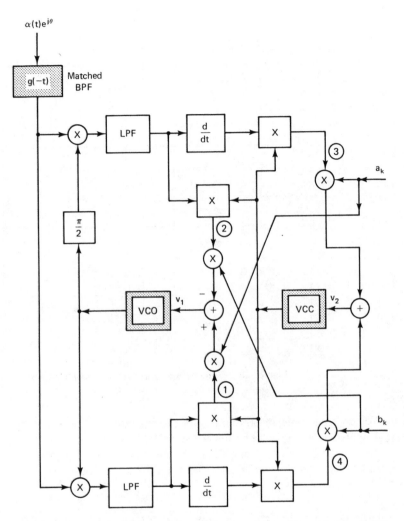

Figure 7.20 Joint carrier phase and symbol timing recovery (data-aided) for QAM (QPSK).

So we see that the VCC control signal is obtained by sampling the time-differentiated demodulated signals, then multiplying these samples at points 3 and 4 by the data values to form the $v_2(kT)$ control signal.

7.4.3 Demodulation–Remodulation Recovery Methods

At this point it is appropriate to discuss briefly another popular phase-recovery circuit for the QPSK format [Lundquist et al., 1974; Weber and Alem, 1980a, b]. It is called a *demod–remod* tracking scheme and the configuration is shown in Fig. 7.21.

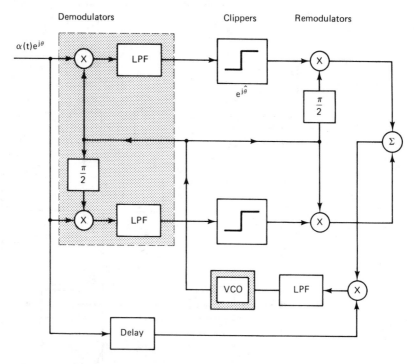

Figure 7.21   Demod–remod phase-recovery loop for QPSK.

We can gain some understanding of the operation of the tracking loop by looking at approximate expressions for the signal $v(t)$ driving the VCO. Let us assume that the arm LPFs and the loop filter do not alter baseband signal components; then if the clipping devices in the arms were removed, we would get

$$v(t) = \tfrac{1}{2} \, \text{Im} \, [(a + jb)^2 e^{-j2\phi}] \tag{7.56}$$

where $\phi = \hat{\theta} - \theta$. This is the same control signal that results in the ordinary Costas loop in Fig. 7.6. On the other hand, we might suppose that under ideal conditions, the phase error is small, so that the clipper outputs correspond to clipped versions of the baseband binary PAM I and Q signals, $a(t)$ and $b(t)$, respectively. Now if we assume that the clippers perform effectively as data detectors [i.e., sgn $a(t) = a_k$ and sgn $b(t) = b_k$ for t in the kth symbol interval], we would get

$$\begin{aligned} v(kT) \simeq \tfrac{1}{4} \, a_k \, b(kT) &- \tfrac{1}{4} \, b_k \, a(kT) \\ &+ \tfrac{1}{4} \, a_k \, \text{Re} \, [je^{j2\hat{\theta}} \{a(kT) + jb(kT)\}] \\ &- \tfrac{1}{4} \, b_k \, \text{Re} \, [e^{j2\hat{\theta}} \{a(kT) + jb(kT)\}] \end{aligned} \tag{7.57}$$

Comparing the $\hat{\theta}$-dependent terms in equation (7.57) with the terms in (7.54), we see that under these idealized conditions, the demod–remod tracking loop should perform like the phase-recovery part of the data-aided QPSK tracking loop. A detailed performance analysis of the demod–remod loop has been published [Weber and Alem, 1980a, b].

PROBLEMS

7.1. Determine the complex envelope relative to f_o of the real signal

$$x(t) = \cos(2\pi f_o t) + \sin(2\pi f_o t + 2\pi f_1 t)$$

Does your answer depend on the size of f_1, compared to f_o? Consider positive and negative values of f_1 separately.

7.2. Determine the (complex) impulse response, $\omega(t)$, of the low-pass equivalent (relative to f_o) for a bandpass filter, $H(f)$, of the following forms:

(a) A rectangular passband, with $H(f) = 1$ for $|(|f| - f_o)| \leq B$ and $H(f) = 0$ otherwise.

(b) A single-resonator BPF, with

$$H(f) = \frac{j(f_o/Q)f}{f_o^2 - f^2 + j(f_o/Q)f}$$

Let us simplify this problem by considering only the narrowband case with a Q factor $\gg 1$. Then show that

$$H(f) \approx \frac{(f_o/2Q)}{j(f - f_o) + (f_o/2Q)} \qquad \text{for } f > 0$$

and consequently, the resulting $\omega(t)$ is approximately a real exponential function.

(c) A BPF that can be approximated, for $f > 0$, by

$$H(f) = e^{-jx}\frac{\sin x}{x} \qquad \text{where } x = \frac{\pi}{B}(f - f_o)$$

7.3. Suppose that each bandpass filter in Problem 7.2 has a white noise input, with a double-sided spectral density of N_o volts²/hertz. Determine the rms value of the noise at the filter output. What is the "noise bandwidth" of each of these filters?

7.4. Consider the product of two statistically independent complex stationary random processes, $\alpha(t)$ and $\beta(t)$, which is filtered by a (complex) low-pass filter, $\omega(t)$, to produce $\gamma(t)$ (i.e., $\gamma = \omega \otimes \alpha\beta$). Verify the following formulas for the mean-square value of γ. (Note the independence on t_o).

$$E[|\gamma(t_o)|^2] = \int\!\!\int_{-\infty}^{\infty} \omega(t)\omega^*(t + \tau)k_{\alpha\alpha}(\tau)k_{\beta\beta}(\tau)\, d\tau\, dt$$

$$= \int\!\!\int_{-\infty}^{\infty} |\Omega(f)|^2 K_{\alpha\alpha}(f - v)K_{\beta\beta}(v)\, dv\, df$$

where $k_{\alpha\alpha}(\tau)$ and $k_{\beta\beta}(\tau)$ are autocorrelation functions as defined in equation (7A.11). Now suppose that $\Omega(f)$ has a substantially narrower bandwidth than either $K_{\alpha\alpha}(f)$ or $K_{\beta\beta}(f)$; then show that

$$E[|\gamma(t_o)|^2] \simeq \left[\int_{-\infty}^{\infty} |\Omega(f)|^2 \, df\right] \left[\int_{-\infty}^{\infty} K_{\alpha\alpha}(-f)K_{\beta\beta}(f) \, df\right]$$

These formulas are very helpful for calculating rms values at the output of the squarer/ BPF timing- or phase-recovery circuits.

7.5. Prove the *Poisson sum formula,* which states that

$$\sum_{k=-\infty}^{\infty} s(t - kT) = \frac{1}{T} \sum_{l=-\infty}^{\infty} S\left(\frac{l}{T}\right) e^{j2\pi l t / T}$$

[*Hint:* The left-hand side of the equation is clearly a periodic function of t (with period T). Make a Fourier series expansion, relating the Fourier coefficients to the Fourier transform of $s(t)$. Use this result to verify the result stated in equation (7.11).]

7.6. In the BPSK carrier recovery scheme using the square-law device followed by the BPF centered at $2f_o$, let us consider a signal having a bandwidth in excess of the Nyquist bandwidth. Specifically, let $g(t) = \text{sinc}^2\,(t/T)$ in equation (7.11) and assume that the input BPF (Fig. 7.4) is rectangular with a bandwidth of $2/T$ (twice the Nyquist bandwidth). Evaluate the mean-squared phase jitter corresponding to equation (7.14) for this case. Compare the squaring-loss terms in the two cases. Note that the mean-squared phase jitter is no longer time independent. What is the implication of this in designing phase recovery circuits?

7.7. Consider the baseband PAM signal in equation (7.19) with independent, zero-mean, data symbols. Calculate the mean-square value of this signal (which is the same as the mean timing wave for the squarer/BPF timing recovery circuit) as a function of the excess bandwidth (roll-off factor, α) for the following types of Nyquist data pulse (assume that $\alpha \leq 1$):

(a) $g(t) = \cos\,(\alpha t/T)\,\text{sinc}\,(t/T)$

(b) $g(t) = \dfrac{\cos\,(\pi\alpha t/T)}{1 - (2\alpha t/T)^2}\,\text{sinc}\,(t/T)$

(c) $g(t) = \text{sinc}\,(\alpha t/T)\,\text{sinc}\,(t/T)$

Show that the mean-square value is a constant plus a sinusoidal term. [*Hint:* Notice that the spectral roll-off shapes corresponding to the pulses above are (a) double-jump (see Fig. 7.13); (b) raised-cosine; (c) trapezoidal.]

7.8. Consider the squarer/BPF timing recovery circuit for baseband PAM timing recovery when $g(t)$ is a T-second duration rectangular pulse. This is sometimes called NRZ (non-return-to-zero) pulse signaling. What is the mean timing wave in this case? Is this a suitable timing-recovery scheme? Suppose that the NRZ signal is differentiated before it enters the squaring device. Describe qualitatively the behavior of the timing recovery, assuming pulses with very short rise times.

7.9. In the situation of Problem 7.8, assume that the data pulses entering the squaring device are rectangular, with a $T/2$-second duration. Assuming no additive noise, determine the approximate value of rms fluctuations in the zero crossings of the timing wave as a function of the BPF transfer function, $H(f)$.

7.10. Demonstrate the validity of equation (7.26) for the jitter variance of data-aided baseband timing recovery by proving the following intermediate steps. Remember that we are assuming independent, zero-mean, data symbols.

(a) $\dot{\Lambda}(\tau) = \sum\limits_{k=0}^{K-1} \sum\limits_{m=-\infty}^{\infty} a_k a_m \dot{r}(mT - kT) + \sum\limits_{k=0}^{K-1} a_k \int_{-\infty}^{\infty} \dot{g}(t - kT - \tau)u(t)\,dt$

(b) $E\dot{\Lambda}(\tau) = 0$

(c) $E\ddot{\Lambda}(\tau) = -K\sigma_a^2 \ddot{r}(0)$

(d) $\dot{r}(0) = 0$

(e) $\dot{r}(-kT) = -\dot{r}(kT)$

(f) $E\dot{\Lambda}^2(\tau) = \sum\limits_{k=0}^{K-1} \sigma_a^4 \left[\sum\limits_{m=-\infty}^{\infty} \dot{r}^2(mT - kT) - \sum\limits_{m=0}^{K-1} \dot{r}^2(mT - kT) \right] - K\sigma_a^2 N_o \ddot{r}(0)$

APPENDIX 7.A

COMPLEX ENVELOPE
REPRESENTATION OF SIGNALS

A straightforward extension of the familiar two-dimensional phasor representation for sinusoidal signals has proven to be a great convenience for dealing with carrier-type data signals where properties of amplitude and phase shift are of special significance. As a supplement to this chapter, only the most basic relationships are presented. More details and the derivations of the formulas can be found in some texts on communication systems or in [Franks, 1969, Chaps. 4 and 7].

An arbitrary signal $x(t)$ can be represented exactly by a complex envelope $\gamma(t)$ relative to a "center" frequency f_o, which for modulated-carrier signals is usually, but not necessarily, taken as the frequency of the unmodulated carrier.

$$x(t) = \text{Re } [\gamma(t) \exp (j2\pi f_o t)] \tag{7A.1}$$

Expressing the complex value $\gamma(t)$ in polar form reveals directly the instantaneous *amplitude* $\rho(t)$ and *phase* $\theta(t)$ of the signal.

$$\gamma(t) = \rho(t) \exp [j\theta(t)] = c_I(t) + jc_Q(t) \tag{7A.2}$$

In some situations, the rectangular form of $\gamma(t)$ in (7A.2) has a more direct bearing on the problem, as it decomposes the signal into its *in-phase* and *quadrature* (*I* and *Q*) components.

$$x(t) = c_I(t) \cos 2\pi f_o t - c_Q(t) \sin 2\pi f_o t \tag{7A.3}$$

328

Equation (7A.1) might be regarded as one part of a transform pair. The other equation [i.e., how to get $\gamma(t)$, given $x(t)$] presents a small problem. Due to the nature of the "real part of" operator Re, there is not a unique $\gamma(t)$ for a given $x(t)$. We solve this problem by making the definition

$$\gamma(t) = [x(t) + j \, \hat{x}(t)] \, \exp \, (-j2\pi f_0 t) \qquad (7A.4)$$

where $\hat{x}(t)$ is the Hilbert transform of $x(t)$. The prescription for getting $\gamma(t)$ from $x(t)$ is especially simple in the frequency domain, and requires no direct evaluation of Hilbert transforms. The Fourier transform $\Gamma(f)$ is obtained by doubling $X(f)$, suppressing all negative-frequency values, and frequency translating the result downward by an amount f_0. Incidentally, using this approach, no narrowband approximations concerning $x(t)$ are necessary, and an arbitrary value of f_0 can be selected.

We now characterize the two most important signal-processing operations, filtering and multiplication, in terms of equivalent operations on complex envelopes. Consider first the time-invariant *bandpass filtering* operation in Fig. 7A.1. We express the bandpass transfer function $H(f)$ in terms of an equivalent low-pass transfer function $\Omega(f)$, according to

$$H(f) = \Omega(f - f_0) + \Omega^*(-f - f_0) \qquad (7A.5)$$

$\Omega(f)$ is not necessarily a physical transfer function. If $H(f)$ exhibits asymmetry about f_0, then $\Omega(f)$ is asymmetric about $f = 0$ and the corresponding impulse response $\omega(t)$ is complex. In fact, $\omega(t)$ is precisely the complex envelope of $2h(t)$, where $h(t)$ is the real impulse response of the bandpass filter.

Straightforward manipulation shows that the input–output relation for complex envelopes is also a time-domain convolution

$$\beta(t) = [\omega \otimes \gamma](t) \qquad (7A.6)$$

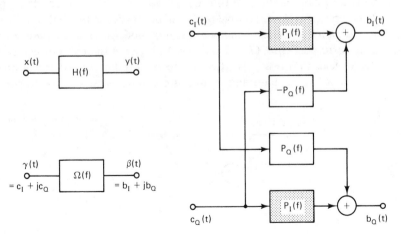

Figure 7A.1 Bandpass filtering and low-pass equivalent operation on complex envelope signals.

and this result is general because of our particular method for defining the complex envelope in (7A.4). If we express $\omega(t)$ in terms of its real and imaginary parts, $\omega(t) = p_I(t) + j p_Q(t)$, then the two-port bandpass filtering operation can be represented by a real four-port filter with separate ports for the I and Q components of input and output. The four-port filter is a lattice configuration involving the transfer functions $P_I(f)$ and $P_Q(f)$, as shown in Fig. 7A.1.

$$P_I(f) = \tfrac{1}{2} \Omega(f) + \tfrac{1}{2} \Omega^*(-f)$$

$$P_Q(f) = \frac{1}{2j} \Omega(f) - \frac{1}{2j} \Omega^*(-f) \qquad (7A.7)$$

Notice that if $H(f)$ is symmetric about f_o, then $P_Q(f) = 0$ (this is the definition of symmetry for a bandpass filter) and there is no cross-coupling of the I and Q components in the filtering operation.

Next we consider the output of a *multiplier circuit*, $z(t) = x(t)y(t)$, when the two inputs are expressed in complex envelope notation. From equation (7A.8), the multiplier output consists of two terms, one representing low-frequency components and the other representing components around $2f_o$.

$$z(t) = \text{Re} \ [\gamma(t) \ \exp \ (j2\pi f_o t)] \ \text{Re} \ [\beta(t) \ \exp \ (j2\pi f_o t)]$$

$$= \tfrac{1}{2} \ \text{Re} \ [\gamma(t)\beta^*(t)] + \tfrac{1}{2} \ \text{Re} \ [\gamma(t)\beta(t) \ \exp \ (j4\pi f_o t)] \qquad (7A.8)$$

In most applications a multiplier is followed by either a low-pass filter (LPF) or a bandpass filter (BPF), as shown in Fig. 7A.2, in order to select either the first or second term in equation (7A.8) and completely reject the other term. In our application, we may regard $y(t)$ as the reference carrier, then the LPF output $z_1(t)$ is the response of a coherent demodulator to $x(t)$. If $y(t) = x(t)$, so that the multiplier is really a squarer circuit, the BPF output $z_2(t)$ can be used for carrier phase recovery. Its complex envelope, relative to $2f_o$, is proportional to $\gamma^2(t)$.

Finally, when the bandpass signal is modeled as a *random process,* we use the same correspondence, (7A.1) and (7A.4), between the real process $x(t)$ and the complex envelope process $\gamma(t)$. It is of interest to relate the statistical properties of $x(t)$ to those of its inphase and quadrature components, relative to some f_o. First we

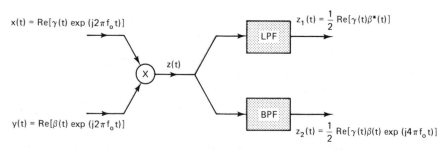

Figure 7A.2 Low-frequency and $2f_o$ terms of the product of two bandpass signals.

note that $E[x(t)] = \text{Re}\{E[\gamma(t)] \exp (j2\pi f_o t)\}$; hence for a wide-sense stationary (WSS) $x(t)$ process, $\gamma(t)$ must be a zero-mean process, in order that $E[x(t)]$ be independent of t. Proceeding to an examination of second-order moments, it is a simple matter to show that $\gamma(t)$ must be a WSS process if $x(t)$ is to be a WSS process. The converse is not true. A WSS $\gamma(t)$ may produce a nonstationary $x(t)$, as indicated below. Rewriting equation (7A.1) as

$$x(t) = \tfrac{1}{2} \gamma(t) \exp (j2\pi f_o t) + \tfrac{1}{2} \gamma^*(t) \exp (-j2\pi f_o t) \tag{7A.9}$$

the autocorrelation for $x(t)$ can be expressed as

$$k_{xx}(t + \tau, t) = E[x(t + \tau)x(t)]$$
$$= \tfrac{1}{2} \text{Re} \left[k_{\gamma\gamma}(\tau) \exp (j2\pi f_o \tau) \right] \tag{7A.10}$$
$$+ \tfrac{1}{2} \text{Re} \left[k_{\gamma\gamma^*}(\tau) \exp (j4\pi f_o t + j2\pi f_o \tau) \right]$$

where, for complex WSS processes, we define the autocorrelation of $\gamma(t)$ as

$$k_{\gamma\gamma}(\tau) = E[\gamma(t + \tau)\gamma^*(t)] \tag{7A.11}$$

The quantity $k_{\gamma\gamma^*}(\tau) = E[\gamma(t + \tau)\gamma(t)]$ in equation (7A.10) can be regarded as the cross-correlation between signal components centered at $+f_o$ and at $-f_o$. If $x(t)$ is WSS, this cross-correlation must vanish in order that the t-dependent term in (7A.10) vanish. Otherwise, $x(t)$ is a cyclostationary process.

If we let $\gamma(t) = u(t) + j\,v(t)$, where the I and Q processes, $u(t)$ and $v(t)$, are jointly WSS, then we have

$$k_{\gamma\gamma^*}(\tau) = k_{uu}(\tau) - k_{vv}(\tau) + j[k_{vu}(\tau) + k_{uv}(\tau)] \tag{7A.12}$$

and the condition for stationarity of $x(t)$ requires that

$$k_{uu}(\tau) = k_{vv}(\tau) \qquad \text{and} \qquad k_{vu}(\tau) = -k_{uv}(\tau) \tag{7A.13}$$

Thus for a WSS bandpass process, the I and Q components are balanced, in the sense that they have the same autocorrelation function. Also, the cross-correlation of the I and Q components must be an odd function, since $k_{vu}(\tau) = k_{uv}(-\tau)$ for any pair of WSS processes. For example, $u(t) = v(t)$ would satisfy the autocorrelation condition in equation (7A.13), but not the cross-correlation condition. The size of $k_{\gamma\gamma^*}(\tau)$ indicates the degree of cyclostationarity of a bandpass process. In the extreme case where either the I or Q component is missing, as in DSB-AM, we would have $k_{\gamma\gamma^*}(\tau) = \pm k_{\gamma\gamma}(\tau)$; for example, for $v(t) = 0$,

$$k_{xx}(t + \tau, t) = \tfrac{1}{2} k_{uu}(\tau)(1 + \cos 4\pi f_o t) = k_{uu}(\tau) \cos^2 (2\pi f_o t) \tag{7A.14}$$

In modeling an additive noise process $n(t)$ on received signals, we often use the white noise assumption, wherein $k_{nn}(\tau) = N_o \delta(\tau)$. If we let $r(t) + js(t)$ be the complex envelope of the process relative to any f_o that is significantly larger than the passband width of the signals, then the white-noise process is equivalently modeled by I and Q processes whose correlation functions are given by

$$k_{rr}(\tau) = k_{ss}(\tau) = 2N_o \delta(\tau) \qquad k_{rs}(\tau) = 0 \tag{7A.15}$$

APPENDIX 7.B

MAXIMUM-LIKELIHOOD ESTIMATION
OF SIGNAL PARAMETERS

There are two strategies commonly regarded as optimal for the estimation of a signal parameter, τ, based on observations of a received signal process, $z(t)$. One is the maximum *a posteriori* probability (MAP) estimator and the other is the maximum-likelihood (ML) estimator. The ML estimator is employed when the parameter τ is *unknown*, but nonrandom. When τ is characterized as a random variable, the MAP strategy is the natural one to use. However, in our applications there is normally a great deal of uncertainty in the *a priori* values of the parameters. If we regarded τ and θ as random variables, we would assign a uniform probability density over intervals $(0, T)$ and $(0, 2\pi)$, respectively. In this case, the MAP and ML strategies are equivalent [Van Trees, 1968].

We can briefly describe the difference between ML and MAP estimation as follows. For this purpose, we suppose that there are only a finite number of true values of the parameter $\tau[\tau = \tau_i; \ i = 1, 2, \ldots, N]$ and we also partition the set of all possible received signals into a finite number of nonoverlapping subsets S_j [i.e., $z(t) \in S_j; \ j = 1, 2, \ldots, M$]. This "discretization" of the parameter set and the signal space allows us to assign meaningful probabilities to the events $[\tau = \tau_i]$, and to the events $[z(t) \in S_j]$. Since N and M can be made arbitrarily large, the description can be made as accurate as desired. Now for a particular realization of the signal process, we identify the subset containing $z(t)$, say $z(t) \in S_m$. Then the MAP strategy is to choose the estimate $\hat{\tau} = \tau_i$ which maximizes the conditional probability function,

$P(\hat{\tau}) = \Pr[\hat{\tau} = \tau_i | z \in S_m]$. The ML strategy maximizes the conditional probability $L(\hat{\tau}) = \Pr[z \in S_m | \hat{\tau} = \tau_i]$, called the *likelihood* function.

In the case that $z(t)$ is a deterministic signal $x(t; \tau)$ depending on τ, plus white, Gaussian noise with a (double-sided) power spectral density of N_o volts²/hertz, then the likelihood function can be shown to be proportional to [Van Trees, 1968]

$$L(\hat{\tau}) = \exp\left\{-\frac{1}{2N_o}\int_{T_o} [z(t) - x(t; \tau)]^2\, dt\right\} \tag{7B.1}$$

where we normally omit multiplicative and additive constants because our interest is in maximization of the function. The parameter T_o is called the *observation interval* and it is an important aspect of the design of an estimator because it characterizes the allowable delay in the estimation process. It is closely related to the loop bandwidth of the parameter tracking schemes described. Notice, from equation (7B.1), that maximization of $L(\tau)$ is equivalent to minimization of the energy of the difference signal, $z(t) - x(t; \hat{\tau})$, over the time interval, T_o.

Let us consider the problem of estimation of the symbol timing parameter in a baseband PAM signal, where

$$x(t; \tau) = \sum_{k=-\infty}^{\infty} a_k g(t - kT - \tau) \tag{7B.2}$$

Expanding the binomial term in equation (7B.1) and observing that $\int_{T_o} z^2(t)\, dt$ is independent of $\hat{\tau}$ and $\int_{T_o} x^2(t; \hat{\tau})\, dt$ is essentially independent of $\hat{\tau}$ if $T_o \gg T$, we see that maximization of $L(\hat{\tau})$ is equivalent to maximization of

$$\Lambda(\hat{\tau}) = \sum_{k=-\infty}^{\infty} \int_{T_o} z(t) g(t - kT - \hat{\tau})\, dt \tag{7B.3}$$

which is sometimes referred to as the "correlation" between $z(t)$ and $x(t; \hat{\tau})$. This corresponds to a data-aided strategy because the a_k values are assumed known. As a practical implementation consideration, we must consider that only a finite number of the a_k (say K) are available. This allows us to approximate $\Lambda(\hat{\tau})$ in equation (7B.3) with an expression involving a K-term sum and an integration from $-\infty$ to $+\infty$ if we let $T_o = KT$. The modified log-likelihood function becomes

$$\Lambda(\hat{\tau}) = \sum_{k=0}^{K-1} a_k q_k(\hat{\tau}) \qquad \text{where } q_k(\hat{\tau}) = \int_{-\infty}^{\infty} z(t) g(t - kT - \hat{\tau})\, dt \tag{7B.4}$$

and the integral can be viewed as a convolution integral. The physical implementation for realizing the $q_k(\hat{\tau})$ quantities is discussed in Section 7.3. In that section it is also pointed out that using the implementable approximation equation (7B.4) instead of (7B.3) introduces a pattern-dependent component of jitter in the evaluation of rms performance.

Extention to the two-parameter estimation case is straightforward. Using complex envelope notation, equation (7B.1) becomes

$$L(\hat{\theta}, \hat{\tau}) = \exp\left[\frac{-1}{4N_o} \int_{T_o} |\alpha(t) - \beta(t - \hat{\tau})e^{j\hat{\theta}}|^2 dt\right] \qquad (7B.5)$$

For the QAM (QPSK) case, we have

$$\beta(t) = \sum_k (a_k + jb_k)g(t - kT) \qquad (7B.6)$$

and the log-likelihood function corresponding to equation (7B.4), including the implementation approximation, becomes

$$\Lambda(\hat{\theta}, \hat{\tau}) = \sum_{k=0}^{K-1} a_k q_k(\hat{\theta}, \hat{\tau}) + b_k p_k(\hat{\theta}, \hat{\tau}) \qquad (7B.7)$$

where

$$q_k(\hat{\theta}, \hat{\tau}) \overset{\Delta}{=} \text{Re}\left[e^{-j\hat{\theta}} \int_{-\infty}^{\infty} \alpha(t)g(t - kT - \hat{\tau})dt\right]$$

$$p_k(\hat{\theta}, \hat{\tau}) \overset{\Delta}{=} \text{Im}\left[e^{-j\hat{\theta}} \int_{-\infty}^{\infty} \alpha(t)g(t - kT - \hat{\tau})dt\right] \qquad (7B.8)$$

For a non-data-aided estimation strategy, the appropriate likelihood function is obtained by averaging the likelihood function (not the log-likelihood function) over the data values. Consider the baseband timing recovery problem with the K-term implementation approximation incorporated, and with multiplicative constants neglected:

$$L(\hat{\tau}) = \exp\left[\frac{1}{N_o}\sum_{k=0}^{K-1} a_k q_k(\hat{\tau}) - \frac{1}{2N_o}\sum_{k=0}^{K-1}\sum_{j=0}^{K-1} a_k a_j r(kT - jT)\right] \qquad (7B.9)$$

where $r(t) \overset{\Delta}{=} \int_{-\infty}^{\infty} g(t + s)g(s)\,ds$. Now if we assume that $r(kT) = 0$ for $k \neq 0$, then $L(\hat{\tau})$ in equation (7B.9) can be expressed as a product of K factors, each factor depending on a single random variable a_k.

$$L(\hat{\tau}) = \prod_{k=0}^{K-1} \exp\left[\frac{a_k q_k}{N_o} - \frac{r(0)}{2N_o} a_k^2\right] \qquad (7B.10)$$

For the binary ± 1 PAM case, we have $a_k^2 = 1$ and letting $\text{Pr}[a_k = 1] = \frac{1}{2}$ we get

$$E \exp\left[\frac{a_k q_k}{N_o} - \frac{r(0)}{2N_o}\right] = \exp\left[\frac{-r(0)}{2N_o}\right]\left[\frac{1}{2}\exp\left(\frac{q_k}{N_o}\right) + \frac{1}{2}\exp\left(\frac{-q_k}{N_o}\right)\right] \qquad (7B.11)$$

Therefore, neglecting the multiplicative constant and taking the natural logarithm, our strategy is to maximize

$$\Lambda(\hat{\tau}) = \sum_{k=0}^{K-1} \ln \cosh \frac{q_k(\hat{\tau})}{N_o} \qquad (7B.12)$$

The expressions for multilevel (nonbinary) data have been derived [Meyers and Franks, 1980]. It is important to recognize that $a_k^2 \neq 1$ in equation (7B.10) for the multilevel case. Similar expressions are also obtained for the two-parameter case.

Another case of theoretical interest is for data values having a Gaussian distribution. Completing the square on each of the arguments of the exponential functions in (7B.10) and then averaging leaves a product of terms each proportional to exp [constant $x \; q_k^2 \, (\hat{\tau})$]. Then taking the logarithm, we want to maximize the quantity

$$\Lambda(\hat{\tau}) = \frac{1}{2} \sum_{k=0}^{K-1} q_k^2(\hat{\tau}) \qquad (7B.13)$$

It is interesting to note that the nonlinear function $\ln \cosh x$ can be approximated by $x^2/2$ for small x; thus the square-law type of nonlinearity is close to optimum, even for binary data, at low signal-to-noise ratios. For multilevel data, the square-law nonlinearity is an even better approximation. In any case, it is interesting to note that the square-law nonlinearity which is so often employed in timing-recovery circuits does actually correspond to the correct type of nonlinearity for ML estimation when the data are Gaussian.

8

TIME-DIVISION MULTIPLE-ACCESS SYSTEMS (TDMA)

DR. S. JOSEPH CAMPANELLA   DR. DANIEL SCHAEFER*

Executive Director,
Communications Technology
COMSAT Laboratories
Clarksburg, Maryland

Manager,
Digital Applications
COMSAT Laboratories
Clarksburg, Maryland

> *In the digital satellite communications community Dr. S. J. Campanella is known as a principal architect of TDMA/DSI technology. He wrote numerous benchmark papers and has made lasting contributions in the areas of signal processing, SCPC, FDMA, and TDMA systems. Dr. D. Schaefer is a well-known digital signal processing and satellite communications research engineer and manager. I wish to thank both of them for this comprehensive chapter on the principles and applications of domestic and international TDMA systems.*
>
> *Dr. K. Feher*

8.1 INTRODUCTION

Time-division multiple-access (TDMA) is a method of time-division-multiplex transmission of digitally modulated carriers used for establishing communications links among the earth stations of a satellite network. In a TDMA system, each participating station transmits one or more traffic bursts, synchronized so that they occupy assigned nonoverlapping epochs in a TDMA frame. This is illustrated in Fig. 8.1. Each station's

---

* In January 1982 he joined the LINKABIT corporation.

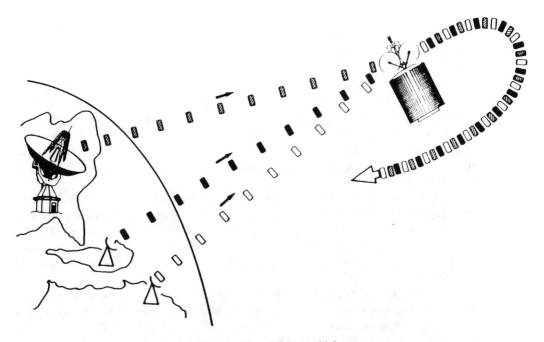

Figure 8.1 Time-division multiple-access system.

burst is synchronized so that at the time of arrival at the satellite it is the only signal present and no collision occurs with the traffic burst of any other station. The traffic bursts are amplified by the satellite transponder and retransmitted in a downlink beam which is received by all of the participating stations. Thus all stations in the network can receive the traffic bursts and select from them the traffic destined for a particular station. The term "multiple access" refers principally to the fact that any number of participating earth stations can enter the network by occupying exclusive traffic burst epochs in the TDMA frame. It is also proper to refer to the system as "multiple destination," since a traffic burst can be received by all stations in the downlink beam and any station can extract its traffic from any of the bursts.

TDMA offers a number of advantages over the frequency-division multiple-access (FDMA) systems which have dominated the first generation of multiple-access satellite communications systems. Perhaps its most significant advantage is the presence of only one carrier at a time in the satellite transponder. FDMA requires simultaneous transmission of a multiplicity of carriers through a common traveling-wave-tube amplifier (TWTA) in the satellite. It is well known that TWTAs are highly nonlinear and the intermodulation products produced by the presence of multiple carriers generate interference which degrades individual channel performance if left uncorrected. To avoid this, it is common practice in FDMA systems to *back-off* the TWTA operating point from maximum power output, consequently forcing a reduction in the amount of traffic capacity that can be realized in that tube. With TDMA, since only one carrier appears at a time, the intermodulation distortion is eliminated and the resulting capacity reduction due to TWTA nonlinearity is significantly reduced. The impact of nonlinearity is, however, not totally eliminated but reappears in the form of nonlinear intersymbol interference which must be minimized by careful transmission path filter function design.

A second important advantage is the use of the time domain rather than frequency domain to achieve selectivity. In an FDMA system, an earth station must transmit and receive on a multiplicity of frequencies to achieve a desired traffic plan and must accordingly provide a large number of frequency-selective up-conversion and down-conversion chains. In a TDMA system, the needed selectivity is accomplished in time rather than frequency, and this is much simpler and less costly to implement.

TDMA is also ideally suited to digital communications since digital signals are naturally acclimated to the storage, rate conversions, and time-domain processing operations used in TDMA terminal implementation. For the same reasons TDMA is ideal for accomplishing satellite on-board processing. Also, TDMA is ideally suited to demand-assigned operation in which the durations of traffic bursts are adjusted to accommodate demand.

Yet another very significant advantage of TDMA compared to FDMA occurs in *multiple-beam* satellite systems. In such systems it is desirable for the stations of each beam to communicate to the stations in all other beams. If FDMA methods are used, then on board the satellite the up-beams must be routed to the down-beams through transponder filter banks and additive combiners which sum the *noise from all uplink beams,* thus aggravating the overall uplink noise problem. Use of TDMA permits the use of a satellite switch which selectively connects individual up-beams to individual down-beams, thus avoiding addition of up-beam noise. Also, by jointly adjusting the locations and durations of individual station traffic bursts and the locations and dwell times of satellite switch beam-to-beam connections, the overall traffic flow among all beams can be accommodated in a simple and optimum manner which cannot be easily matched by FDMA methods.

TDMA operating at a rate of 48-Mb/s on digital carriers at 14/11 GHz has been adopted by Satellite Business Systems (SBS) [Goode, 1978], a continental U.S. domestic business satellite communications company to provide high-speed digital data and teleconferencing services on a demand basis from company site to company site using small earth terminals. Many other business communications networks in the United States, Canada, and Europe are adopting similar schemes.

INTELSAT is about to introduce a TDMA system operating on 120-Mb/s digital carriers at 6/4 GHz and using digital speech interpolation (DSI) to conservatively achieve over 3000 channels per 80 MHz of transponder frequency assignment in its international multibeam INTELSAT-V system. It also plans to introduce a satellite-switched TDMA (SS-TDMA) system using the same 120-Mb/s TDMA terminals in its INTELSAT-VI system to be introduced in the late 1980s. INTELSAT will become the major digital link between the growing digital terrestrial networks of Europe and North America. This will significantly spur the growth of the *integrated services digital network* (ISDN) in the transoceanic international arena.

Some of the system advantages and the efficiencies resulting from the use of satellites in a TDMA mode are discussed in Chapters 1 and 2. Here in this chapter, the techniques for implementing these powerful system concepts are presented.

8.2 BASIC TDMA ARCHITECTURE

TDMA traffic bursts are organized in a TDMA frame as illustrated in Fig. 8.2(a). The frame in this case begins with a reference burst RB_1 and there may be a second reference burst RB_2 for reasons of reliability. The locations of traffic bursts are assumed to be referenced to the time of occurrence of reference burst RB_1 in the case illustrated. Each traffic burst originates from a participating earth station and carries the traffic from that station to *all* destination stations in a digital transmission format. In the example shown, the start of traffic burst from station A occurs a time T_A after the reference burst, that from station B at T_B, and so on. The position and duration of each traffic burst relative to the reference burst is assigned according to a protocol established for network operation. This may be a *preassignment protocol,* in which case the position and duration assigned is changed infrequently and only for overall network rearrangement, or a *demand-assignment* protocol, in which case the positions and duration of the bursts may be adjusted almost continuously to meet traffic demand.

The TDMA frame duration may range from as small as 125 μs established by the Nyquist sampling period of 4 kHz voiceband signals (i.e., $f_s = 8$ kHz), typical of T-carrier and CEPT digital PCM transmission standards, and extend to as much as 25 ms for systems that use demand assignment protocols. The digital transmission rates may be as low as or as high as desired. For example, a TDMA system used

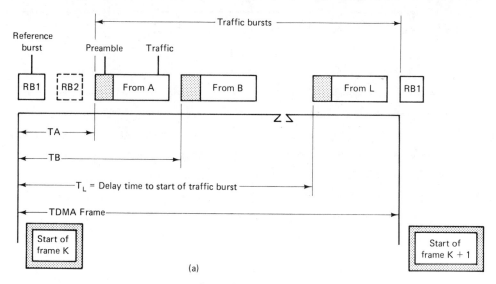

Figure 8.2(a)   TDMA frame organization.

for signaling distribution in the INTELSAT SPADE system operates at a transmission rate of only 128 kb/s. [Cacciamani, 1971] The TDMA system being adopted in the INTELSAT system for high volume international traffic operates at a rate of 120.832 Mb/s. In general, bit rate is determined by the bandwidth and power available in the satellite transponder, and the receive station's antenna gain-to-noise temperature ratio. The duration of traffic bursts depends strictly on the amount of traffic carried and the duration of the preamble needed to accommodate reception processing, network control, and service signaling.

Frequently in this chapter, examples from a TDMA system having a TDMA frame period of 2 ms and a bit rate of 120.832 Mb/s will be used. For QPSK modulation, this corresponds to a symbol rate of 60.416 Msym/s. These values represent a choice based on compatibility with CCITT-*recommended* CEPT and T-carrier standards and other considerations. The 2-ms frame period is long enough to result in high frame efficiency, considering the *overhead* due to reference bursts, traffic burst preambles, and guard time.

The 2-ms frame period also corresponds to the CEPT primary multiplex signaling

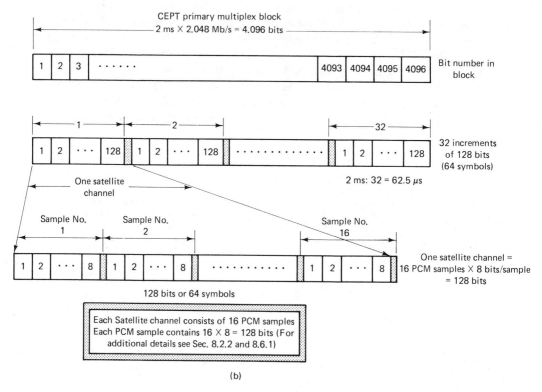

(b)

Figure 8.2 (*continued*) **(b)**   Illustration of Satellite channels, SC, in the INTELSAT-V-120.832 Mb/s TDMA system. The bit rate of 120.832 Mb/s is selected to be the 59th multiple of the CEPT 2.048 Mb/s primary multiplex rate.

and alignment multiframe which contains 16 Nyquist frames (125 μs). The bit rate of 120.832 Mb/s is selected to be the 59th multiple of the 2.048 Mb/s CEPT primary multiplex rate and results in 241,664 bits or 120,832 QPSK symbols per 2-ms TDMA frame. During the 2-ms TDMA frame, a 2.048-Mb/s CEPT digital stream is carried as a block of 4096 bits divided into 32 128-bit channel increments, each comprising 16 8-bit PCM samples and called a satellite channel (SC). Hence SCs will occur in 64 QPSK symbol increments following the last symbol of the unique word (UW) in a traffic burst, see Fig. 8.2(b) and Fig. 8.2(c).

The primary multiplex for T-carrier consists of a time-division multiplex of 24 8-bit PCM channel samples plus one additional frame alignment bit per 125-μs frame. This yields 193 bits per frame and a bit rate of 1.544 Mb/s. In the 2-ms TDMA frame, a 1.544-Mb/s T-carrier can be carried as a block of 3088 bits containing sixteen 193-bit primary multiplex frames. Thus the selection made is compatible with T-carrier as well as CEPT primary multiplex transmission standards. Also, individual

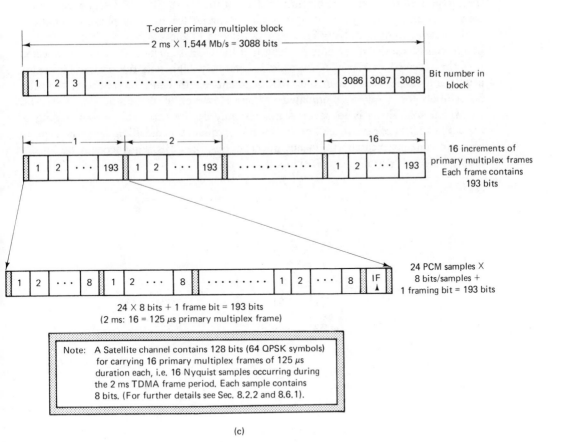

(c)

Figure 8.2 (*continued*) **(c)**   Illustration of the INTELSAT-V TDMA system satellite channels, SC, for T-1–1.544 Mb/s carrier primary multiplex systems.

channels may be carried in 64-symbol increments in the same way as described above for CEPT. In the latter case, the extra alignment bit used in T-carrier could be carried in additional space at the end of the TDMA traffic burst. In certain cases where the information of the extra bit is not used, it may simply be discarded. *Solve Problems 8.1 and 8.2.*

8.2.1 Reference Burst

Reference bursts are emitted by a reference station and, as previously indicated, constitute the basis for synchronizing all other stations in a network. The structure of a typical reference burst is shown in Fig. 8.3(a). This burst contains information necessary for other stations to derive the precise location of their bursts in the frame. The reference bursts consist of three parts. First, there is a *carrier and bit timing recovery* (CBR) sequence which serves the purpose of locking a receive station to the carrier frequency and the bit timing clock of the burst. The CBR sequence usually consists of an initial segment of unmodulated carrier followed by phase alternations of the carrier frequency between 0 and π radians at the symbol clock rate. System design considerations determine the length of the CBR sequence. If it can be expected that the received carrier-to-noise ratio is relatively high and the carrier frequency acquisition range small, then the CBR segment length can also be short, typically 30 symbols for a QPSK modulated carrier. However, if the carrier-to-noise ratio for which acquisition is to occur is low, as may be the case when severe fading of the RF signal is expected and/or the carrier frequency acquisition range is large to accommodate large carrier frequency uncertainties, the CBR sequence may be considerably longer, for example, 300 symbols.

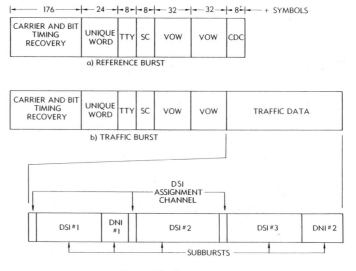

Figure 8.3 Burst format.

The CBR sequence is followed by the unique word (UW). The UW is a sequence of ones and zeros on both the P and Q phases of the carrier selected to exhibit good correlation properties. UWs vary in length and may be as short as 10 QPSK symbols or as long as 24. At a receiver, the UW is supplied to a UW correlator, where it is correlated with a stored pattern of itself. A typical UW correlator is shown in Fig. 8.4. If all bits of the received pattern correspond to those of the stored pattern, the UW is received with error $E = 0$. This, of course, occurs at the instant of reception of the last bit of the UW and constitutes an accurate indication of the instant of arrival of the burst. For this reason, *the output of the UW correlator is always used to reference the time of occurrence of a burst.* It also constitutes the time reference for timing the demultiplexing of the channels carried in the traffic data portion of the burst. The UW correlator may suffer two types of errors, miss and false alarm. The mechanism of these errors is discussed later, and provision must be made in the receive processor to counter them.

The UW serves other important purposes in addition to marking the time of occurrence of a burst. The instant of occurrence of the UW correlation spike marks the symbol time reference for decoding information in the traffic part of a burst. Different UWs can be used to distinguish between two reference bursts or between reference bursts and traffic bursts. For example, simple inversion of the unique word causes a negative-going rather than a positive-going correlation spike which may be used to distinguish between bursts. Another important use of the UW is resolution of the ambiguity in the phase of the recovered carrier when coherent QPSK modulation is used. This is discussed in a later section.

The control and delay channel (CDC) of the reference burst serves to communicate information to control the burst positions of stations in the network. In systems where traffic stations are controlled by a central reference station (or stations) which is permanently assigned and in which the traffic station is not able to see its own return emissions from the satellite, the information carried on the control and delay

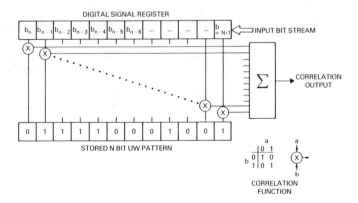

Figure 8.4   Typical UW correlator. Typical unique word correlator outputs are shown in Figs. 8.27 and 8.28.

channel is critically important to bringing a station's traffic burst into the network from a cold start (a function called initial acquisition) and maintaining its position in the frame with high precision (a function called synchronization). In demand-assigned systems in which traffic burst positions and durations must be continuously adjusted to optimumly allocate the TDMA frame capacity, the traffic control information may be carried over the control channel. For systems that operate using simple preassigned global beam or regional beam networks and in which stations control burst position from direct observation of their position in the frame, there may be no need for a control channel at all. In such systems, the traffic burst of one of the traffic stations may serve as the reference burst. In this case, a service channel (SC) may be used to carry information important to transfer of the reference station responsibility among participating stations. Several examples are given later in this chapter which illustrate various possible network control schemes.

8.2.2 Traffic Burst Structure

Traffic bursts are emitted by the traffic stations and are synchronized relative to the reference bursts to occupy assigned positions in the TDMA frame. The structure of a typical traffic burst is shown in Fig. 8.3(b). The first two parts of a traffic burst consist of a CBR sequence and UW and are the same as the corresponding parts of the reference burst. It is reasonable to expect that the CBR sequence will be the same for both types of bursts since the traffic bursts must operate in the same environment as the reference burst, suffering the same degradations due to noise, interference, and distortion. The UW should be common to both since this *permits the same UW detector to be used for detecting both reference and traffic bursts*. Discrimination of different types of bursts may be accomplished by using coded patterns in terms of the UW and its inverse UW. Also, periodic inversions of the UW may be used to signal multiframes or superframes used in some acquisition and synchronization protocols and also for traffic control.

Following the UW, the traffic burst may also contain a *service channel (SC)* and order wires which are used for supporting system operating protocols and for utility *teletype* (TTY) and voice (VOW) communications among the stations.

The next part of the traffic burst is the traffic data field, which carries the **payload** of customer service. Organization of the traffic data field can vary significantly depending on the traffic data rates carried, the preestablished structure of the traffic as it is supplied to the system at the transmit side, and of course delivered at the receive side, and the TDMA frame period. The simplest traffic format carried is that of a continuous digital data stream. If such a source stream has a rate R_o bits/s and the TDMA frame period is T_F in seconds, then the number of bits in the traffic burst data field must be $B_{TR} = R_o T_F$. Furthermore, if the TDMA transmission bit rate is R_T, the *duration of the traffic data field* is

$$T_{TR} = \frac{R_o T_F}{R_T} \qquad (8.1)$$

Example 8.1(a)

If a terrestrial link having a rate of $R_o = 2.048$ Mb/s is carried over a TDMA system having a frame period of 2 ms, the number of bits in the data field will be 4096. If the transmission rate were 120.832 Mb/s, the duration of the traffic field part of the burst would be 33.9 μs.

■

To obtain the total length of a traffic burst, the duration of the preamble must be added. If the preamble contains S_p symbol periods and QPSK transmission in which each symbol carries 2 bits of traffic information is assumed, then the *total length* of the *traffic burst measured in QPSK symbols* is

$$S_T = \frac{R_o T_F}{2} + S_P \tag{8.2}$$

and the burst time duration is

$$T_T = \frac{2 S_T}{R_T} \tag{8.3}$$

Example 8.1(b)

At a transmission rate of 120.832 Mb/s corresponding to a symbol rate of 60.416 Msym/s, and assuming that the preamble length is $S_P = 300$ symbols, the total number of symbols contained in the traffic burst for carrying 2.048 Mb/s is 2348. This is to be compared with a total of 120,832 symbols in the entire 2-ms TDMA frame. The combined traffic and preamble burst duration is 38.86 μs.

■

Thus far, the discussion has pertained to traffic carried with no attention paid to alignment between the TDMA frame and any frame structure existing in the traffic. To incorporate concepts such as digital speech interpolation, demand assignment, and plesiochronous interworking among networks when telephony traffic is involved, it is most desirable that an alignment discipline exist between the frame structure of the terrestrial plant signals and the space segment's TDMA frame.

Alignment refers to the fact that the TDMA frame period is blocked in terms of channels as illustrated in Fig. 8.5. Positions occupied by the blocks are fixed relative

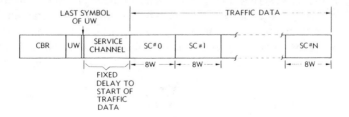

Figure 8.5 Alignment of satellite channels in the TDMA frame.

to the instant of occurrence of the last symbol of the UW. These channels are called *satellite channels* (SCs). Assume that information on a *terrestrial channel* (TC) is grouped in 8-bit PCM samples which occur at a rate of 8000 samples/s, yielding a primary multiplex frame period of 125 μs, and the TDMA frame period contains an integer multiple W of these primary multiplex frames. Then each SC in the TDMA frame is a block of $8W$ bits which carries the most recent W samples of the TC. For example, in the INTELSAT-V system a satellite channel corresponds to $8 \times 16 = 128$ bits, as there are 16 primary multiplex frames in one TDMA frame, that is, 2 ms : 125 μs $= 16$. More will be said concerning the organization of the SCs in the section on interfaces. If the number of such channels carried is N, the total number of symbols contained in the traffic burst, including the preamble, is

$$S_T = \frac{8NW}{2} + S_P \tag{8.4}$$

and the duration is given by equation (8.3).

8.2.3 Guard Time

It is the function of the burst position control strategy to maintain a burst at its assigned target position in the TDMA frame. As previously described, this position is measured by the time of occurrence of the last symbol of the UW of the burst. Due to propagation time uncertainties and clocking granularities encountered in making observations of burst positions, the position of a burst cannot be maintained precisely. For this reason, a *guard time, expressed in terms of the number of symbols, S_G,* is associated with each burst. *Guard time is the width of the time interval in which the unique word correlation spike will be found with a high probability of success.* More will be said about the factors determining guard time in Section 8.3.6. Let it suffice for the present to say that assuming a 60.416-Msym/s transmission rate, S_G can be as small as 10 symbols for tightly controlled direct satellite loop-back to as high as 60 symbols for multiple-beam systems in which position control is established from a remote station by feedback methods.

8.2.4 Frame Efficiency

Frame efficiency of a TDMA system is defined as the *ratio* of the number of symbols available for carrying *traffic* to the *total* number of symbols available in the TDMA frame. The total number of QPSK symbols in a TDMA frame is $R_F T_F/2$. The total number of symbols available for carrying traffic is this value less the overhead due to reference bursts, the preambles of the traffic bursts, and the guard time. The following expression is easily developed for the TDMA frame efficiency:

$$\eta = \frac{R_F T_F/2 - NS_P - KS_{\mathrm{RB}} - (N+K)S_G}{R_F T_F/2} \times 100\% \tag{8.5}$$

where N is the number of traffic bursts in the frame, K the number of reference bursts, S_P the number of symbols in the traffic burst preamble, and S_{RB} the number of symbols in each reference burst and S_G is the guard time between bursts expressed in number of symbols.

Example 8.2

A typical system used for multibeam service may have the following parameters:

$$R_F = 120.832 \text{ Mb/s}$$
$$T_F = 2 \text{ ms}$$
$$N = 16$$
$$K = 2$$
$$S_P = 280 \text{ symbols}$$
$$S_{RB} = 288 \text{ symbols}$$
$$S_G = 60 \text{ symbols}$$

Substituting these values into the expression for frame efficiency yields a value of 95%. It is interesting to consider the relationship of frame efficiency to the number of traffic bursts in the frame with TDMA frame period constant, and the relationship of frame efficiency to the TDMA frame period with the number of bursts constant. These are plotted in Fig. 8.6 using the system parameters given above. ■

Note that *shorter TDMA frames are less efficient and larger numbers of traffic bursts are also less efficient.* This is because in both instances the number of symbols devoted to overhead functions increases at the expense of the capacity available to traffic. *Solve Problem 8.3.*

8.2.5 Transmit Side Burst Processing

The basic elements comprising the transmit side burst processor of a TDMA terminal are shown in Fig. 8.7(a). The digital data stream to be transmitted occurs as continuous input to a pair of buffer memories A and $\overline{A}$. When one of these buffer memories is filling, the other is emptying. Filling and emptying alternate from frame to frame by alternating application to the buffers of a continuous low rate clock which controls filling and a burst high rate clock that controls emptying. The alternating action is accomplished by the switches shown in the diagram. This buffer memory arrangement causes the continuous low data rate input to be compressed to short bursts at the high data rate. For this reason it is called the *compression buffer.*

The continuous clock used to fill the compression buffer may be operated in either a synchronous or asynchronous manner. If operated *synchronously,* its rate is a precise integer multiple of the TDMA frame rate so that the same number of bits are stored in the buffer memory during each TDMA frame. The rate of the high-bit-rate clock may also be an integer multiple of the TDMA frame, in which case there will be a constant number of bits contained in each TDMA frame. The high-rate clock is applied as a burst having a duration sufficient to empty the contents of the buffer during each TDMA frame. The time of application of this clock burst,

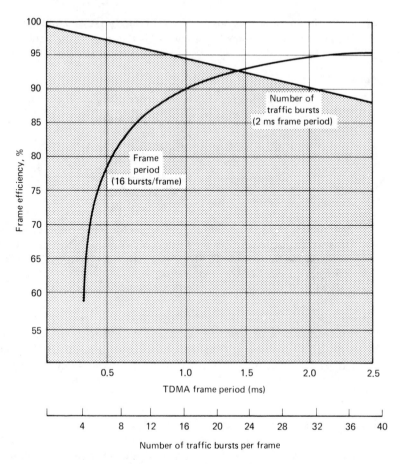

Figure 8.6 TDMA frame efficiency as a function of number of traffic bursts per frame and frame period for 120.832-Mb/s TDMA rate.

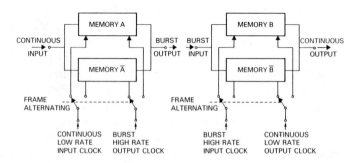

Figure 8.7 Burst compression and expansion buffers. (a) Transmit side compression buffer; (b) receive side expansion buffer.

which is controlled by the *common TDMA terminal equipment* (**CTTE**), causes the traffic burst to be transmitted in the proper time interval to arrive at the satellite in its assigned position in the TDMA frame.

The continuous clock may also be operated *asynchronously* relative to the data stream applied to the input. In this case its rate is not an integer multiple of the TDMA frame rate, which causes the number of bits stored in the buffer to vary from frame to frame depending on the timing relationship existing. The length of the resulting traffic burst may vary by a few symbols. Also, the high-bit-rate clock may not necessarily be an integer multiple of the TDMA frame rate, in which case the duration of the TDMA traffic burst will correspondingly vary. In general, it is preferable that the high-bit-rate clock be a multiple of the continuous clock to simplify design of the compression buffer. Both the continuous and high-bit-rate clocks may be derived from a local crystal oscillator operating at the nominal bit or symbol rate frequency of the TDMA system. This clock should have an accuracy sufficient to count the number of symbols in a TDMA frame to within a symbol at the end of the frame. Thus for a 120.832-Mb/s system with a QPSK symbol period of 16.67 ns and frame period of 2 ms, a clock with an accuracy of $\pm 10^{-6}$ will be able to position the last symbol in the frame with an accuracy of ± 2 ns. The transmit and receive side clock rates will differ because of satellite motion and the attending Doppler. Ideally, in either case, the clock rate should be adjusted so that the number of symbols per TDMA frame period is equal to the specified value. The receive side frame period is the time between receptions of the reference bursts and the transmit side. TDMA *frame period* is the time between the *starts of the transmit side TDMA frame* (SOTF). The SOTF is continuously adjusted to compensate for the changing range to the satellite. This adjustment will be discussed in greater detail later in the chapter.

8.2.6 TDMA Receive Side Processing

A TDMA terminal is designed to receive traffic bursts by the means now discussed. First, the radio frequency carrier is down converted to an intermediate frequency carrier and then supplied to a QPSK demodulator. The output of the demodulator is the traffic burst information in digital form. The CBR portion of the TDMA burst is used within the demodulator to lock up its carrier recovery and bit timing-recovery loops. This is done for *each* individual burst received by the TDMA terminal. The demodulator delivers the demodulated digital information and the recovered clock to the common TDMA terminal equipment (CTTE) for processing to recover the traffic destined to the terminal.

A key factor in the recovery process is UW correlation. Assuming that the UW word is received with sufficiently few errors (less than the error threshold E), the UW correlator generates a correlation spike at the instant of occurrence of its last symbol. This spike initializes a clock running at the recovered bit timing clock rate which counts the number of bits elapsed since the time of occurrence of the correlation spike and permits any *desired portion* of the traffic data field contained in the burst to be routed into a buffer memory as shown in Fig. 8.7(b). The input to the buffer memory occurs at the TDMA transmission rate as a burst. This buffer stores the desired contents of the traffic data portion of the burst occurring in a given TDMA frame. The contents of the buffer are supplied as output data at the desired continuous output rate during the next TDMA frame. Since the buffer is used on one frame to store the burst and on the next to exhaust its contents, two buffers operating in an alternating (ping-pong) manner are typically used to achieve continuous data flow. Since this **ping-pong buffer** converts the received short-duration TDMA burst to continuous data flow, it is referred to as the *expansion buffer*.

As noted, the UW is the key factor involved in successfully demultiplexing a burst, and its instant of occurrence must be accurately detected for each burst at the proper time. Also, it must not be mistakenly detected elsewhere in the frame. To maximize assurance of proper UW detection, the TDMA terminal's receive side processor opens a **window** in the vicinity of the expected time of occurrence of the UW correlation spike for each traffic burst it intends to receive in each TDMA frame. Only if the correlation spike occurs in this window will the system recognize its occurrence. When such a window is used the error threshold E of the UW correlator is increased from zero to improve the probability of detection. This is possible because the action of the window significantly reduces the probability of false UW detection, which might otherwise become excessive. This window must be of sufficient width to accommodate traffic burst position uncertainty, and this is equal to the guard time. During the acquisition process, when burst position may not be accurately known, the window is either disabled or widened to allow detection of the correlation spike. Also, in this case, the error threshold of the correlator may be made equal to zero to improve resistance to false alarm declarations.

8.3 TDMA CONTROL ARCHITECTURES

TDMA requires a method for precise timing of the epochs of burst transmission to prevent burst overlapping in the satellite. TDMA as originally conceived existed in a single wide-coverage global or regional beam, as illustrated in Fig. 8.8, and relatively

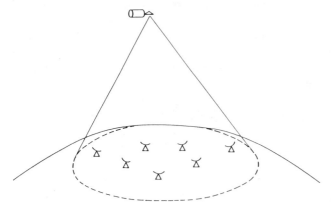

Figure 8.8 Global or regional beam network.

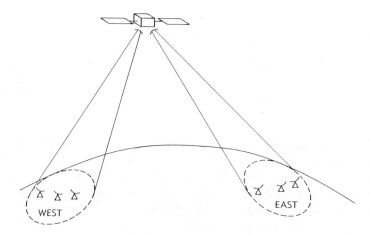

Figure 8.9   Multiple-beam network.

simple acquisition and synchronization methods were possible [Schmidt, 1973; Campanella et al., 1979]. With the evolution of multiple-beam satellites, as illustrated in Fig. 8.9, different procedures have had to be developed, such as those in the new TDMA specification which will be used for INTELSAT-V [Pontano et al., 1981]. Continued evolution of TDMA will lead to the introduction of *satellite-switched* TDMA (SS/TDMA) [Perillan and Rowbotham, 1981] [Campanella et al., 1980] in the satellite generation following INTELSAT-V. This requires that *TDMA bursts be synchronized with satellite switch epochs* so that traffic bursts are routed from up-beam to down-beam without collision with a switching boundary. Such operation imposes further constraints on accomplishing acquisition and synchronization among the earth stations due to the need to synchronize to the satellite switch [Campanella and Colby, 1983].

8.3.1 Acquisition and Synchronization

Acquisition refers to the process of **entry** of a TDMA burst into its assigned location in the TDMA frame, and synchronization refers to *precision maintenance* of a burst at its assigned location. Methods for accomplishing TDMA network acquisition and synchronization can be categorized as *satellite loop-back, open loop,* and *cooperative feedback.* Each of these is described and analyzed for its ability to meet the needs of both nonswitched and satellite switched network operations.

8.3.2 Satellite Loop-Back Control

Satellite loop-back control was introduced in the first TDMA system designed for use in global and regional beams. This control method relies on the ability of any station to see reference and "own" burst retransmissions from the satellite, thus enabling the station to adjust its "own" burst timing to occupy an assigned epoch in the TDMA frame. Global and regional beams inherently allow each station to see all TDMA bursts, including its own. This **cannot** be applied to **multiple-beam** systems because they do not permit direct satellite loop-back. To establish a time reference, one station's burst may be designated as the network reference and all transmissions are timed relative to it. This reference function may be taken over by another station for system reliability reasons. In some systems, reference-only stations may serve the reference function.

Loop-back acquisition and synchronization.   The method of loop-back acquisition and synchronization is illustrated in Fig. 8.10. In the *acquisition phase,* the station first synchronizes its receive side to the *reference bursts R,* thus establishing a local timing reference. It next transmits an *acquisition burst A* at a time ΔT after the instant of reception of each reference burst. The time ΔT is a coarse estimate obtained by methods discussed later. This burst arrives at the satellite in the position indicated, which is displaced an amount ϵ from the desired target location in the frame. This is the burst position error. The earth station does not see this error until the retransmitted burst returns to the station at a time equal to twice the one-way propagation time to the satellite. The station is then able to observe the error e and appropriately adjust its delay value from ΔT to $\Delta T - e$. This correction may be applied on the next frame, or one or two frames later to allow for processing time. Application of the corrected delay value will cause the burst to reside at its

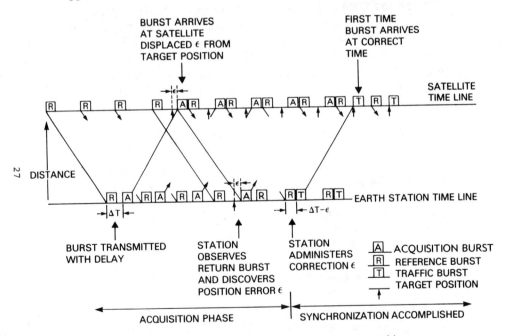

Figure 8.10   Direct loop-back correction of traffic burst position.

assigned position in the TDMA frame at the satellite. Note that the time delay ΔT between the time of reference burst reception and the time of acquisition burst transmission is not equal to the delay between the reference and the acquisition bursts at the satellite for reasons to be described in the section on open-loop control. For the present discussion, the difference does not matter.

At this point, the station has completed the acquisition phase and may be considered to now enter the *synchronization phase,* wherein it simply maintains synchronization by continued loopback error observation and correction cycles. It will also stop transmitting the acquisition burst A and commence transmission of its full traffic burst T. The *acquisition burst usually consists of only the preamble* of the traffic burst, and therefore is a short burst. This minimizes the possibility of collision with other traffic bursts during acquisition.

Before the loop-back correction acquisition method can be initiated, a means of obtaining a *coarse estimate of the delay* ΔT must be provided. This is also part of the acquisition phase. Several methods are possible. In one method, a long-duration, **low-power burst** of **unmodulated** carrier is used to search for the desired time slot. The peak power level is maintained at 25 dB or more below the peak power level of the normal traffic bursts so as not to interfere with other stations' traffic bursts transmissions. Since this low power burst is of long duration (typically equal to the entire length of the assigned traffic burst slot in the frame), a narrowband filter can be used to enhance its detection. This burst is first scanned through the TDMA frame by applying a stepping sawtooth function to control its transmission time delay ΔT relative to the reference burst. Once the burst is observed to fall in the middle of the assigned traffic slot, the corresponding value of ΔT is used to start the burst position loopback correction procedure described previously.

A modified version of the above method for obtaining an initial estimate of ΔT is to use a PN sequence modulated rather than an unmodulated carrier burst. This permits the use of a correlation demodulator to locate accurately the position of the *low-power search burst.*

Another method used to obtain an initial value of ΔT is to compute it from the estimated distance or range between the earth station and the satellite. This can be obtained from knowledge of the coordinates of the earth station and the satellite. The former is available from an accurate calibration of the earth station location and the latter from satellite tracking data obtained from the satellite's *telemetry, tracking, and control* (TT&C) system. The method for computing ΔT from satellite range is given later in the section on open-loop control and is frequently referred to as *open-loop acquisition.* The distance accuracy needed to position a burst to a time accuracy of ΔT in the frame is $\frac{1}{2} \Delta Tc$, where c is the speed of light. Thus, to achieve a burst position accuracy of ± 100 μs requires a satellite range accuracy of ± 30 km. This level of accuracy is easily possible from the satellite TT&C systems that are currently used, and is sufficient to enter a 2-ms TDMA frame. Open-loop entry into systems with shorter-duration frames will require correspondingly higher accuracies.

Satellite range extrapolation.   A TDMA terminal effectively continuously ranges on the satellite, smoothes the observed data, and predicts the future range in the process of performing stationkeeping of its burst position by the loop-back method. It may use zero-order or first-order prediction, as described below.

Figure 8.11 illustrates the satellite ranging principle. A station may obtain satellite *range estimates at intervals T,* where *T* is longer than the round-trip propagation time to the satellite. Any attempt to measure range in less time than the round-trip time will produce an erroneous observation and result in system instability. Thus, as illustrated in the figure, individual range observations $\overset{*}{d}_n$ are made at intervals *T*. If *zero-order prediction* is used, a station applies the last **observed value** $\overset{*}{d}_{n-1}$ to the next interval. Hence the **prediction** $\hat{d}_n$ is given by the relation

$$\hat{d}_n = \overset{*}{d}_{n-1} \tag{8.6}$$

As illustrated in Fig. 8.11, if the change in range per interval *T* is Δ_n, this method suffers an error Δ_n at the beginning of the interval to which it is applied, and $2\Delta_n$ at the end.

First-order prediction is obtained by determining the slope of the range change and using it to predict the next value to be used. This requires that two observation

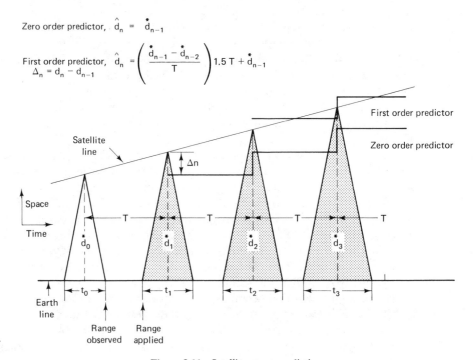

Zero order predictor, $\hat{d}_n = \overset{*}{d}_{n-1}$

First order predictor, $\hat{d}_n = \left(\dfrac{\overset{*}{d}_{n-1} - \overset{*}{d}_{n-2}}{T} \right) 1.5\,T + \overset{*}{d}_{n-1}$
$\Delta_n = d_n - d_{n-1}$

Figure 8.11   Satellite range prediction.

intervals be included in the computation. Hence the slope is estimated by

$$\dot{\Delta}_n = \frac{\overset{*}{d}_{n-1} - \overset{*}{d}_{n-2}}{T} \tag{8.7}$$

This must be extrapolated to the middle of the next interval and summed with the most recent range observation to obtain the predicted range. The predicted range is given by

$$\hat{d}_n = \overset{*}{d}_{n-1} + \frac{\overset{*}{d}_{n-1} - \overset{*}{d}_{n-2}}{T} 1.5T \tag{8.8}$$

As illustrated in Fig. 8.11, this results in a prediction function which is a staircase with zero error at the center of each interval and an error of $\Delta_n/2$ at the extremes.

In actual application, a terminal does not observe ranges; rather, it observes *differences in range.* The expressions above can be converted to differences (i.e., burst position errors) simply by making the transformations

$$\hat{e}_n = \hat{d}_n - \hat{d}_{n-1} \tag{8.9}$$

$$\overset{*}{e}_n = \overset{*}{d}_n - \overset{*}{d}_{n-1} \tag{8.10}$$

When this is done, the zero-order range expression becomes

$$\hat{e}_n = \overset{*}{e}_{n-1} \tag{8.11}$$

and the first-order expression becomes

$$\hat{e}_n = \overset{*}{e}_{n-1} + \frac{\overset{*}{e}_{n-1} - \overset{*}{e}_{n-2}}{T} 1.5T \tag{8.12}$$

Solve Problem 8.4.

8.3.3 Open-Loop Control

Open-loop control [Campanella et al., 1979] refers to control of traffic burst position based on knowledge of the *propagation time, which may be viewed as the distance between the satellite and the earth station.* It requires accurate measurement of satellite position and precise knowledge of earth station locations. This method can be used in multiple-beam systems because it does not require visibility of own station bursts.

Time delay, D_N.   Open-loop control of traffic burst position at an earth station requires the introduction of a time delay between reception instant of the TDMA reference burst and transmission instant of the station's own burst. *The value of time delay (D_N) introduced is selected so that the time elapsed between departure of the burst from the satellite and the return of a response to the satellite from an earth terminal is an integer number of TDMA frame periods.* The instant of arrival of the reference burst at a station is called the **"start of receive frame"** (SORF). Elapse of the delay, D_N, from the SORF marks the instant of the **"start of transmit frame"** (SOTF). Consequently, traffic bursts adjusted to assigned positions relative to the SOTF at the earth station fall at their assigned positions in the TDMA frame at the satellite. The relationship between SORF and SOTF is shown in Fig. 8.12.

For the INTELSAT-V-TDMA system specification, the TDMA frame period is $T_F = 2$ ms. For traffic management purposes, a multiframe (T_M) consisting of 16 TDMA frames and having a duration of 32 ms has been introduced. For this reason, the relations in the following text are expressed in terms of a **multiframe** T_M which has the duration $16T_F$, that is, $T_M = 16T_F$. The method for determining the time delay, D_N, for any station N is illustrated by the space-time graph in Fig. 8.13. The time line at the top of the figure shows the timing of events at the satellite. The pulses designated R are the reference bursts, which occur with a TDMA multiframe period, T_M. All traffic bursts in the satellite are timed relative to these pulses. The bottom dashed line designated as the range limit represents the maximum distance between the satellite and an earth station. It is selected to result in a round-trip propagation time of MT_M, where M is an integer. MT_M must be greater than the round-trip propagation time for the farthest station from the satellite and T_M is the multiframe period. For a stationary orbit satellite, the greatest round trip propagation delay that occurs is 283 ms. The minimum integer value of M that causes the value MT_M to exceed this amount is 9. Hence $MT_M = 288$ ms. A station located on this time line would need to introduce zero delay relative to the reception instant of the reference burst (SORF) in order to align its local start of TDMA frame (SOTF) instant.

For an earth station, N, located nearer to the satellite, a delay D_N equal to the difference between the actual round-trip propagation time and MT_M, must be introduced to mark the SOTF instant and thereby achieve proper frame synchronization at the satellite. Thus

$$D_N = MT_M - 2\frac{d_N}{c} \tag{8.13}$$

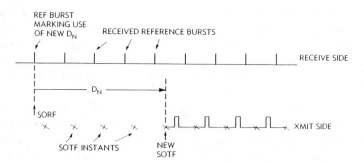

NOTE: EACH REFERENCE BURST RECEPTION MARKS AN SORF. VALUES OF D_N ARE INVOKED OVER GROUPS OF TDMA FRAMES DURING WHICH SOTFs OCCUR ON PREDICTED FRAME BOUNDARIES.

Figure 8.12 Transmit and receive side TDMA frame timing.

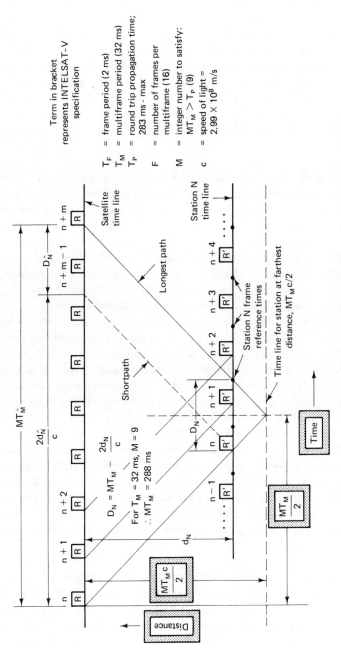

Figure 8.13 Time/space graph for determination of synchronization delay D_N for traffic station N-open loop control.

Term in bracket represents INTELSAT-V specification

T_F = frame period (2 ms)
T_M = multiframe period (32 ms)
T_P = round trip propagation time; 283 ms - max
F = number of frames per multiframe (16)
M = integer number to satisfy: $MT_M > T_P$ (9)
c = speed of light = 2.99×10^8 m/s

Satellite time line

Station N time line

Longest path

Shortpath

Station N frame reference times

Time line for station at farthest distance, $MT_M c/2$

$MT_{\hat{M}}$

$\dfrac{2d_{\hat{N}}}{c}$

$D_{\hat{N}}$

$D_N = MT_M - \dfrac{2d_N}{c}$

For $T_M = 32$ ms, $M = 9$
∴ $MT_M = 288$ ms

D_N

d_N

$\dfrac{MT_M c}{2}$

Distance

$\dfrac{MT_M}{2}$

Time

357

where d_N is the distance between the satellite and the earth station designated by subscript N. A traffic burst transmitted at this instant will fall at the beginning of the TDMA frame at the satellite. To position a traffic burst at its assigned location in the frame, its delay relative to the SOTF must be added to D_N.* D_N may also be augmented by an integer multiple of TDMA frames to allow for processing time without changing burst positions in the frame. For the INTELSAT TDMA system specification [INTELSAT, 1980] three multiframes have been added to allow for processing time, making the value of M equal to 12. Also, D_N must be corrected for propagation delays within the earth station caused by waveguides, cables, and filters.

Ranging for open-loop control with direct satellite loop-back.   Open-loop control requires precise determination of satellite position relative to all participating earth stations. In a global or regional beam system where the position of traffic bursts relative to the reference burst can be observed directly by the controlling reference station, corrections to satellite range can be determined by observing the time difference between a traffic burst's actual position and the target position and multiplying by the speed of light. This is a consequence of the fact that any difference between an estimated value of D_N and the actual value is equal to the difference between actual position of the traffic burst and its target position in the TDMA frame.

This method of range determination has an *ambiguity interval* equal to T_M in time or $cT_M/2$ in distance. For a $T_M = 16$ ms, the ambiguity interval is 2400 km. Ranging must be performed by at least *three widely separated earth stations,* preferably located on the periphery of the region occupied by traffic terminals to achieve precision burst position control. The station which is responsible for the control (i.e., assigning target values of D_N and observing the traffic burst position error) must be able to see the position of the controlled traffic burst relative to the common reference burst. This is obviously satisfied in a global or regional beam system.

Ranging accomplished by the direct loop-back method can be sufficiently accurate to support both the acquisition and synchronized traffic operating phases of TDMA systems provided that the stations using the computed values of D_N for burst position control lie within or near to the triangle formed by the ranging stations as vertices. Burst position precision of $\pm$ 10 symbols (120 Mb/s) are easily possible. Further details on this type of operation are given in [Campanella and Hodson, 1978].

Ranging for open-loop control in multiple-beam systems.   In multibeam systems where a reference station cannot observe traffic bursts relative to its own burst, an additional complication arises which is best illustrated by the example shown in Fig. 8.14. Consider two separate coverage areas, East and West. Traffic stations in the West coverage area use the East-to-West reference burst to time their transmissions,

---

* The delay ΔT used in the previous discussion on open-loop acquisition is the sum of D_N and the time delay corresponding to burst position in the TDMA frame.

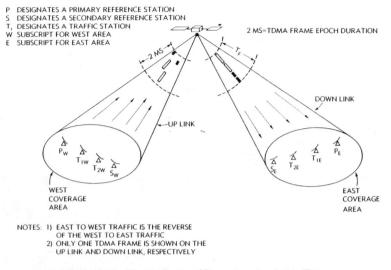

P DESIGNATES A PRIMARY REFERENCE STATION
S DESIGNATES A SECONDARY REFERENCE STATION
T, DESIGNATES A TRAFFIC STATION
W SUBSCRIPT FOR WEST AREA
E SUBSCRIPT FOR EAST AREA

2 MS=TDMA FRAME EPOCH DURATION

DOWN LINK

UP LINK

P_W

T_{1W} T_{2W} S_W

S_E T_{2E} T_{1E} P_E

WEST
COVERAGE
AREA

EAST
COVERAGE
AREA

NOTES: 1) EAST TO WEST TRAFFIC IS THE REVERSE
           OF THE WEST TO EAST TRAFFIC
        2) ONLY ONE TDMA FRAME IS SHOWN ON THE
           UP LINK AND DOWN LINK, RESPECTIVELY

Figure 8.14   West-to-East multibeam transponder traffic.

while the controlling station in the East coverage area uses the West-to-East reference
burst to observe the timing variations. The *difference between the instants of occurrence
of the reference bursts at the statellite in the West-to-East and East-to-West transponders
must be known to accomplish ranging* by the method given earlier in this section. It
may not be possible to satisfy this requirement at system startup because of the
uncertainty in satellite position, and hence in a multibeam system, it may be necessary
to adopt alternative methods for satellite position determination.

One alternative is **double-hop ranging** between earth stations located in the
different beam coverage regions. Double-hop ranging is accomplished by measuring
the delay time between the transmission instant of a burst from a station in one
beam (e.g., West) and the reception instant at that same station of the response to
that burst from a station in the opposite beam (e.g., East) and observed at the original
transmitting station. Using a pair of stations in each beam, it is possible to obtain
four independent double-hop range measurements, as illustrated in Fig. 8.15.

Only three measurements are needed to accomplish a position solution for the
location of the satellite using the intersection of three ellipsoids of rotation. A complete
analysis of this method is given in [Lunsford, 1981] for earth station configurations
that may be encountered in the INTELSAT-V system. The stations used for ranging
would most likely be the reference stations also needed to control the TDMA network.
It has been determined that the method can locate the satellite sufficiently accurately
to allow positioning of short bursts (preamble only) from traffic terminals to accomplish
initial acquisition in the INTELSAT-V within a 4000-symbol interval. This is acceptable for TDMA initial acquisition but not for synchronized traffic operation.

A second alternative is to use predicted satellite position as estimated from
telemetery, tracking and control (TT&C) data. Presently, this source of information

P_W, P_E ARE WEST AND EAST PRIMARY REFERENCE STATIONS
S_W, S_E ARE WEST AND EAST SECONDARY REFERENCE STATIONS

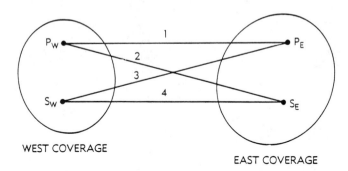

WEST COVERAGE

EAST COVERAGE

a. EACH STATION CAN MEASURE THE DOUBLE HOP RANGE TO THE PAIR OF
STATIONS IN THE OPPOSITE BEAM. ONLY THREE INDEPENDENT RANGES ARE
REQUIRED FOR A SOLUTION.
b. IT IS POSSIBLE THAT A TRAFFIC STATION CAN ALSO SERVE AS A RANGING
REPEATER

Figure 8.15   Double-hop ranging between pairs of reference stations.

is not intended to be used for a critical application such as TDMA traffic burst
position control, but even so, when the solution has stabilized, it appears to be sufficiently accurate to accomplish initial acquisition. However, during and after maneuvers
of the satellite, its accuracy is presently insufficient for this purpose. To achieve a
burst position accuracy of 200 symbols at a symbol rate of 60 Msym/s a satellite
position accuracy of ±2500 m is required.

8.3.4 Cooperative Feedback Control

For accomplishing synchronized traffic burst operation, the precision needed for burst
position control must be considerably better than that needed for acquisition. The
cooperative feedback method now discussed provides the needed precision. The cooperative feedback method of traffic burst position control is suited to synchronization
in systems where traffic stations cannot see their own burst. It uses values of delay,
D_N, as described above, but corrects them by cooperative feedback as illustrated in
Fig. 8.16. It is particularly applicable to multibeam systems. Open-loop computation
by double-hop ranging may be used to initialize values of D_N for acquisition, with
synchronization accomplished by observation of burst position error and feedback
of corrections. The following simplified example demonstrates the salient features
involved in the technique. Two beam regions, West and East, are defined. The West
region includes a reference station and one traffic station. The East region has only
a reference station. Figure 8.16 is a time/space graph showing time lines for the
reference and traffic stations in the West region, and a reference station in the East.

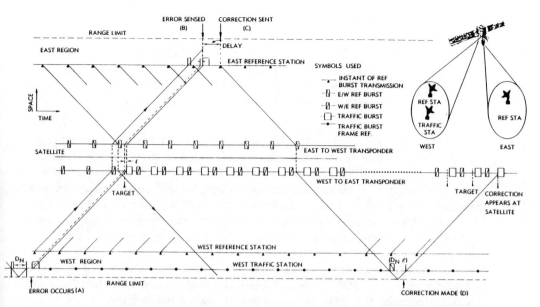

Figure 8.16   Cooperative feedback method for synchronization of traffic bursts.

Two transponder time lines are also shown, one for West-to-East and the other East-to-West traffic. The reference stations establish reference bursts in their respective transponders as shown in the diagram. It is assumed that these bursts are synchronized but not necessarily time coincident in the satellite. (The limitations imposed by this condition are discussed later.)

Traffic burst position control.   To illustrate the cooperative feedback synchronization principle, a traffic burst is followed from the traffic station to the controlling reference station. At point A on the left of Fig. 8.16, the West traffic station receives a value of delay, D_N, from the reference burst sent by the East reference station and applies it to mark its local SOTF. The initial value of D_N is obtained by the open-loop method using double-hop ranging. In the figure, it is assumed that D_N is in error by an amount e, causing the traffic burst to miss its target position at the satellite by an amount e. This error is sensed at the East reference station (point B) by observation relative to the West reference burst. The East reference station decrements (or increments) the previously transmitted value of D_N by the estimated error e and sends it (point C) to the traffic station via its reference burst. The west station decodes the new value ($D_N - e$) and invokes it on its next transmission. This results in a correction, e, to the time phase of the local TDMA frame and consequently removes the error in traffic burst position at the satellite.

The burst position *accuracy* of this method is *limited* principally by the **timing granularity** existing at the cooperating station and at the traffic station making the correction. If the clocks operate at the symbol rate, and are synchronized to the instant of clock pulse occurrence, one symbol of granularity can be encountered at each location, giving a total uncertainty due to granularity of two symbols. The granularity error will be doubled if the clock used for measurement is not synchronized to the instant of reception of the reference burst at both the reference and traffic stations. Also, the correction must be time-extrapolated to compensate for satellite motion between the instant of observation and the instant of application.

Reference burst alignment.   The position of a traffic burst is observed relative to the reference burst of the reference station located in the originating beam, but *burst timing is applied relative to the reference burst arriving from the opposite beam.* For synchronization of the cooperative traffic burst, any difference between the occurrence of reference bursts at the satellite will be automatically corrected by the cooperative feedback process described above. However, considerations of initial acquisition and transponder hopping reveal problems caused by the noncoincidence of reference bursts at the satellite.

Acquisition is accomplished by estimating a value of D_N to locate approximately a traffic burst in the TDMA frame using methods described in Section 8.3.3. Any difference between the times of occurrence of the reference bursts at the satellite will appear initially as a difference between the actual traffic burst position and the target position. If the difference is known, it can be compensated for, or, if the reference bursts are made to coincide at the satellite, the need for compensation is eliminated. Any unknown variation in the relative positions of reference bursts appears as an error in the initial acquisition process and results in initial burst position uncertainty.

When transponder hopping is performed to a third beam by stations in both the West and East beam regions, only one reference burst can be used as the reference. Any difference in the time of occurrence for the two reference bursts will appear as an equal difference in the time of occurrence of traffic bursts from each of the beam regions into the third beam. For this reason, *time differences between the instants of occurrence of reference bursts at the satellite in the transponders involved in transponder hopping should be reduced to zero.*

One method of aligning reference bursts in different transponders is to monitor their locations at a station which can see both bursts and send corrections back to one using the cooperative feedback approach. This is possible for systems involving transponder hopping between collocated beams. Another method is to achieve precision control by very accurate knowledge of satellite position and consequently of D_N value. The latter is expected to be the case for satellite systems equipped with precision ranging capabilities for open-loop control.

8.3.5 Transponder Hopping

Transponder hopping refers to the capability of an earth station either to *send traffic bursts to more than one transponder, receive traffic bursts from more than one transponder, or a combination of both.* Transmit-side transponder hopping requires that a station send two or more bursts per frame on different uplink frequencies or polariza-tions. Its most important use is to serve two or more down-beams by selectively illuminating the transponders serving those beams. Receive-side hopping requires a down-converter able to switch among the downlink frequencies or polarizations of various transponders whose down-beams illuminate a station. Its principal use is to accomplish cross-transponder operation, and thereby achieve traffic management flexi-bility among a multiplicity of transponders serving a common area. Transponder hopping is practical only when the reference bursts in all transponders used in the hopping exercise are synchronized to a common time reference.

8.3.6 Guard Time Constitutents

In this section the elementary timing uncertainties that contribute to traffic burst position uncertainty are identified and quantified. This is done for the case of *cooperative feedback control* as administered in the INTELSAT BG-42-65 TDMA system specifi-cation for multiple-beam application of the INTELSAT-V system.

The elementary timing uncertainties are identified by following the route of a traffic burst from its originating station to the remote cooperative feedback station and its return via a satellite transponder. Assume that a traffic burst is transmitted in response to an impulse generated by a local transmitting station clock. The fact that this *clock* periodically marks discrete time instants *causes a timing granularity error* having a minimum of zero and a maximum of one clock period magnitude. This will be treated as an uncertainty of ± 0.5 symbol on a 0.5-symbol bias. In a feedback system the bias can be ignored. This component is identified as E_Q. In general, the clock may be considered to operate at the TDMA symbol rate, which for a 120-Mb/s system using QPSK modulation results in a clock period of 16.67 ns. All estimates in terms of symbols used in this discussion are based on this symbol period.

The next elementary uncertainty encountered is that due to variations in the propagation time within the station between the TDMA terminal and the antenna feed caused by propagation-time variations through the cables, filters, amplifiers, con-verters, and so on, which make up the transmission path in the station. It is called the *earth station path-length variation* and is identified as E_E. It is assigned an estimated value of ± 2 symbols individually on each the transmission (up-chain) and reception (down-chain) sides of a station.

Once the traffic burst leaves the antenna it traverses the path to the satellite. A small part of this path intercepts the earth's atmosphere and is subject to a variation of approximately ± 0.5 symbol due principally to variation in moisture content. This is called the *atmospheric path elementary uncertainty* and is identified as E_A. This uncertainty is also encountered on the down-path from the satellite to the earth.

At the satellite, propagation delay variations occur due to on board transmission path components and principally to the traveling-wave-tube amplifiers when different tubes are encountered in the two directions of transmission (i.e., East to West and West to East). This component, called the *satellite path-length elementary uncertainty,* is designated as E_S and is assigned an estimated value of ±0.5 symbol which is encountered separately in each direction of transmission.

A second satellite originated elementary uncertainty is the path-length variation caused by satellite motion and the consequent displacement in satellite position between the instant the traffic burst passes through the satellite in one direction and the instant the reference burst carrying the burst correction passes through the satellite in the opposite direction. This time difference is approximately 288 ms and results in an uncertainty of approximately ±0.5 symbol. This component is called the *satellite range rate induced error* and is designated as E_R.

The last component to be accounted for is that due to internal TDMA terminal implementation and may be caused by internal reclocking. This component, called the *TDMA equipment implementation quantization,* is designated as E_I, and may have a minimum of 0 and a maximum of 1 symbol. It is treated as a variation of ±0.5 symbol with a 0.5-symbol bias.

All of the components described above are tabulated in Table 8.1 and the accumulated result of all elemental components involved in the cooperative feedback control path is shown in Fig. 8.17. From this figure it is seen that the total peak-to-peak uncertainty is

$$E_{\text{total}} = 2E_I + 2E_Q + 4E_E + 4E_A + 2E_S + E_R \tag{8.14}$$

For the estimated magnitudes of the elements given above, assuming that the clocking components E_Q and E_I which take on values 0 or 1 are each equivalent to a variation of ±0.5 symbol, the cumulative uncertainty is ±13.5 symbols. Each traffic burst will exhibit this uncertainty independently, and to assure that a pair of neighboring bursts

TABLE 8.1 ELEMENTARY BURST TIMING UNCERTAINTIES

Designation	Definition	Magnitude in symbols at 60 Msym/s
E_Q	Clocking quantization error assuming symbol rate clock	0, 1
E_I	TDMA equipment implentation quantization	0, 1
E_E	Earth station path-length variation; occurs on both up- and down-chains	±2
E_S	Satellite path-length variation; encountered when different transponders are involved	±0.5
E_A	Atmospheric path-length variation; due principally to variations in atmospheric moisture content	±0.5
E_R	Range rate induced error; due to difference in satellite position between passage of go and return bursts	±0.5

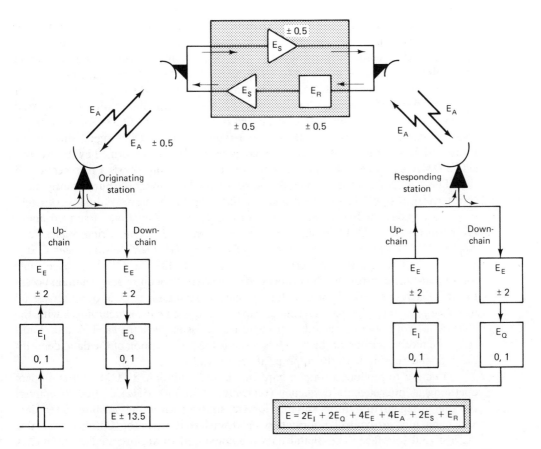

Figure 8.17 Accumulation of elementary uncertainties encountered in cooperative burst position feedback control.

do not overlap, they must be assigned a separation equal to twice the peak-to-peak variation for each burst (i.e., 54 symbols). This is referred to as the *guard time*. In the 120-Mb/s INTELSAT system, the guard time has been specified to be 60 symbols.

For global or regional beam systems that use direct loop-back burst position control, the uncertainty elements E_S and E_R can be ignored and all other components are encountered only half as many times, giving a total peak-to-peak uncertainty of

$$E_{total} = E_I + E_Q + 2E_E + 2E_A \tag{8.15}$$

This results in a peak-to-peak uncertainty for each burst of 12 symbols and a guard time of 24 symbols.

8.4 TDMA TERMINAL IMPLEMENTATION

A TDMA terminal must be capable of processing a variety of terrestrial signal formats, both analog and digital. Since TDMA is inherently a digital transmission method, analog baseband signals must be converted into a digital format prior to transmission over the TDMA link. Similarly, digital baseband signals may also require additional processing before actual transmission to reorganize channels or modify channel coding and multiplex signals to conform to recommended digital link transmission formats. This is discussed in Section 8.6.1. The transmit side of the TDMA terminal equipment must provide interface processing necessary to achieve compatibility with the terrestrial network, process the input signals as needed to format them suitably for burst mode transmission, and then convert them into a radio-frequency (RF) signal suitable for transmission over the TDMA link. An analogous function is performed by the receive-side TDMA equipment. In this case, the receive processing consists of reception of the TDMA burst from the RF link, digital processing to convert the burst transmission to a terrestrial signal format, and finally interfacing to the terrestrial network itself.

As depicted in Fig. 8.18, and also in Fig. 1.12, the functions described above are performed by TDMA terminal equipment composed of four primary elements: *common TDMA terminal equipment* (CTTE), *terrestrial interface equipment* (TIE), *burst modems* (BM), and *RF terminal equipment* (RFT). The CTTE is responsible for all real-time control and data processing necessary to acquire and maintain access to the TDMA network. Typically, this equipment maintains control of all terrestrial interface port multiplexing, determines capacity requirements, synchronizes with the TDMA network, and formats data prior to transmission over the satellite link. Similarly, the receive portion of the CTTE is responsible for reception of the data, demultiplexing, and routing it to the appropriate terrestrial port.

In order to produce a simpler hardware implementation, a TDMA burst modem operating at intermediate frequency (typically 70 to 140 MHz) is used to convert the digital signal to a form suitable for transmission over the satellite. User data and burst modem control information are transferred from the CTTE to the BM via a digital interface. The digital data are converted to an intermediate frequency (IF) signal using a selected form of digital modulation, typically quadrature phase-shift keying (QPSK). The IF signal is then converted by the RFT to the appropriate frequency for transmission through the satellite transponder.

From a satellite system user's viewpoint, the terrestrial interface equipment

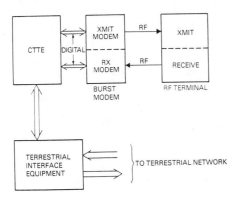

Figure 8.18 TDMA terminal simplified block diagram.

represents the most important part of the TDMA terminal equipment, since it provides the physical interface to the terrestrial equipment. In addition to maintaining a specified set of interface standards, the TIE may provide the compression and expansion buffering necessary to convert the continuous mode terrestrial data port into a burst format suitable for TDMA transmission. If necessary, the TIE is also responsible for any standard conversion to convert signaling or data transmission formats to those desired for TDMA network operation. A good example of this is the conversion of 1.544-Mb/s T-carrier formatted multiplex groupings to 2.048-Mb/s CEPT-32 formatted multiplex groupings. Hence the TIE enables the TDMA network to be transparent to the user while providing enhanced transmission capability.

8.4.1 Common TDMA Terminal Equipment (CTTE)

The common TDMA terminal equipment (CTTE) consists of all digital equipment responsible for terminal operator interface, control of terrestrial interface equipment, burst modems, and where applicable, RFT hardware. The CTTE controls the following functions: transmit start time, preamble generation, forward error correction coding, data scrambling, and terminal frequency hopping if used. It generates apertures for detecting unique words, detects receive bursts, descrambles received data, and processes common signaling channels used for network control. The CTTE is thus a central control system for the TDMA terminal which determines overall performance in terms of ease of operation, ability to achieve and maintain synchronization, accept new TDMA assignment structures, and interface to terrestrial equipment.

Even though the CTTE is physically placed between the TDMA burst modems and the terrestrial equipment, it is functionally implemented using a *bus-type architecture,* as shown in Fig. 8.19. This architecture allows the centralization of critical TDMA control functions in the CTTE while enabling simple addition of a wide variety of terrestrial interface options without change to basic TDMA hardware. Specialized interface requirements are accommodated in the TIE, thereby allowing a common CTTE design for most applications. Figure 8.19 depicts a variety of terrestrial interfaces provided in this manner. Specifics regarding these are provided later in this section.

CTTE transmit functions.   The CTTE can be thought of as having three major functional blocks associated with:

(1) maintenance of the basic TDMA frame synchronization and operator functions,
(2) control of transmit timing and data manipulation, and
(3) receive-side synchronization or data processing.

A more detailed functional block diagram of these functions is provided in Fig. 8.20. Each of the three major CTTE functions are reviewed, beginning with CTTE transmit operation.

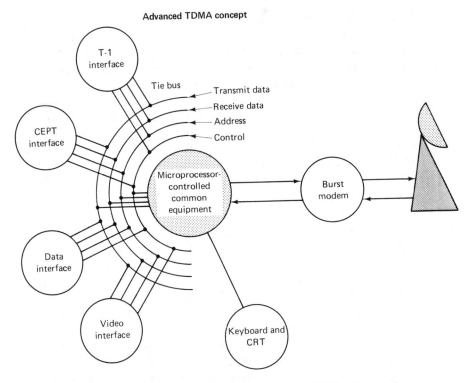

Figure 8.19 TIE/CTTE bus architecture. TIE, terrestrial interface equipment; CTTE, common TDMA terminal equipment.

Referring to Fig. 8.20, CTTE transmit operation involves collection of data from the terrestrial interface equipment buffers via the CTTE/TIE bus, preparation of data for transmission, and finally, actual transmission at the appropriate time epoch which will allow it to arrive at the satellite in its assigned TDMA frame position relative to the frame reference burst. The transmit control processor (TXC) is responsible for sequencing these functions. Satellite range information is obtained from the central controller to determine the transmit time offset necessary to compensate for propagation time to the satellite. The TXC then sequences all transmit functions relative to this offset requirement. This sequencing is best described by example. Timing control signals are generated within the resolution accuracy of one TDMA symbol and are locked to the start of the TDMA frame. The TX timing control is responsible for initiating a transmit burst. The starting locations of multiple TDMA transmit bursts are stored in the TX timing map. These maps are easily modified under microprocessor control in the event of burst-time plan changes. These changes may occur infrequently, as for a fixed assigned network, or take place dynamically in the event of full *demand assignment.* Once it is determined a transmit burst is to be transmitted, a series of well-defined events must take place. First, a TDMA preamble consisting of a *carrier* and *symbol clock recovery sequence* (CBR) is generated

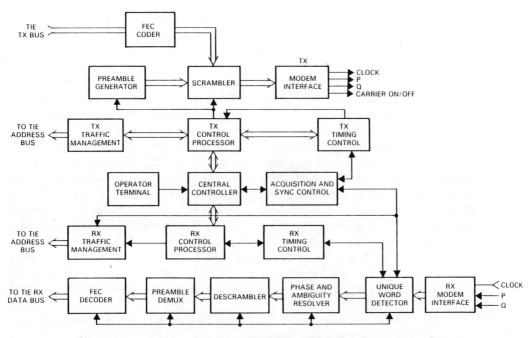

Figure 8.20   CTTE block diagram.

The CBR is followed by a unique word to indicate the start of data. Since the CBR and UW provide the basic burst synchronization capability, they are not scrambled. Only the traffic data of a burst are scrambled. Also included in the preamble may be additional fixed format data blocks used for support of system order wire and TTY communication channels. These channels provide a convenient method for TDMA terminal operators to communicate between terminals. These data, together with any following data obtained from the TIE, are scrambled before transmission. Scrambling consists of modulo-2 adding of a PN sequence to both P and Q data before transmitting in order to distribute the transmitted frequency spectrum more evenly across the satellite transponder. The length and other characteristics of the PN sequence used depend on the specific application and data rate. Scramblers for PN sequence are described in Chapter 2.

Once the preamble is generated and the scrambler functioning, the remaining portion of the burst is formed from data obtained via the TIE bus. Depending on satellite link characteristics and bit-error requirements, forward error correction (FEC) may be added to TIE data before transmission. Alternatively, FEC may be added in the CTTE. FEC is discussed in Chapter 6. Typically, some form of BCH block code is used which will, as a minimum, correct single-bit errors and detect double-bit errors. Examples of FEC applied to TDMA are given later in this chapter. Implementation of a FEC encoder may employ either serial or parallel techniques. The serial approach operates on a serial bit stream at the TDMA rate. For the INTELSAT

TDMA system, this requires serial processing at a 120-Mb/s bit rate. Because of the speeds involved, the timing is very critical and requires careful design to ensure synchronous operation at worst-case specifications. The parallel approach operates on the input data in N-bit parallel blocks after a serial-to-parallel conversion. For example, N equal to 8 in the INTELSAT case reduces the processing speed to 15 Mb/s, thereby reducing potential timing difficulties. This technique does not have a significant impact on the amount of actual hardware since lower speeds allow more liberal use of LSI components available for the slower processing rates.

Transmit data for a particular TDMA burst may come from one or more specific TIE modules. All data to be transmitted flow over the TIE transmit bus as shown in Fig. 8.20. Selection of specific TIE modules to be used for the particular TDMA transmit burst is determined by the TX Traffic Management hardware. The TX Traffic Management hardware performs all the mapping functions which match specific terrestrial data ports with selected TDMA bursts. The TIE/TDMA burst mapping is part of the TDMA terminal initialization and is updated via the TX Control Processor/ Operator interface. Physical selection of a specific TIE module is achieved by placing the appropriate range of a bus address on the TIE address bus. The resulting data, after addition of FEC if applicable, and scrambling, are transferred from the CTTE via the modem interface.

The modem interface converts the signal formats into a type suitable for use with the TDMA burst modem. For higher-data-rate TDMA applications in excess of 25 Mb/s, control and data interfaces are typically differential emitter-coupled logic (ECL) with delay compensation to allow for propagation delays over the cable interface. For a QPSK signaling format, the transmit burst modem requires a set of four signals for operation: P and Q data, TDMA symbol clock, and transmit burst enable. The signals are derived from the CTTE as shown in Fig. 8.20. Care must be taken to ensure alignment of the P and Q data lines with the transmit clock. Transmit carrier on/off is usually derived from the transmit timing control circuitry used to control the transmit preamble generator, scrambler, and FEC coder. The carrier on/off is enabled with the first bit of the TDMA burst preamble and remains so until the last data bit from the TIE used to assemble that particular TDMA burst.

CTTE receive functions.   The receive side CTTE performs an analogous function to that previously described for transmit CTTE operation. The receive modem interface connects to the TDMA burst demodulator and recovers P and Q data streams together with a receive-side clock. The receive TDMA clock represents the transmit clock of the distant TDMA terminal perturbed by Doppler due to the satellite motion and transmission path uncertainties as described in Section 8.3.6. Since in all probability the receive TDMA clock does not exactly match the internal clock used by the receiving CTTE, some compensation is required to convert the incoming data to a local synchronous clock. This is discussed later in the section.

Data and clock signals are typically available on a continuous basis from the burst demodulator whether or not a TDMA burst is present. As a result, the CTTE must provide a means of discriminating between actual TDMA transmissions and false data produced by random noise. This is accomplished through the combined effects of the receive unique word detector, receive timing control, and CTTE acquisition and synchronization processing.

Each TDMA burst contains, as part of its preamble, a unique word to mark the beginning of transmitted data channels. It is the purpose of the unique word (UW) detector to identify when these UWs occur, and in the case of multiple UWs, which one it is. This is done through a correlation process in which data segments of a length equal to the UW length are compared against the known unique word pattern. When *the total number of symbol disagreements is less than a threshold, the UW is considered found.* Since the UW correlation process itself tends to result in a large number of false detections due to random noise, an additional level of gate control is included in the UW detector. This gate, commonly referred to as **receive aperture,** is generated by the CTTE receive timing control circuitry whenever a receive TDMA burst is expected by the CTTE. The expected **time of arrival** (TOA) of a TDMA burst is determined by knowledge of the TDMA traffic burst plan relative to the TDMA frame reference burst. The frame reference burst is identified as part of the receive acquisition process to be described later. The receive timing map residing in the RX control processor referenced in Fig. 8.20 contains a map of the complete TDMA frame and can be updated via software control to accommodate burst time plan changes. If desired, a background memory can be used to store new timing information and switched on-line at a time that will ensure no receive bursts are missed.

The receive timing maps, together with TIE port/TDMA burst mappings, are updated via the receive control processor. This processor is responsible for monitoring and control of the basic receive-side TDMA functions. It provides an interface to the main CTTE central controller, to be discussed shortly.

Once a UW has been detected within the desired aperture, receive processing may begin. This consists of the inverse operation described for the transmit portion of the CTTE. Before the receive processing can be initiated, incoming data must be retimed to ensure full synchronization with the local CTTE clock. This retiming requires a single-stage buffer which can be alternately written and read.

The process of carrier recovery within the burst modem results in a phase ambiguity of as many as four different states in the case of QPSK. Since this would result in erroneous data being decoded, the receive phase ambiguity must be resolved before further processing can occur. This is usually accomplished using the TDMA UW and comparing the P and Q data of the UW pattern. Since the phase error results in specific UW states, it is a simple process to determine the correct receive phase. In the 120-Mb/s INTELSAT TDMA system four unique words, UW0, UW1, UW2, UW3, are used in the preamble.

Having correctly resolved any receive-phase ambiguity, the data can be descrambled by again modulo-2 adding a PN sequence to the received P and Q data. Since this is a modulo-2 process duplicated at both the TDMA transmitter and receiver, the original data are recovered. Sequentially, the first data following the UW are associated with the TDMA preamble voice-order wires and TTY. These data blocks are now demultiplexed from the burst. FEC decoding is next performed in the CTTE and the desired data transferred to specific TIE modules. Alternatively, FEC decoding can be performed in individual TIEs. This is performed through the combined use of the receive TIE data and address bus under control of the RX traffic management maps.

CTTE central control.   The CTTE central control processes involve maintenance of overall TDMA terminal synchronization, operator interfaces, and control of TDMA frame burst-time plan information. Terminal synchronization is a two-step process which first requires *receive and then transmit synchronization.* Receive synchronization requires reliable reception of the TDMA reference burst. Hence the first step in the CTTE synchronization process is to search for, identify, and continually track the TDMA frame reference burst. The second step of the synchronization process is to achieve transmit synchronization. In other words, the CTTE must place its transmit burst at its assigned frame location within an acceptable tolerance, often plus or minus one symbol interval. Transmit acquisition can be achieved in a variety of ways. Some systems, SBS for example, transmit an estimated range to each traffic terminal for open-loop acquisition purposes. This information is included as part of the TDMA frame reference burst. The CTTE then employs this range estimate to determine the approximate start of transmit TDMA frame time reference. Alternatively, some systems such as INTELSAT directly transmit a time delay offset to be applied relative to the instant of reception of the reference burst to establish the start of TDMA frame time reference. A short TDMA burst centered in the traffic assignment is then transmitted. This is referred to as **open-loop acquisition.** The actual terminal transmit position relative to the frame reference burst is then measured by observing the return downlink signals. CTTE transmission offset is *refined* to within required specifications using the **closed-loop feedback** information obtained from the downlink observations. Once proper transmit position is achieved, the TDMA terminal may begin normal traffic operation. After initial acquisition, transmit and receive synchronization is updated on a frame to frame basis. The actual rate at which adjustments may be made may not exceed the minimum round-trip link propagation time for the feedback control, since otherwise corrections would be made before their results could be observed at the CTTE.

The CTTE central control is also responsible for other processing related to the overall TDMA terminal performance. Typical examples are terminal operator displays and data base control. Operator entry or modification of burst-time plan information, additional TDMA terminal performance monitoring, demand assignment information processing, and maintenance of remote terrestrial control links are others. *Solve Problem 8.5.*

8.4.2 Terrestrial Interface Equipment (TIE)

The TDMA terrestrial interface equipment (TIE) establishes the link between the TDMA terminal equipment and terrestrial interface ports. A wide variety of terrestrial interface characteristics may be required by the TDMA equipment depending on the type of service; voice, video, data, or some mixture thereof. A common bus architecture has been proposed earlier to accommodate this range of interface require-ments without requiring modification to the remainder of the TDMA terminal equip-ment. This basic TIE/CTTE architecture, consisting of data, address, and control lines, was shown in Fig. 8.19. This section describes in a general manner the primary functions of a TIE module. The generalized block diagram of the interface module, consisting of compression expansion buffers, CTTE bus interface, terrestrial interface conditioner, and interface control processor appears in Fig. 8.21. The interface condi-tioner is responsible for performing any standard conversion necessary to adapt the TIE module hardware to the terrestrial interface lines.

Requirements in this area vary. For example, voice applications with analog interfaces require transformer coupled inputs and outputs with some means of process-ing associated signaling lines, if present. Once the analog signal is coupled to the interface conditioner, it must be converted from an analog signal to digital word format whose length varies with the application. Any associated line signaling must also be converted to a suitable format for processing.

In the case of a digital interface, a significantly different set of interface require-ments are present. A T-1 multiplex carrier, for example, requires an analog bi-sync interface having ±1 values. Once the bi-sync is converted to standard logic levels, a T-1 multiplex frame must be recovered, signaling and frame bits stripped off for processing, and the remaining data sent to the compression/expansion buffers.

The interface control processor is used to process any signaling and control interface information associated with the terrestrial link equipment. In the T-1 case, this would constitute frame recovery and processing of other control bits contained in the multiplex structure. The analog interface would require only processing of the control signaling lines for out-of-band or E-M lead applications.

Referring to Fig. 8.21 the purpose and operation of the TIE module compression and expansion buffers were described in Sections 8.2.5 and 8.2.6, respectively. Their primary purpose is to store and receive burst-mode data from the CTTE while provid-ing the terrestrial link with a continuous mode data port.

The CTTE bus interface controller and bus adapter provide a means of accessing the CTTE/TIE bus. The controller is responsible for local control and interface module diagnostics. It adapts specific CTTE/TIE bus addressing into specific local control commands.

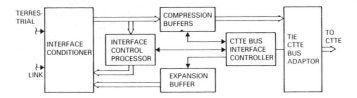

Figure 8.21 Generalized TIE module block diagram.

8.4.3 Burst Modem

The TDMA burst modem converts the all-digital data and control signals into a modulated intermediate-frequency carrier suitable for upconversion to radio frequencies and transmission over the satellite link. Quadrature phase-shift keying (QPSK) is a commonly used modulation method for TDMA applications. QPSK transmission simultaneously encodes biphase data onto in-phase (P) and quadrature (Q) carriers. Modem input and output consist of a P and Q data channel and companion data clock. These signals are shown in the modem block diagram of Fig. 8.22.

The transmit side of the burst modem consists of an interface, modulator, and output filters. The modem interface to the CTTE typically consists of a set of differential digital lines for the P and Q data channels, clock, and carrier on/off control. The interface performs the function of line receiver and includes delay units to compensate for any propagation delay differentials which might occur as a result of the modem-CTTE cabling. Using the P–Q data and clock, a QPSK signal at the selected intermediate frequency (IF) typically 70 or 140 MHz, is constructed. The carrier on-off control enables the actual modem transmit function. Filtering is included at the modem output to limit the transmitted spectrum bandwidth. Typically, this filtering consists of a Nyquist filter adjusted for the transmitted symbol period and having some form of cosine frequency-response character. As shown in Chapter 4, the modem filters may be replaced by premodulation and postdemodulation low-pass filters.

The receive side of the burst modem shown in Fig. 8.22 converts the QPSK IF signal into digital information suitable for use by the CTTE. An input filter may be included in the receive-side modem to further limit the overall bandwidth and equivalent receive noise bandwidth. An automatic gain control (AGC) is included

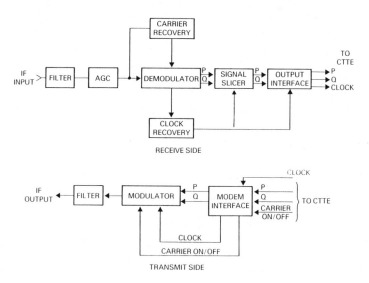

Figure 8.22 Modem block diagram.

in the receive modem to maintain a more constant signal level at the demodulator input. Before the QPSK signal can be fully demodulated, the transmitted carrier frequency and symbol clock must be recovered. A variety of techniques have been employed for carrier and bit timing recovery. A detailed analysis of these techniques is given in Chapter 7. Briefly, since a QPSK transmission inherently has no carrier frequency spectral component, it must first be regenerated and then recovered. One method commonly used for this is referred to as a times-four multiplication, in which the receive signal is twice squared. This results in the generation of the desired carrier component, which can then be filtered for receovery.

A burst symbol clock may be recovered from the received signal by detecting the signal's zero crossings or other suitable method. The QPSK demodulator, using the recovered carrier, converts the received signal into two quadrature analog channels, P and Q. The signal slicer samples the analog baseband waveforms using the recovered symbol clock. Depending on the analog signal value at the sampling instants, a one or zero digital bit is generated. These data are sent to the CTTE via the P and Q data lines together with the recovered TDMA symbol clock.

TDMA burst modems may also include a variety of performance monitoring capabilities, ranging from simple self-diagnostics to capabilities that include link performance monitors. Link bit-error-rate estimates are commonly done using monitoring facilities included in the burst modem. The SBS burst modem contains extensive self-diagnostics, together with a *link performance indicator* (LPI) which estimates link bit-error rate using pseudo-error techniques operating on the P and Q channel analog signals [Feher, K., Chapter 11, 1981].

8.5 ANCILLARY TDMA PROCESSING

8.5.1 Unique Word (UW) Miss and False Alarm Error Rates

The use of the UW in a TDMA system was described previously in this chapter. Successful detection of the UW is necessary to mark the time of occurrence of the SORF and to establish the time reference for decoding the received channels. It must be detected reliably to maintain successful TDMA terminal operation. Here, the two types of error, miss and false alarm, which can influence UW performance, are discussed.

A **miss** refers to the failure of the UW detector to produce a correlation spike that attains or exceeds a preassigned correlation threshold. For an N-bit UW, the peak amplitude of the correlation spike is N. This is true because the correlation spike is the summation of the number of correct bit coincidences which occur when the received UW is aligned with the stored UW in the UW correlator, and each coincidence is given a weight of 1 as illustrated in Fig. 8.4. If the received UW has a number of errors which result in failures to correlate, then if each failure yields a zero, the amplitude of the correlation spike is decreased by an amount equal to the number of errors. The number of such *errors allowed* in detecting the UW correlation spike is called the **correlation threshold** *E*. If the correlation threshold is set to zero, the UW must be received perfectly before it is recognized. If it is set to a value of *I*, then *I* errors anywhere in the UW are allowed in recognizing the UW.

The *probability of a miss* in detecting the UW is the probability of *more than E errors* in the UW. This probability is a function of *the number of bits N in the UW, the bit error probability p of the received data, and the threshold value E.* Let *P* represent the probability of successful detection and correspond to the event that the number of errors is equal to or less than $\dot{E}$. This is sometimes referred to as a UW **hit** (correct detection). The probability of a hit is given by the sum of the probabilities of 0, 1, 2, ... , *E* errors occurring in the UW. For any given number of errors, the number of combinations of a sequence of *N* bits out of which *I* bits are in error ($I \leq E$) and $N - I$ are correct is given by the binomial coefficient

$$\binom{N}{I} = \frac{N!}{I!(N-I)!} \tag{8.16}$$

For example, in an $N = 6$ bit-long UW, there are 15 combinations of two error bits ($I = 2$), 20 combinations of three error bits ($I = 3$), and so on. The probability of a particular sequence of *I* errors and $N - I$ nonerrors is

$$p^I(1-p)^{N-I} \tag{8.17}$$

and the probability of any arrangement of *I* errors and $N - I$ nonerrors is

$$\binom{N}{I} p^I(1-p)^{N-I} \tag{8.18}$$

The total probability of a UW hit is the sum of the above for $0 < I \leq$ E. Thus

$$P = \sum_{I=0}^{E} \binom{N}{I} p^I(1-p)^{N-I} \tag{8.19}$$

Since the probability of hits (correct detections) plus misses must be unity, the *probability of a miss* is simply

$$Q = 1 - P \tag{8.20}$$

The analysis above is illustrated in Table 8.2 for the case where the length of the UW is $N = 4$ and the correlation threshold is $E = 1$. It can be determined easily from this table that *Q* can be expressed also as

$$Q = \sum_{I=E+1}^{N} \binom{N}{I} p^I(1-p)^{N-I} \tag{8.21}$$

Graphs of the miss probability, *Q*, versus the probability of bit error, *p*, are given in Fig. 8.23 for several values of the word length *N*. It is seen that UW miss rates can be made quite low, even for fairly high bit-error rates, especially for large values of *N*. Figure 8.24 gives the plot of miss probability versus UW length. It is

TABLE 8.2   UW HIT AND MISS PROBABILITIES FOR A 4-BIT UNIQUE WORD

Number of bits in error	Number of correct bits	Number of different sequences	Probability of each individual sequence	Probability of obtaining a sequence with I errors and $N - I$ correct bits
I	$N - I \cdots$	$n = \dfrac{N!}{I!(N-I)!}$	$p^I(1-p)^{N-I}$	$n[p^I(1-p)^{N-I}]$
0	4	1	$(1-p)^4$	$(1-p)^4$
P				
1	3	4	$p(1-p)^3$	$4p(1-p)^3$
2	2	6	$p^2(1-p)^2$	$6p^2(1-p)^2$
Q				
3	1	4	$p^3(1-p)$	$4p^3(1-p)$
4	0	1		p^4
	Total	16		1.000

p = bit error probability

P = probability of UW detection = $\displaystyle\sum_{I=0}^{E} \binom{N}{I} p^I(1-p)^{N-I}$

$Q = 1 - P$ = UW miss probability = $\displaystyle\sum_{I=E+1}^{N} \binom{N}{I} p^I(1-p)^{N-I}$

$\dbinom{N}{I}$ = binomial coefficient = $\dfrac{N!}{I!(N-I)!}$

E = error threshold (number of errors allowed)

N = length of UW

seen to vary slowly and become asymptotic as the word length increases. Occasional UW misses can be tolerated provided that a "flywheeling" clock recovery circuit which replaces missing UWs is used. This can be done only in systems in which the burst being detected is predictable. In systems for which this is not true a UW miss constitutes a total miss of the traffic carried by the burst involved.

8.5.2 UW False Alarm

UW false alarm refers to the occurrence of a UW hit from the correlator when the UW is not actually present. If allowed to occur, it may be very disruptive to TDMA terminal performance because it will result in *storage of a garbled message* in the TDMA expansion buffer.

A UW false alarm can occur when bit patterns due to random noise or segments of the data portion of the traffic bursts are allowed to enter the UW correlator. There is a small but finite probability that random patterns will accidentally correspond

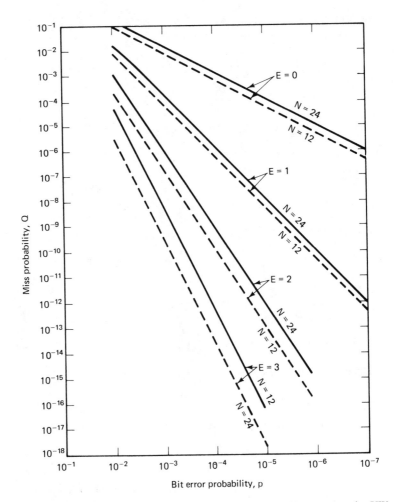

Figure 8.23 Miss probability versus bit-error probability. N represents the UW length; $N - E$ = correlation threshold; E = number of errors in UW correlation threshold.

to the UW and cause the erroneous production of the correlation spike. The *probability of a false alarm, F,* is given by the probability of accidental occurrence of the UW in the wrong location within the number of errors E established by the UW threshold. If $E = 0$, the UW must occur perfectly. Assuming that the UW has N bits, the total number of combinations which can occur is 2^N, and assuming that they are all equally likely to occur, the probability of occurrence of one that precisely corresponds to the UW is 2^{-N}. Thus for $N = 12$, the probability of a false alarm when $E = 0$ is 2.4×10^{-4}. If the number of allowed errors E is set to a nonzero value, the total number of combinations in which E or less than E errors can occur is

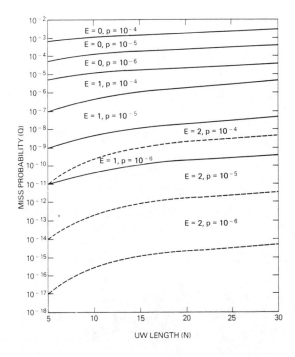

Figure 8.24 Miss probability versus UW length.

$$\text{Probability of } E \text{ or less than } E \text{ errors} = \sum_{I=0}^{E} \binom{N}{I} \qquad (8.22)$$

and since each occurs with probability 2^{-N}, the probability of a false alarm is

$$F = \frac{1}{2^N}\left[\sum_{I=0}^{E} \binom{N}{I}\right] \qquad (8.23)$$

Graphs of the false alarm rate as a function of N for various values of E are given in Fig. 8.25. The false alarm rate decreases as N increases, and increases as E increases. Considering that a separate correlation event occurs for every symbol interval of the TDMA frame, and that a large number of symbols occur in a frame, the rate of occurrence can become quite high.

Example 8.3. If the TDMA symbol rate is 60 Msymbols/s for $N = 12$ and $E = 0$, false alarms can be expected approximately once every 10000 symbols, or once every 0.167 ms; however for $N = 24$ a false alarm can be expected only once every 34×10^6 symbols or once every 559 ms.

During the acquisition process, this false alarm rate can be tolerated because no traffic is carried and special precautions can be taken to discriminate against spurious false alarms by using the expected periodicity of the true UW. However, during operation with actual traffic, additional provisions can be made to decrease the false alarm rate. One very successful means to *reduce the false alarm rate* is to establish a *time window* at the expected time of occurrence of the UW correlation

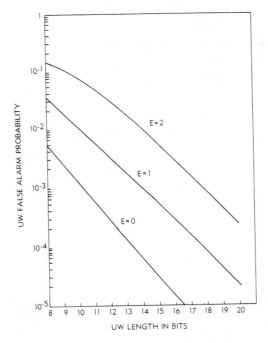

Figure 8.25 Unique word false alarm for random data signal.

spike and permit only those correlation spikes which fall in the window to pass. This can easily be done once acquisition has been accomplished and a terminal has established synchronization with the reference burst. The method is illustrated in Fig. 8.26. After every successful detection of the UW correlation spike, a window centered at the next expected instant of UW occurrence is opened. This can be done by positioning the window one TDMA frame later than the current UW hit.

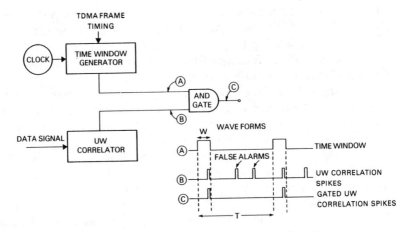

Figure 8.26 Time aperture gating of correlation spike.

The window must have a width sufficient to accommodate the expected variation in burst position due to timing anomalies encountered in the system. Typical window locations relative to the correlation instant are shown in Fig. 8.27. Part (a) shows the window at the most advanced position in which it can intercept the correlation spike. Also illustrated is the span of bits of the input data stream which are involved in the correlation. Note that the span consists of the N bits of the UW and ($W -$ 1) bits of the carrier and bit recovery (CBR) pattern. If the window is advanced any further, the correlation spike will be missed. When it is retarded, the number of bits of the CBR pattern included in the correlation processing will be reduced one bit for each symbol of retardation. These conditions are shown in Fig. 8.27(b) and (c). Since the occurrence of the correlation spike terminates any further correlation processing, none of the bits following the UW become involved.

The UW pattern is selected so as not to exhibit strong correlation when only a portion of the CBR pattern is included as part of the pattern residing in the correlation detector. Figure 8.28 illustrates the performance of one particularly well-chosen 12-bit UW when it is combined with an alternating 0, 1 pattern for the CBR. The figure shows the correlator output for various displacements of the combination of the alternating 0, 1 CBR pattern, and the 12-bit UW pattern 011110001001. The UW pattern immediately follows the 010101 . . . pattern of the CBR. The CBR pattern is used to sychronize the receive bit timing clock in preparation for decoding the UW and the following traffic data. A 1111 . . . pattern may preceed the 010101

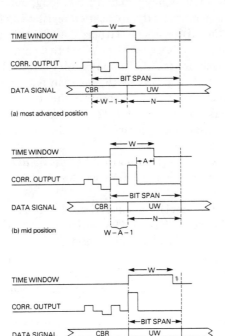

Figure 8.27 Various positions of time window relative to correlation instant. (a) Most advanced position; (b) midposition; (c) most retarded position.

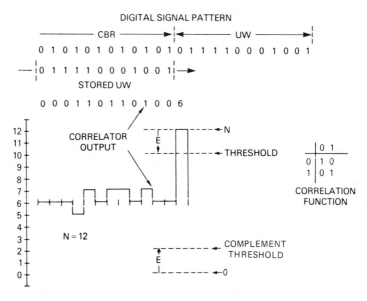

Figure 8.28   UW correlator output for CBR followed by UW.

. . . bit pattern in the CBR for the purpose of initializing the carrier recovery process. Assuming that the time window is more than 12 bits long and resides in the most advanced position, the pattern entering the correlation register can have a span including more than 12 bits of the CBR pattern. Hence, initially only an alternating 0, 1 CBR sequence resides in the correlator register. When this pattern is correlated with the UW pattern, the net correlation is $N/2$. Note that $N/2$ is the value of the correlator output for equal numbers of 1's and 0's. As the bit span pattern passes through the correlator, an increasing fraction of the UW appears in the correlation register, causing the correlator output to take on values of $N/2 + 1$ or $N/2 - 1$ occasionally. At zero advance, the UW correlator output jumps to the value N, exceeding the detection threshold E and declaring the UW hit.

Under the circumstances described above, the *probability of false alarm* may now be estimated. Consider first that the UW window is 1 bit wide and coincides with the instant of UW correlation. Then, only the N bits of the UW are admitted to the correlator register, the correlation spike is produced, and the possibility of a false alarm is zero. Next, consider that the UW window is 2 bits long and is aligned so that it terminates at the instant of the correlation. Then, during the bit period just prior to the instant of correlation, the register will contain 1 bit of the CBR plus $N - 1$ bits of the UW. For the UW illustrated in Fig. 8.28 this will yield a correlation output of $N/2$ and, of course, the correlation threshold is not exceeded. However, if this pattern should contain between $N/2 - E$ and $N/2$ errors, the correlation threshold would be reached and a false alarm would occur. For $N/2$-I errors, patterns yielding correlation hits occur in as many ways as $N/2$ items can be divided into sets of $N/2$-I and I items. Consequently, the probability of false alarm is

$$F = \sum_{I=0}^{E} \binom{N/2}{I} p^{N/2+I}(1-p)^{N/2-I} \tag{8.24}$$

Next, consider that the window is widened to 3 bits. Then two opportunities for false alarm exist and we must sum their individual probabilities of occurrence to obtain the total probability. The probability that the first group of N to enter the correlation register results in a false alarm is the value F given by equation (8.24). The probability that the second grouping of 12 bits obtained advancing the data signal by one bit produces a false alarm is the joint probability that the first bit was correct, which is $1 - p$, and that the second grouping of N bits contains $N/2 - E$, or more errors. This is given by $(1 - p)F$. The total probability of false alarm is the sum of the two cases and is $F + (1 - p)F$. Continuing this same rationale, the probability of a false alarm for a window of width W aligned so that it terminates at the instant of correlation is given by the expression

$$F_W = F + (1-p)F + (1-p)^2 F \cdots (1-p)^{W-1}F \tag{8.25}$$

or

$$F_W = F \frac{1-(1-p)^W}{p} \tag{8.26}$$

A limiting expression that is quite accurate for values of $p < 0.001$ is

$$\lim_{p \to 0} F_W = WF \tag{8.27}$$

The expressions above give the probability of false alarm for the worst possible positioning of the window relative to the instant of unique word correlation since it admits the maximum number of opportunities for generating false alarms (i.e., $W - 1$). More retarded positions reduce this number, as illustrated in Fig. 8.27. However, an additional modification must be made in certain cases. Since certain combinations of the CBR and the UW produce patterns yielding a correlator output that differs by ± 1 from the mean random value of $N/2$, the range of summation for a shift of C used in equation (8.24) should be $0 \gtrless I \gtrless (E + C)$ rather than $0 \gtrless I \gtrless E$. This change can be made selectively bit for bit once the correlation output function for the UW selected is determined.

Figure 8.29 gives graphs of the false alarm rate for $W = 2$ for the UW given in Fig. 8.28 for $E = 0$, 1, and 2. Note that the false alarm rate increases dramatically with increasing E, indicating that under normal synchronized operating conditions, the value of E should be kept small.

Figure 8.30 gives a graph of increase in the probability of false alarm as a function of window width W. The increase is not rapid, but nonetheless it is significant and W should be kept as small as possible. Values of W of 60 may be used in the 120 Mb/s TDMA system for the INTELSTAT V satellite. *Solve Problem 8.6.*

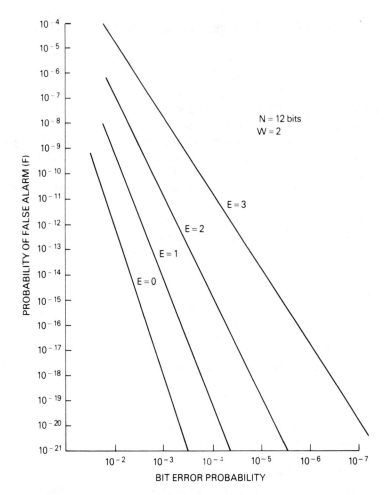

Figure 8.29   Probability of false alarm for CBR plus UW for 12-bit UW.

8.5.3 Multiple Unique Words

In some cases, several unique words may be used in a TDMA burst. The probabilities of miss and false alarm for such multiple UWs are now discussed. It is assumed that all of the UWs have the same bit pattern, although some may be the complement of the pattern. Each UW is considered to be correlated and detected separately. Hence the probabilities of miss and false alarm discussed in the preceding section apply for the individual UWs. A successful declaration of the UW hit requires that all K UWs of a given TDMA burst individually achieve their correlation thresholds. The probability of individual hits P is given by equation (8.19). Using the relation that $Q = 1 - P$, the probability of miss for K UWs can be shown to be

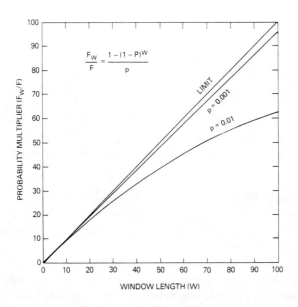

Figure 8.30   Increase in the probability of false alarm as a function of window width W.

$$Q_K = 1 - (1 - Q)^K \approx KQ \qquad (8.28)$$

where Q is given by equation (8.21). The approximation, which is valid for $Q <$ 0.01, indicates that the probability of miss is increased by K times.

Consider next the influence on the multiple UWs on the false alarm probability. *A false alarm will occur only when all K UWs are mistakenly declared present.* This is the joint probability of occurrence of K false hits in one burst. Thus if the probability of false alarm for one UW is F, as given in equation (8.22), then the probability of false alarm for K UWs is

$$F_K = F^K \qquad (8.29)$$

This indicates that the false alarm probability is significantly decreased by the use of multiple UWs.

It should be pointed out that if the capacity of the K UWs is used to form only one UW, the false alarm probability will be equal to or less than that given by equation (8.29).

8.5.4 Ambiguity Removal by UWs

In the carrier recovery circuit of a QPSK modem, the carrier component is recovered by double squaring (quadrupling) the received modulated signal, filtering this to recover a frequency which is four times the carrier frequency, and dividing the latter by 4 to recover the actual carrier frequency. The process results in a four-phase ambiguity for the recovered carrier; that is, any one of the phases $\phi = 0$, $\pi/2$, π, or $3\pi/2$, may result for the carrier. This ambiguity may be resolved by use of the four-element UW shown in Fig. 8.31. The four-element UW consists of four UWs carried on the QPSK modulated carrier. Two (UW followed by UW) are carried on the p channel, and another two (UW followed by $\overline{\text{UW}}$ where the bar indicates the complemented UW), are carried on the q channel. These are decoded individually to determine their correlation with a stored UW. The correlation spike is positive when UW is correlated with UW, and negative when UW is correlated with $\overline{\text{UW}}$. The result of the four correlations for the four different carrier phases is shown in Fig. 8.31. It is seen that each carrier phase results in a unique pattern of correlation values 1 and -1 and indicate carrier phase. This result can be used to resolve the carrier phase and permit the p and q channels to be properly decoded for the remainder of the traffic burst. *Solve Problem 8.7.*

8.5.5 Forward Error Correction

As stated previously, it is not uncommon for some form of forward error correction (FEC) to be associated with a digital transmission, particularly TDMA. Typically, FEC is implemented using linear block codes which may be realized in hardware using a simple shift register with feedback connections. Since this type of linear code generally has a well defined algebraic structure, decoding can be achieved using table look-up techniques which are readily implemented using read-only-memory (ROM) technology. A complete description of error correcting codes is given in Chapter 6. This section discusses implementation associated with TDMA applications.

An (n, k) block code consists of a word n bits in length, $n > k$, where the first k bits of a word consist of unaltered information bits. The remaining $n - k$

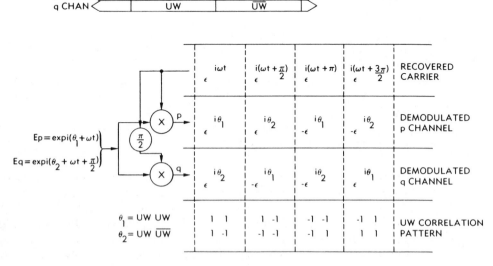

Figure 8.31 Use of UWs to resolve carrier phase ambiguity.

bits consist of parity check bits used for error correction of the first k digits. This is demonstrated below.

one block

total n bits/block

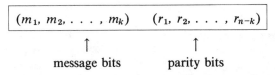

$$\begin{array}{cc} (m_1, m_2, \ldots, m_k) & (r_1, r_2, \ldots, r_{n-k}) \\ \uparrow & \uparrow \\ \text{message bits} & \text{parity bits} \end{array}$$

In the INTELSAT 120 Mb/s TDMA system a (128, 112) BCH code is used for FEC on selected TDMA traffic bursts. The specific cyclic code implementation is defined by a generator polynomial of degree $n - k$.

$$g(x) = 1 + g_1 x + g_2 x^2 + \cdots g_{n-k-1} x^{n-k-1} + x^{n-k} \tag{8.30}$$

Encoding a message of k digits requires the determination of the $n - k$ parity bits associated with the specific message. This can be accomplished using a $(n - k)$ feedback shift register with taps defined by the generator polynomial. This is demonstrated in Fig. 8.32.

Referring to Fig. 8.32, operation of the FEC encoder consists of transmitting the desired message bits while simultaneously loading the $n - k - 1$ stages of the shift register with the message bits. During the loading of the shift register, the feedback gate is enabled and the FEC output switch, SW, is in position B so that the original message data are sent from the FEC encoder. At the end of k, bits, the shift register is fully loaded with the desired parity bits ready for transmission. The feedback gate is disabled to inhibit further changes of the parity bits as they are clocked from the shift register. Output switch SW, is placed in position A and

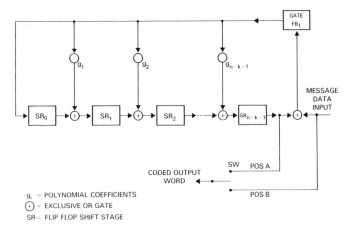

Figure 8.32   FEC encoder block diagram.

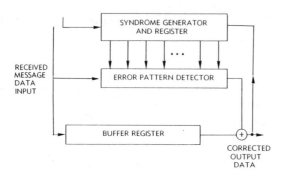

Figure 8.33   Block diagram of a generalized decoder for a (n, k) cyclic code.

the $n - k$ parity bits clocked from the register to the output channel. Note at this point a total of n bits equal to the block code length have been transmitted. This process is repeated for every set of k data bits.

     The decoding of a block code requires first determination of which bits in the received data block may be in error, and second, a correction process that corrects the received errors to the limit of the code's ability to do so. The error detection process is accomplished through the *generation of a code* **syndrome.** The syndrome is obtained by calculating the modulo-2 sum of the received parity check bits and the parity check bits generated directly from the received information bits. If the syndrome is zero, the received data block is error free. Otherwise, an error has occurred at the digit indicated by the nonzero syndrome bit. The local receiver-side estimate of the parity bits is obtained using a encoder of the form given in Fig. 8.32, but operating on the receive message bits rather than the data to be transmitted. A block diagram of a generalized decoder for a (n, k) cyclic code is given in Fig. 8.33. Using this decoder, a syndrome is generated by loading k data bits into the syndrome generator circuit. The syndrome is tested for an error pattern. When an error is detected, the error pattern detector outputs a "1" which is modulo-2 added to the current bit at the buffer register output. This addition corrects that erroneous bit. This technique is limited by the realizability of the error pattern detector which can be implemented using combinatorial logic. With the availability of high-speed read-only memory (ROM), this logic has been replaced with simple *table-lookup* procedures, thereby simplifying decoder design. A typical 1/2 rate coding example

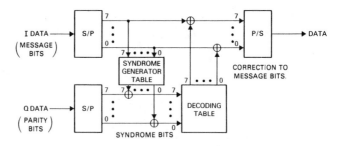

Figure 8.34   Table-lookup FEC decoder.

is shown in Fig. 8.34 for a QPSK transmission in which data bits are solely contained in the I channel and parity bits in the Q channel. The serial data are converted into parallel form and used to address lookup-table ROMs containing all possible data combinations. The syndrome generator is a table-lookup ROM. The syndrome itself is generated by parallel modulo-2 additions with the result used to address a ROM-based error-decoding table. As shown in Fig. 8.34, the decoding table directly drives a set of exclusive-OR gates which correct any erroneous data.

8.6 TERRESTRIAL INTERFACES

8.6.1 CEPT and T-Carrier Interfaces

In modern worldwide digital telephony practice, two hierarchies are commonly used. These are the CEPT standards practiced largely in European countries and a number of other nations around the world, and the T-carrier standards practiced largely in North America and Japan, and also adopted by a number of other nations around the world. Both of these standards, discussed in Chapter 2, have been recognized by the CCITT and are described in detail in CCITT Recommendations G732 and G733. Both have in common a frame rate of 8000 frames/s corresponding to a primary multiplex frame duration of 125 μs. This is because they both recognize the same 8000-sample/s Nyquist sampling rate to preserve the information content of the nominal 4000-Hz bandwidth telephone channel. Both also recognize the use of 8 bits per sample for PCM encoding of amplitude information, giving a basic rate of 64 kb/s for carrying a telephone channel, but they use *different companding laws* which prevent the signals encoded by one from being directly decodable by the other. They also differ in the structure of their primary multiplex configuration. In CEPT, a primary multiplex group comprises 32 time slots, each having 8 bits, as shown in Fig. 8.35(a), yielding a frame having 256 bits and occurring in a time period of 125 μs and having a transmission rate of 2.048 Mb/s. In T-carrier, a primary multiplex group comprises 24 time slots, each having 8 bits, as shown in Fig. 8.35(b). However, one extra bit is appended to the frame, giving 193 bits in a period of 125 μs, yielding a transmission rate of 1.544 Mb/s.

The traffic data part of a TDMA burst structured to carry T-carrier or CEPT-32 primary multiplex traffic is shown in Fig. 8.36. It is assumed that the TDMA frame has a duration equal to W primary multiplex frames (i.e., $T_F = WT_{\text{PRI}}$, where T_{PRI} is the multiplex frame period). For Fig. 8.36, $T_{\text{PRI}} = 125$ μs and $W = 16$, yielding $T_F = 2$ ms. Each satellite channel contains 128 bits (64 QPSK symbols) for carrying the 16 Nyquist samples occurring during the TDMA frame period. Each **satellite channel** can be assigned to a different destination if desired, or alternatively, blocks of channels can be assigned to different destinations. The number of channels carried depends on the traffic capacity of the burst. If all of the traffic channels appearing at the interface to the system are carried continuously over the system, the number of channels in the burst will equal the number appearing at the interface.

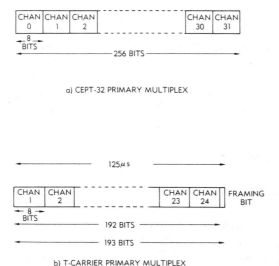

a) CEPT-32 PRIMARY MULTIPLEX

Figure 8.35   CEPT-32 and T-carrier primary multiplexes.

However, if a technique such as digital speech interpolation is used at the interface in which satellite channels are shared among the terrestrial channels in accordance with voice spurt activity, the number of satellite channels will be less than the number of terrestrial channels.

8.6.2 Digital Speech Interpolated (DSI), and Digital Noninterpolated (DNI) Interfaces

TDMA systems are inherently compatible with the digital signal format characteristics of the CEPT and T-carrier structures, as illustrated in the preceding section. An interface to a TDMA system which is organized to carry individual primary multiplex channels in permanently assigned satellite channels located in a traffic burst is known as a digital noninterpolated (DNI) interface. The designation DNI is used to distinguish it from the digital speech interpolated (DSI) interface, which is described below. The structure of the traffic data part of a burst for carrying either DNI or DSI channels is shown in Fig. 8.36. Typically, DNI interfaces can be arranged to carry selected individual primary multiplex channels, or complete primary multiplex groups

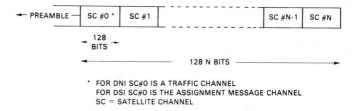

* FOR DNI SC#0 IS A TRAFFIC CHANNEL
FOR DSI SC#0 IS THE ASSIGNMENT MESSAGE CHANNEL
SC = SATELLITE CHANNEL

Figure 8.36   Traffic data portion of TDMA burst for DNI and DSI use.

of channels such as a CEPT-32. Carrying complete 32-channel groups of CEPT-32 has the merit of retaining the association between the signaling and supervisory information carried in time slots 0 and 16 and the 30 telephone channels of the group.

Digital speech interpolation (DSI) interfaces assign *satellite channels* (SCs) to *terrestrial channels* (TCs) only when a speech spurt is declared present on the terrestrial channel. The DSI interface unit contains a digital speech detector which detects the presence of the speech signal while discriminating against line noise. Whenever the speech detector declares speech to be present, a TC is assigned to an SC. The SC assigned is a purely arbitrary choice and may change from one speech spurt to the next for the same TC. Assignments must be indicated to the destination terminal. This is done by means of an assignment message carried on an assignment message channel which identifies the TC/SC associations generated by the DSI transmit side. Because the SCs are used only during active speech intervals and may serve any of the TCs on demand, the ratio of the number of TCs served to the SCs serving is typically $2:1$ or more when the number of TCs is greater than 60. This ratio is referred to as the DSI **gain.**

A curve of DSI gain for various numbers of TCs served, assuming a speech spurt activity of 40% on each TC, is shown in Fig. 8.37. DSI gain is limited by the quality degradation caused by a phenomenon called *competitive clipping.* A competitive clip occurs whenever a TC on which a speech spurt is declared present cannot immediately be assigned to an SC. Any time spent waiting causes loss of a corresponding fragment of speech, and this is called a competitive clip. The DSI gain curves given in the figure are based on occurrence of a competitive clip of 50-ms duration or less occurring for 2% of the time or less. A 50-ms clip constitutes an event that may occasionally be just noticeable, but not perturbing.

To improve further the quality of the DSI interface unit a technique called **bit stealing** may be introduced to generate overload channels. Whenever a TC cannot be served because all normal SCs are busy, a situation that would otherwise result in a competitive clip, an **overload channel** is generated by stealing the least significant bits (LSBs) of seven normal SCs. This reduces the number of bits carried on the overload channel and the SCs from which the LSBs are stolen, slightly increasing the quantizing noise on these channels. The overall decrease in quality due to quantizing noise increase is small as long as the DSI gain is not increased significantly over that acceptable for competitive clipping operation, while the degradation caused by competitive clipping is essentially eliminated. A curve of DSI gain versus the number of SCs served for overload channel operation with 1 dB of average quantizing noise increase on the channels influenced is also given in Fig. 8.37.

8.6.3 Interfacing Satellite Links to Digital Terrestrial Facilities

When satellite links are used as part of an overall end-to-end connection, their peculiarities need to be accounted for. Almost all satellites that are used in the public telephone and other private networks reside in the *geostationary orbit.* The geostationary orbit

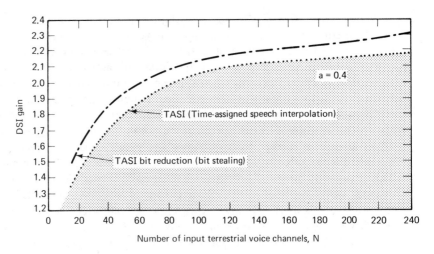

Figure 8.37 DSI gain with and without overload channels.

has a radius from the earth's geocenter of 41,600 km, which places it at an altitude of 35,200 km. In this orbit the satellite moves at the same angular rate as the rotation of the earth, and if it is placed in the equatorial plane, the satellite appears to remain stationary over the earth. Actually, the satellite does not remain truly stationary. Inaccuracies in adjusting the orbital parameters, combined with the influence of non-uniformities in the earth's gravitational field, and the perturbations caused by the moon and other celestial bodies cause the satellite to depart from the true stationary orbit. The departure is characterized by oscillations about a mean position which have the periodicity of the **sidereal day** [the time it takes the earth to rotate back to the same constellation (86,163.3 s)]. In addition, there is an East/West drift in the mean position which is compensated once every thirty to forty sidereal days by orbital maneuvers. This motion causes the propagation path distance between the satellite and points on the earth to vary with a magnitude sufficient to require compensation at earth stations which use satellite links for transmission of digital data. In the following text, details regarding the magnitude of the variation, the impact it has on digital clocks recovered at various earth stations, and the means for compensating for clock differences thereby generated are discussed.

Elements of satellite motion. A geostationary orbit satellite never remains truly stationary; instead, it moves about in a box centered at its nominal location. A satellite orbit can be specified in terms of its inclination angle relative to the equatorial plane and its orbital eccentricity. These elements are illustrated in Fig. 8.38(a) and (b). In addition, the nominal position of the satellite can drift slowly in the East/West direction over a period of many sidereal days, as illustrated in Fig. 8.38(c). Orbital plane inclination angles can vary over a range of as much as ± 0.5° relative to the equatorial plane, and the orbit eccentricity can typically have a value as great

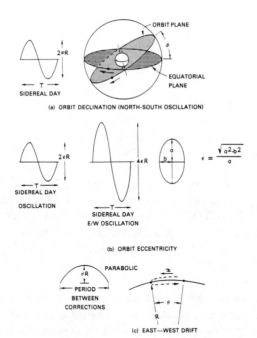

Figure 8.38 Satellite position deviation components.

as $\epsilon = 5 \times 10^{-4}$. For the radius of the stationary orbit (41,600 km), the peak-to-peak East/West variation due to eccentricity is 84 km, while the peak-to-peak altitude variation due to eccentricity is 42 km. The maximum peak-to-peak North/South position variation due to a 0.5° orbital inclination angle is 728 km. All of these variations exhibit a sidereal day period of 86,160 s. In addition, the long-term East/West drift can be as much as 146 km. The amount of motion seen by an earth station depends on the position of the earth station on the surface of the earth. For the orbital parameter variations just cited, the maximum variation in one-way propagation time between satellite and earth is 550 μs peak to peak. If the variation in the satellite orbital plane inclination is reduced to 0.1°, the maximum variation in one-way propagation time is 250 μs. All of the values thus far given are for peak-to-peak sinusoidal variations having a sidereal day period. These variations produce an associated Doppler shift determined by the peak-to-peak variation and the sidereal day period. For the 0.5° orbital inclination, the maximum Doppler is 4×10^{-8}. If the orbital plan inclination is reduced to 0.1°, the maximum Doppler is reduced to 2×10^{-8}.

It is important to note that the TDMA *frame rate is different* on the up- and downlinks to the satellite by twice the Doppler experienced on either link separately. This is because the TDMA *frame reference is really on the satellite.* Hence, if a Doppler Δf is experienced on the downlink, the frame rate transmitted from a terminal must be shifted by $-\Delta f$ so that upon arrival at the satellite it occurs at the on-board frequency.

Digital interface structures.   The interfaces to the TDMA compression and expansion buffers may operate in either a synchronous or asynchronous fashion. When operated synchronously, the clock rate of the continuous digital stream, which is the input to the compression buffer and the output from the expansion buffer, must be an integer multiple of the TDMA frame frequency. When operated asynchronously, the clock rate is not an integer multiple of the TDMA frame frequency. Each of these methods of operation is influenced by the path-length variation and associated Doppler characteristic of satellite links and must be compensated appropriately to achieve continuity of data flow.

1. *Synchronous operation—terrestrial clock locked to TDMA frame rate.* The simplest method to achieve synchronous operation is to derive the terrestrial system clock as a multiple of the TDMA frame rate. This must be done separately on the transmit and receive sides of the terminal. It can be practically applied when the terrestrial digital streams terminate directly into analog-to-digital or digital-to-analog converters which can easily absorb the rate variations induced by the satellite motion.

2. *Synchronous operation—terrestrial clock not locked to TDMA frame rate.* This form of synchronous operation is achieved by placing a buffer at each interface to smooth the Doppler variation. It must be able to store the number of bits that occur in an interval equal to the peak-to-peak propagation time variation over the satellite path.

Because of the satellite motion, the length of the propagation path between any two earth stations of a network will change. As indicated above, the worst-case one-way propagation time change between an earth station and a satellite is 550 µs. Such a time change could be encountered on both the up- and downlinks of a satellite, thus doubling the change for the entire path to 1.1 ms. To absorb the rate differences caused by satellite motion, an interface such as that shown in Fig. 8.39 must exist between a digital terrestrial network and a satellite network. An aligner and Doppler buffer unit are included in the transmit and/or receive sides. The function of the Doppler buffer can be simply understood by thinking of it as a variable extension

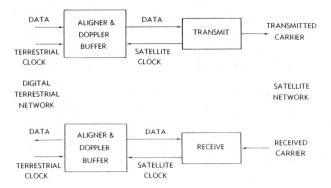

Figure 8.39   Interface between satellite and digital terrestrial links.

of the satellite propagation path between the earth stations which maintains a constant propagation delay. Thus if a path between the transmitter of one station and the receiver of another undergoes a propagation variation of ΔT, the Doppler buffer introduces a compensating delay of ΔT. It does this by storing as many bits of the data message as occur during the interval equal to ΔT.

To allow Doppler buffer operation to be initiated arbitrarily, it is necessary to give it a capacity of $2\Delta T$. In this way it can be initiated at the center of its range and compensate fully in either the $+\Delta T$ or $-\Delta T$ directions. For example, consider a 2.048-Mb/s primary multiplex terrestrial link. If $\Delta T = 1.1$ ms, the capacity of the Doppler buffer is 2×2.048 Mb/s $\times 1.1$ ms $= 4096$ bits. The alignment function is related to accomplishing plesiochronous operation and is described later.

A very important property of the Doppler buffers is that they compensate for satellite motion and result in perfect synchronization of the data stream at the time-varying interface between buffers and the TDMA terminal while providing continuous data rates at their terrestrial interfaces. This synchronization is absolutely necessary if the individual channels of the primary multiplexes are to be aligned in the TDMA frame, as discussed previously in this chapter. This is achieved only if separate Doppler buffers are used on the transmit and receive sides of all terminals in the network.

Asynchronous operation.   When the symbol rate of a digital transmission signal entering the compression buffer is precisely a multiple of the TDMA frame frequency, the interface is said to be synchronous. Otherwise, it is asynchronous. If the nearest multiple of the TDMA frame rate is slightly lower than the synchronous rate, the buffer will occasionally transmit an extra symbol in the TDMA traffic burst to maintain continuity of information flow, and if it is slightly higher, the buffer will occasionally transmit one less symbol. This action is illustrated in Fig. 8.40. The addition or subtraction of a symbol to the transmitted burst will occur only occasionally, as determined by the necessity to conserve all of the symbols in the data stream.

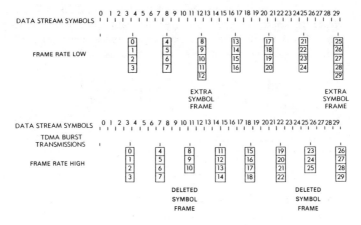

Figure 8.40   Asynchronous operation at a TDMA direct digital interface.

The following analysis provides the relationships for calculating the number of TDMA frames C between these buffer corrections. If the TDMA frame period T_F departs by an amount ΔT_F from a period that is the nearest integer multiple m of the symbol period T_p of the terrestrial data stream, the number of TDMA frames between buffer corrections is given by

$$C = \frac{T_p}{\Delta T_F} \qquad (8.31)$$

ΔT_F is given by the expression

$$\Delta T_F = T_F - mT_P \qquad (8.32)$$

where m is selected to be an integer that minimizes the absolute value of ΔT_F. The value of m is given by the relation

$$m = \frac{T_F}{T_p} \text{ nearest integer value} \qquad (8.33)$$

Example 8.4

Consider now the application of the foregoing relations to a typical situation encountered in a TDMA satellite system. Assume that the TDMA frame period T_F is 2 ms and that a direct digital terrestrial link having a rate of 1.544 Mb/s is to be carried over the system. A simple calculation (given by $T_F R_o$) shows that each TDMA frame should carry 3088 bits, or 1544 QPSK symbols, since each symbol carries 2 bits of the data stream. Thus the symbol period T_p is the duration of two bit periods and hence equal to $2 \times (1.544 \times 10^6)^{-1} = (0.772)^{-1}$ μs. Performing the calculation for the integer m using equation (8.33) yields an integer value of 1544. From equation (8.32), ΔT_F is found to be equal to zero. This situation would prevail only if the satellite remained perfectly stationary relative to points on the earth. Even though the satellite is said to be in a stationary orbit, it only approximates such an orbit. Analysis of the departures from stationary orbit reveals that for worst-case circumstances, the one-way path between the satellite and the earth can exhibit a Doppler of 20 ns/s. Since the reference burst from the reference station traverses this path twice before arriving at a traffic station, the TDMA frame period can exhibit a Doppler contribution of $d(T_F)/dt = 40$ ns/s. Thus the value of ΔT_F given by equation (8.32) will become

$$\boxed{\Delta T_F = T_F \frac{d(T_F)}{dt}} \qquad (8.34)$$

For $T_F = 2$ ms, ΔT_F is 0.08 ns per TDMA frame. Using equation (8.31), the number of TDMA frames between buffer corrections is found to be $C = 25,000$. The calculation above is for an extreme case, and buffer corrections should be expected far less frequently. ∎

When the TDMA bursts are received, they must be supplied to an expansion buffer, and for asynchronous operation, it must be designed to allow for the occasional variation in the number of bits contained in the burst. Assuming the expansion buffer's output data stream is at the same rate as the input data stream to the compression buffer at the far-end transmitter, the data will flow over the TDMA link without loss.

Plesiochronous interface operation.   Plesiochronous operation at an interface is said to prevail when the clocks governing the networks on either side are derived from sources exhibiting uncertainty of $\pm 10^{-11}$ or less. This will result in a 125-μs primary multiplex frame slip once every 72 days. Such operation is advised by CCITT in its Recommendation G.811 [Yellow Book, Vol. III, 1981] at interfaces between digital networks. In the buffer designs that are considered for the interface to the satellite system, the combination of the Doppler correction and the plesiochronous alignment can be combined. Hence the buffers are referred to as *alignment and Doppler buffers*.

To absorb the rate differences caused by satellite motion, an interface such as that shown in Fig. 8.39 must exist between a digital terrestrial network and a satellite network. An aligner and Doppler buffer unit are included in the transmit and/or receive sides. The Doppler function of the buffer was described previously and here its alignment function is described. The fill-state diagrams of Fig. 8.41 illustrate operation of a simple plesiochronous buffer (aligner). A continuous RAM buffer is assumed, with read–write addresses progressing clockwise as shown in Fig. 8.41(a). If the output clock is faster than the input clock, the buffer will empty to some threshold point, as illustrated in Fig. 8.41(b). At this point the aligner will repeat a frame. When the output clock is slower than the input clock, the buffer will fill to a threshold point, as shown in Fig. 8.41(c), and delete a frame. If the size of the Doppler portion

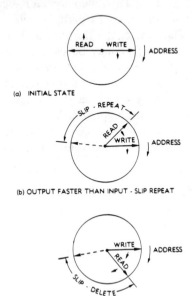

(a) INITIAL STATE

(b) OUTPUT FASTER THAN INPUT - SLIP REPEAT

(c) OUTPUT SLOWER THAN INPUT - SLIP DELETE

Figure 8.41 Fill state diagrams for the multiplex frame alignment portion of the Doppler/alignment buffer.

of the buffer is sufficient to accommodate the satellite position changes, and the consequent changes in propagation time and the clocks are controlled to the accuracies specified in CCITT Recommendation G.811, the interfaces to the satellite system will exhibit a performance consistent with the plesiochronous network requirements.

Alignment/Doppler buffers for various system configurations.   TDMA transmission links in a global/regional beam are illustrated in Fig. 8.42. Station 1 is assumed to establish the TDMA system frame reference period by relating it to an appropriate multiple of the period of f_o. For example, the TDMA frame period may be an integer multiple of the 125-μs frame period used in both T-carrier and CEPT digital transmission rates, and the rate f_o could be either the 1.544-Mb/s T-carrier rate or the 2.048-MB/s CEPT-32 rate. The reference station sends a reference burst to the satellite, and as a result, establishes a network-wide timing discipline observed by all other stations. This timing discipline is the direct result of the fact that each station's transmission burst is assigned a specific epoch in the TDMA frame relative to the reference burst which it must maintain by some means of control. In the following discussion, for simplicity rates in the satellite are not multiplied by the TDMA burst rate expansion ratio, but are expressed as equivalent continuous transmission rates. This has no impact on the validity of the results.

As a consequence of the reference stations' action, the transmission rate at the satellite is $f_o + \Delta f_1$, where Δf_1 represents the Doppler shift on the uplink to the satellite. At traffic station 2, as a consequence of the action of placing its burst at its assigned location in the frame, and of the Doppler shift Δf_2 on the path between the station and the satellite, the receive side rate must be $f_o + \Delta f_1 + \Delta f_2$, and transmit side rate $f_o + \Delta f_1 - \Delta f_2$. The difference $2\Delta f_2$ between the transmit and receive sides is due to the satellite-to-earth path Doppler. All earth stations will exhibit a similar property. The sum of the transmit and receive side rates is twice the rate on the satellite, and this fact may be used to recover the satellite clock rate. Note also that at the reference station, its receive side rate is $f_o + 2\Delta f_1$. At all interfaces in this example, except for the transmit side of the reference station, a Doppler buffer capable of absorbing a path-length change of 1.1 ms corresponding to one up- and one down-path should be provided.

In *multibeam* networks, stations in a beam are timed relative to a reference burst from a reference station in the same beam. The control of timing may be

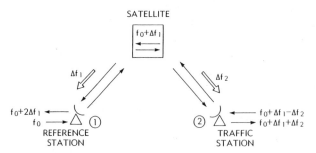

Figure 8.42 Clock rates for TDMA global/regional beam.

accomplished either by cooperative feedback from a station in the opposite beam or open-loop control. In this discussion, two beams are considered, one designated East and the other West. The beams are assumed to be connected by transponders in the satellite. The clock rates appearing at various terminals are influenced by the method used to synchronize the reference stations. Three cases are considered: independent reference station clocks, single clock with uncompensated feedback control, and single clock with compensated feedback control.

The situation pertaining to independent reference station clocks is illustrated in Fig. 8.43. Reference station 1 in the West and 2 in the East each derive timing from *separate clocks* at a frequency f_0. The clocks used could be high-accuracy references, meeting CCITT Recommendation G811 (i.e., exhibiting an uncertainty of $\pm 1 \times 10^{-11}$). Note that the clock frequency appearing in the satellite is different in each direction of transmission because of the different Doppler frequencies on each up-path. Thus the TDMA frame periods at the satellite would also differ slightly. This would constitute a *problem for frequency hopping* in systems using additional beams. The influence on clocks at traffic stations is illustrated at station 3 in the West and station 4 in the East. Each station experiences the influence of the path length change of two satellite-to-earth paths, and should therefore be equipped with Doppler buffers able to absorb 1.1 ms of propagation time change. Note also that on the transmit sides of each traffic station, the net Doppler is the difference between the uplink Doppler for the reference station and the uplink Doppler for the traffic station. For example, the net Doppler is $\Delta f_1 - \Delta f_3$ for traffic station 3 in the West. If this station is near the reference station, Δf_1 and Δf_3 differ by a small amount and the influence of Doppler tends to be canceled, with the consequence that the net path-length change experienced would be correspondingly small. Hence the Doppler buffer could be correspondingly smaller.

Figure 8.44 shows the situation prevailing when the West reference station is the system clock and uncompensated loop-back control of reference burst timing is used at the East reference station. Uncompensated loop-back control refers to the fact that the East reference station synchronizes its reference burst transmission to the reception of the reference burst from the West reference station by adding a *constant delay* adjusted to compensate for the *mean path-length correction* needed to align the reference bursts in the satellite. In this case, only the West reference

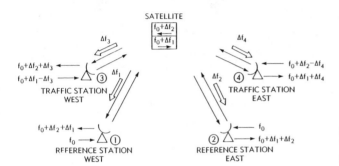

Figure 8.43 Clock rates for multibeam TDMA system with independent reference stations.

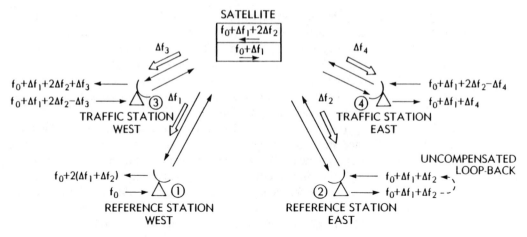

Figure 8.44 Clock rates for multibeam TDMA with master reference and uncompensated cooperating reference station.

station synchronizes its reference burst transmission to the high-accuracy clock f_0. Study of the situation shown in Fig. 8.44 illustrates that the clocks received at the traffic stations are influenced by as many as four satellite-to-earth path components, and require Doppler buffers able to absorb 2.2 ms of propagation time change. Note also that a difference of Δf_2 exists between the two directions of transmission in the satellite, indicating that the TDMA frames would also differ, causing difficulties if transponder hopping were introduced to another beam.

Figure 8.45 shows the situation prevailing when the West reference station is

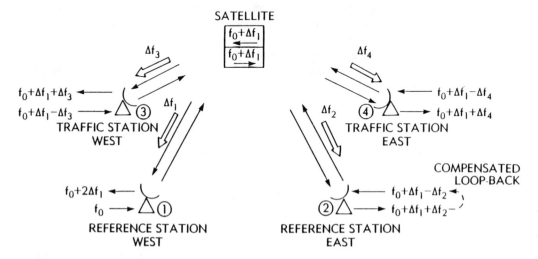

Figure 8.45 Clock rates for multibeam TDMA with master reference and compensated cooperating reference station.

the system clock and compensated loop-back control is used at the East reference station. Compensated loop-back control refers to the fact that the East reference station synchronizes its reference burst transmission to the reception of the reference burst from the West reference station by adding an *adjustable time delay* that is selected to maintain *coincidence of the reference bursts at the satellite* at all times. Examination of the clock rates received at all traffic stations in both the East and West shows that at most, only two satellite-to-earth paths contribute to the total path delay variation. Thus Doppler buffers able to absorb 1.1 ms of propagation time variation are sufficient at all locations. Also for stations near the reference station with the clock f_0, the transmit-side Doppler and net path-length variation are small due to the cancellation effect previously noted in the discussion pertaining to Fig. 8.44. Since the rates at the satellite in both directions are the same, the situation is suited to transponder hopping to other beams.

8.7 TDMA SYSTEM EXAMPLES

TDMA system designs have evolved from simple fixed-assignment, low-data-rate global network configurations to network architectures providing fully variable demand assignment capabilities and data rates in excess of 120 Mb/s. A variety of TDMA network designs have either been implemented or are in the process of implementation. Table 8.3 summarizes the basic parameters of four such TDMA networks: two INTELSAT systems, a Satellite Business Systems network, and the French TELCOM I system. The INTELSAT BG-1–18 system featured a 60-Mb/s data rate with

TABLE 8.3 HIGHLIGHTS OF ILLUSTRATIVE INTELSAT AND OF DOMESTIC TDMA SYSTEMS

	INTELSAT BG-1–18	INTELSAT BG-42–65	SBS	TELECOM I
Frame period	750 μs	2 ms	15 ms	20 ms
Transmission bit rate	60 Mb/s	120.832 Mb/s	48 Mb/s	25 Mbs
Demand assignment	No	No	Yes	Yes
Central reference station control	No	Yes	No[a]	Yes
Acquisition method	Low power burst	Open-loop ranging	Open-loop ranging	Open-loop ranging
Synchronization method	Direct loop-back	Cooperative loop-back	Direct loop-back	Cooperative loop-back
Modulation	QPSK	QPSK	QPSK	DPSK
Transmission coding	Diff.	Nondiff.	Nondiff.	Nondiff.

[a] Central reference station control is planned for mid-1980s.

Figure 8.46 Photograph of an advanced low-cost TDMA terminal developed by COMSAT Laboratories. This 60 Mb/s terminal was field-tested in the Pacific Ocean region.

fixed TDMA frame architecture and was the forerunner of today's 120-Mb/s (BG-42–65) INTELSAT TDMA. It was intended for global operation and was implemented only on a field trial basis. The BG-42–65 system is intended for multiple-beam operation and utilizes a more sophisticated centralized network control concept. The Satellite Business Systems (SBS) network is intended for commercial corporate applications with small customer premise earth stations. The network is capable of supporting a variety of communications links, including both voice and data applications, with fully variable demand assignment of network capacity. The TELCOM I system is intended to provide service similar to that of the SBS network.

PROBLEMS

8.1. Explain the advantages of TDMA systems in comparison to FDMA systems. How many times do we increase the satellite capacity by using TDMA/DSI technology? (Why does this increase occur?)

8.2. What is the duration of the frame period of the INTELSAT 120.832 Mb/s TDMA system. Why is this frame period specified? How many bits are transmitted per TDMA frame for

a) a 2.048 MB/s (CEPT) primary multiplex rate?

b) a 1.544 Mb/s (T-1 carrier) primary multiplex rate?

8.3. Assume that the number of symbols in each reference burst is 330, the number of symbols in the traffic burst preamble is 300, the transmission rate is 180 Mb/s, the TDMA frame is 2 ms and there are 16 traffic bursts in the frame. The system has two reference bursts and the guard time between bursts, expressed in number of symbols is 75 symbols. How much is the frame efficiency of this TDMA system? How does your result compare to the result given in Example 8.2? How could you increase the frame efficiency of your system? What does an increased frame efficiency mean for the end user? How does an increased efficiency impact on the system hardware specifications? Which parts of the TDMA hardware will be changed?

8.4. What are the advantages of first order satellite range prediction compared to zero order range prediction? Are there any disadvantages?

8.5. List the main functions of the common TDMA terminal equipment (CTTE). Try to list these functions, without referring to your book. What are the functions of the receive aperture gate?

8.6. Explain the relationship between the height of the unique word correlation spike and the probability of false alarm. Are these quantities related? How? Is the occurence of a miss (in the unique word detection) or of a false alarm more disturbing? How is the practical system operation modified by the error rates of these quantities? Specify a somewhat marginal unique word miss and false alarm error rate. Could your choice lead to a lower cost TDMA earth station implementation?

8.7. The four-phase ambiguity introduced by conventional carrier recovery circuits of QPSK demodulators can be resolved by the use of four element unique words or by differential decoding (assuming the modulator contains a differential encoder—see Chapter 4). What are the advantages/disadvantages of these methods?

9

REGENERATIVE (ON-BOARD PROCESSING)
SATELLITE SYSTEMS

9.1 INTRODUCTION TO REGENERATIVE (ON-BOARD PROCESSING) SATELLITE SYSTEMS

Time-division multiple-access (TDMA) systems, described in Chapter 8, have been utilized for providing high-capacity flexible digital satellite communication services. For example, the INTELSAT-V satellites, in an all-digital TDMA mode, could transmit over 1 billion bits of information a second (1 Gb/s)—enough capacity to allow the *Encyclopedia Britannica* to be sent from, say, the United States to Italy six times a minute. The INTELSAT-VI satellite is planned to attain its highest capacity in the *satellite-switched* (SS-TDMA) mode. In this mode it uses a pair of 6×6 dynamic RF switches and is expected to have an operational capacity of 35,000 digitized voice circuits or the equivalent of 2.5 to 3.0 Gb/s (10^9 bits/s) [Astrain, 1981].

The phenomenal growth rate in the demand of satellite circuits and services, as described in Chapter 1, together with the increase in the number of digital communication satellites and their increased capacity, necessitates a detailed review of transmission and modulation procedures. One of the major factors contributing to the cost of TDMA and of related satellite systems is the generation of high transmitted power at both the earth station and the satellite. To achieve this high output power, which offsets the large propagation losses, large transmit and receive antennas are also required for economic reasons. For a cost-effective system, the radio-frequency (RF) amplifying devices must be operated in a nonlinear mode (i.e., close to saturation). These nonlinearities degrade the performance of digitally modulated bandlimited signals.

The nonlinear operation of the power amplifier spreads the spectrum of the previously filtered modulated signal. To reduce spillover into adjacent channels, an RF filter is needed. This bandlimited AM/AM and AM/PM converted (distorted) signal is further degraded by various satellite components, including the transponder input multiplex filter, the output TWT, and the output multiplex filter. The main satellite degradation is caused by interaction caused by *cascading* two bandlimited nonlinearities. *Intersatellite links* may have more than two cascaded nonlinearities, in which case even more degradation can be expected. In a number of operational high-speed digital satellite systems (operating in the range of 100 Mb/s per transponder), a significant degradation from theoretical performance is measured (e.g., 5 dB). This has a drastic economic impact if one considers that several dB in additional EIRP (effective isotropic radiated power) amounts to a considerable expenditure in satellite or earth stations.

More spectrally efficient modulation methods, such as those described in Chapter 5, may be employed in order to utilize the spectrum better. These factors aggravate the problem of uplink power. Spectrum conservation methods also include frequency reuse and multiple-spot-beam antennas. With these sophisticated changes, co-channel interference is envisioned as becoming an important performance-limiting factor [Cuccia et al. 1977; Koga et al., 1977].

An innovative solution to these problems is the use of on-board signal processing-regenerative satellite repeaters instead of the conventional translating repeaters.* The "regenerative satellite" demodulates the incoming uplink signals into baseband data and then remodulates them for retransmission. By splitting the total satellite link into two distinct parts in this way, on-board regeneration provides the same performance with reduced satellite and earth station power levels. It also allows system capabilities beyond those achievable with simple translating satellites. These include considerable interference protection, and interconnection between different types of terminals.

Regeneration prevents the accumulation of noise, co-channel, and adjacent-channel interference. Any effect of noise or distortion in the uplink is removed at the satellite and only bit stream *errors* are propagated on to the downlink; hence instead of the noise in the *two* links added up, only the bit errors of each sector are added (double errors and compensating errors are assumed negligible). We have, therefore,

$$P_{et} = P_{eu} + P_{ed} \qquad (9.1)$$

where P_{eu}, P_{ed} and P_{et} is the uplink, downlink, and total probability of error, respectively.

---

* The material contained in this chapter is adopted predominantly from the report by [Feher et al., 1981], with permission.

For identical uplinks and downlinks, an approximately 2.8-dB improvement is obtained by using a regenerative repeater in a linear channel. *Solve Problem 9.1.*

In nonlinear channels, however, the regenerative satellite system may well offer more than 3 dB gain in E_b/N_o over the conventional nonregenerative one. The isolation of the uplink and downlink sectors leads to an efficient use of the available EIRP by allowing the optimization of each link separately. For instance, each link may be individually equalized to reduce intersymbol-interference (ISI), and different modulation methods may be used for the two links.

Regeneration also provides versatility in the system. It is, for example, compatible with a store-and-forward concept for use with a single antenna beam. This facilitates bit rate conversion at the satellite, which provides system flexibility as outlined above. The availability in the satellite repeater of the detected baseband data stream leads to flexibility in interconnection for SS-TDMA systems and added conveniences to users, such as insert and drop, which are not possible without regenerative satellites. Electronic on-board switching, as performed in the baseband of the demodulated system using LSI techniques, ensures miniature and very light circuitry, as opposed to the heavier microwave switches needed if switching is to be done at RF as in conventional systems.

In this chapter the principles and the advantages of using on-board regenerative satellites are described. We limit our presentation to the study of performance characteristics of earth station–satellite *transmission* systems. As an example we compare the P_e performance of both 120-Mb/s QPSK and DQPSK conventional systems with regenerative systems.

Traffic routing, data storage, data processing, and packet- and message-switching flexibility all increase dramatically with the use of regenerative satellites. The true *"switchboard-in-the-sky"* concept of regenerative satellites further enhances the operational flexibility of satellite systems. The proceedings of the Fifth International Conference on Digital Satellite Communications, Genoa, Italy, 1981, and of IEEE International Conferences on Communications (1978–1982) contain numerous papers which describe the traffic routing advantages of regenerative systems.

9.2 PERFORMANCE COMPARISON OF REGENERATIVE AND CONVENTIONAL QPSK SATELLITE SYSTEMS

The majority of operational and planned digital satellite systems use QPSK modems. To describe the advantages in the P_e performance of QPSK regenerative satellites over conventional QPSK satellites, an illustrative example of a 120-Mb/s system having cascaded nonlinearities (including AM/AM and AM/PM) is used. The parameters used in this system study are those specified in the INTELSAT-V specifications described earlier. Although these specifications are for a conventional translating TDMA system, it is expected that a number of future-generation regenerative satellites will have similar requirements.

9.2.1 An Illustrative QPSK System Model

System models or networks for the conventional and regenerative satellite circuits are given in Figs. 9.1 and 9.2, respectively.* In Fig. 9.1, the satellite link considered consists of a transmitting earth station, a satellite transponder, and a receiving earth station. At the transmitting earth station the transmit bandpass filter (F_1) is used to bandlimit the spectrum. The high-power amplifier (HPA), operated near saturation, creates both AM/AM and AM/PM conversion of the modulated carrier. The satellite input and output multiplex filters $(F_2$ and $F_3)$, which are used to bandlimit the signal and thereby reduce the spectral spreading caused by the TWT, may induce ISI. At the receiving earth station, the receive filter, F_4, used to bandlimit the thermal noise and reduce the adjacent channel interference, may also degrade the system performance. For computer simulation purposes, the filter F_2' is usually lumped with the on-board filter F_2. For the regenerative system simulation reported in this chapter, we restrict our attention to a class of systems where the uplink E_b/N_0 is much higher than the downlink E_b/N_0. This is true in the majority systems. Hence we can assume, as a first approximation, that the uplink $E_b/N_0 \approx \infty$. This is equivalent to bypassing the HPA and the filters F_1, F_2, and F_4 (i.e., the blocks in the shaded area in Fig. 9.2 are omitted).

In Fig. 9.1, the data source in an equiprobable pseudo-random binary NRZ sequence, having a rate of 60 Mbaud, corresponding to 120 Mb/s. The modulator transmit filter, F_1, is simulated as a filter whose transfer function has a square root of raised-cosine ($\sqrt{\alpha} = 0.4$) shape with $x/\sin x$ amplitude equalization, while the receive filter, F_4, has a transfer function shape of the square root of raised-cosine ($\sqrt{\alpha} = 0.4$). The transponder filters, F_2 and F_3, termed input multiplex (or MUX) and output MUX filters, are simulated as steep filters with a modified raised-cosine shape ($\alpha = 0.1$ and $\alpha = 0.2$), however, with a 3-dB cutoff frequency = 40 MHz. (*Note:* The bandwidth of this filter is wider than a conventional raised-cosine filter.) The **group delays** of all the filters are assumed to be **parabolic**, $\tau = af^2$, where a is in ns/MHz² and f is in MHz. The coefficient a for transmit (F_1) and receive (F_4) filters is equal to 0.002 ns/MHz². The amplitude and group delay characteristics are chosen to fit the INTELSAT-V modem filter masks, as shown in Fig. 4.28 for F_1 and F_4, respectively. Two cases are considered for the MUX filters:

> *Case 1:* $a = 0.011$ ns/MHz² for F_2 and 0.0075 ns/MHz² for F_3.
> *Case 2:* $a = 0.015$ ns/MHz² for F_2 and 0.01 ns/MHz² for F_3.

Case 2 is the worst-case group delay and the chosen characteristics just fit the specified mask [Figs. 9.3(b) and 9.4(b)]. For case 1, as seen in the figures, the group-delay curve passes through the middle of the specified mask.

The characteristics for the earth station HPA and satellite TWTA obtained

---

* A number of figures in this chapter and the corresponding text are after [Feher et al., 1981], with permission from the CRC.

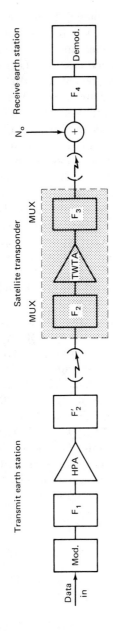

Figure 9.1 Conventional communication satellite system model.

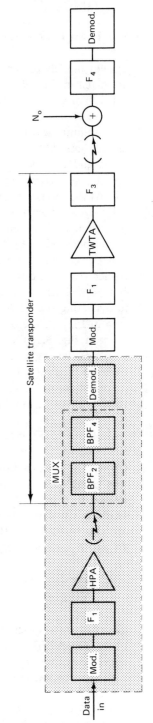

Figure 9.2 Regenerative satellite system model. (After [Feher et al., 1981], with permission of the CRC.)

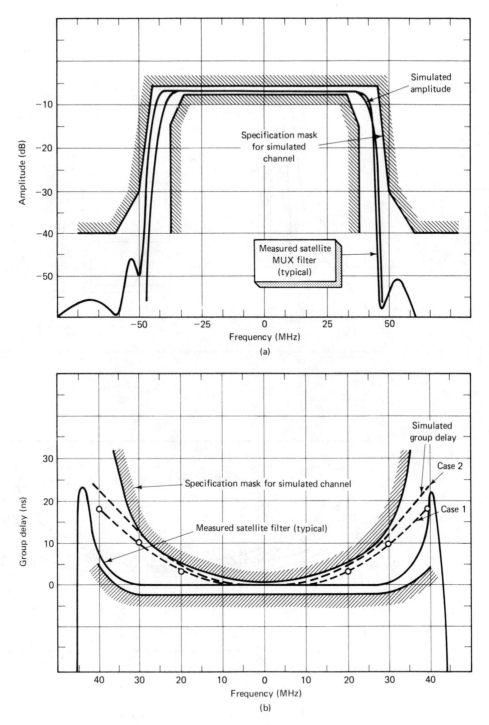

Figure 9.3 Satellite input MUX filter (F_2). (a) Amplitude characteristic; (b) group delay characteristic.

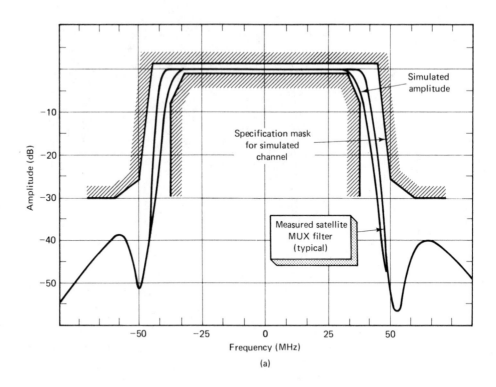

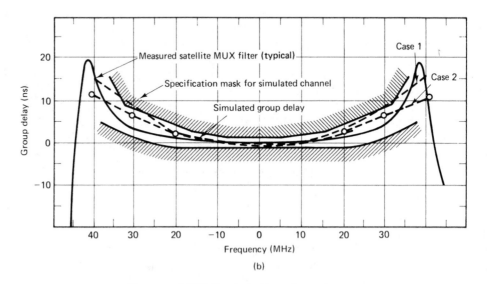

Figure 9.4 Satellite output MUX filter (F_3). (a) amplitude characteristic; (b) group delay characteristic.

from the recommended INTELSAT-V TDMA/DSI system specifications are used
to model the nonlinear devices (Fig. 1.16). Nonlinearities can be described by two
polynomials, $p(r)$ and $q(r)$ [Eric, 1972], so that if the input modulated signal is
written as

$$Z_i(t) = [x(t) + jy(t)]e^{jw_c t} \tag{9.2}$$

where w_c is the carrier frequency, then the output signal of the nonlinear device is

$$Z_o(t) = [x'(t) + jy'(t)]e^{jw_c t} \tag{9.3}$$

where

$$x'(t) = x(t)p[r(t)] - y(t)q[r(t)] \tag{9.4}$$

$$y'(t) = x(t)q[r(t)] + y(t)p[r(t)] \tag{9.5}$$

and

$$r(t) = \sqrt{x^2(t) + y^2(t)} = \text{envelope of the incoming signal} \tag{9.6}$$

The **AM/AM characteristic curve** is represented by

$$\text{AM/AM} = \sqrt{[rp(r)]^2 + [rq(r)]^2} \tag{9.7}$$

and the **AM/PM characteristic curve** is given by

$$\text{AM/PM} = \tan^{-1}\frac{q(r)}{p(r)} \tag{9.8}$$

The polynomials $rp(r)$ and $rq(r)$ may be expressed as

$$rp(r) = \sum_{i=1}^{M} a_{2i-1} r^{2i-1} \tag{9.9}$$

and

$$rq(r) = \sum_{i=1}^{M} b_{2i-1} r^{2i-1} \tag{9.10}$$

where M is the order of the polynomial, and all even coefficients are zero.

9.2.2 Simulation Results

The simulation results for the 120-Mb/s system for QPSK are shown in Figs. 9.5
and 9.6. In all cases, in accordance with INTELSAT-V specifications, the satellite
TWTA is assumed to operate at 2.0 dB input backoff (corresponding to an output
backoff of about 0.2 dB; i.e., the TWTA is run at saturation). In the conventional
mode, results are shown for the earth station HPA operating at different input backoffs
from 0 to 12 dB.

In Fig. 9.5, the results for the group delay of case 1 are shown. For a probability

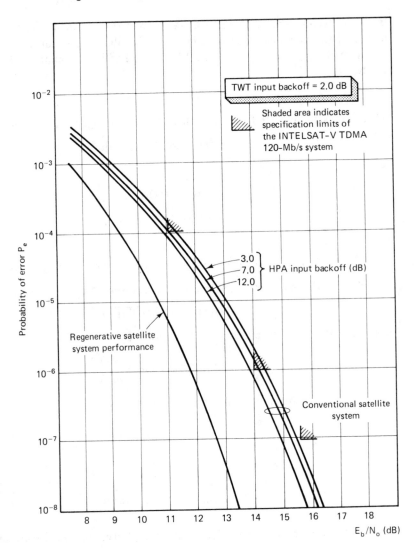

Figure 9.5 P_e performance of regenerative and conventional QPSK systems (case 1 filtering—see Fig. 9.3).

of error of 10^{-4}, conventional-mode satellite systems do not meet the INTELSAT E_b/N_o requirement of less than 11.0 dB, unless the HPA input backoff is greater than 7.0 dB. Comparing conventional and regenerative systems, the latter offers a 2.0-dB gain in E_b/N_o for a P_e of 10^{-4}. At a probability of error of 10^{-6}, the corresponding gain in E_b/N_o is 2.6 dB.

Figure 9.6 shows the results for the worst-case parabolic group delay of the MUX filters. In this case, the group delay is seen to have a considerable effect on

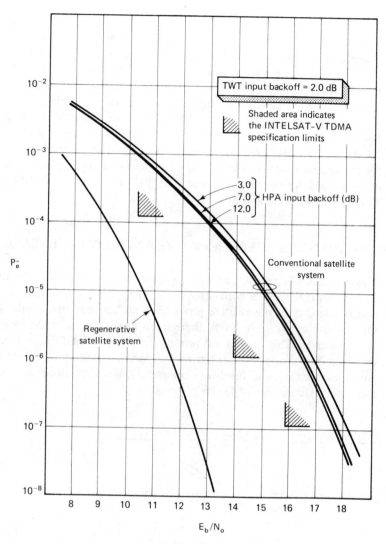

Figure 9.6 P_e performance of regenerative and of conventional QPSK satellite systems—case 2—worst-case parabolic group delay. See Fig. 9.3.

the performance of the conventional system; the performance falls far short of the INTELSAT-V 120-Mb/s system E_b/N_o requirements. At a probability of error of 10^{-4}, a 4-dB improvement results from the use of regenerative systems compared to a translating conventional system. At a probability of error of 10^{-7}, the **gain in E_b/N_o from regeneration is over** 5 dB. Regeneration thus enables the satellite system to meet E_b/N_o requirements that were otherwise impossible. From these results we conclude that more gain is obtained by using a regenerative system when the group delay of the filters is higher.

The effect of filter positioning on the regenerative system was also evaluated. In Fig. 9.7, the equivalent regenerative satellite system is shown. In Fig. 9.7(a) the shaping filter precedes the TWTA at the satellite repeater, whereas in Fig. 9.7(b) the shaping filter is placed after the TWTA. In the latter case, the output MUX filter is replaced by filter F_1, since the bandwidth of the shaping filter is narrower than that of the MUX filter. Figure 9.8 shows the probability of error performance for the two cases considered for QPSK. It is seen that if the filter is placed after the TWTA, a 0.7-dB gain in E_b/N_o results at a probability of error of 10^{-4} and 1.0 dB at $P_e = 10^{-6}$. An improvement in the performance of the system is expected for the system model of Fig. 9.7(b) since the input signal to the TWTA is constant-envelope and hence AM/AM and AM/PM degradations will be greatly reduced. *Solve Problems 9.2 and 9.3.*

9.3 ON-BOARD DQPSK REGENERATIVE SATELLITE SYSTEMS

The principle of operation and the performance characteristics of differential phase-shift keying (DPSK) has been described in Chapter 4. In summary, the digital information is encoded in the relative phase changes between two symbols rather than in the absolute changes. As such, there is no need to acquire the synchronous carrier at the receiver. This has several resultant effects when DPSK is used in regenerative satellites. The omission of the carrier recovery circuitry on board the satellite not only means a saving in hardware complexity, but also, since there is less hardware, there is a reduced risk of failure.

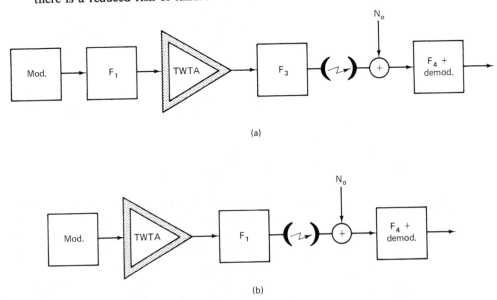

(a)

(b)

Figure 9.7   Effect of different filter positioning on QPSK satellite systems.

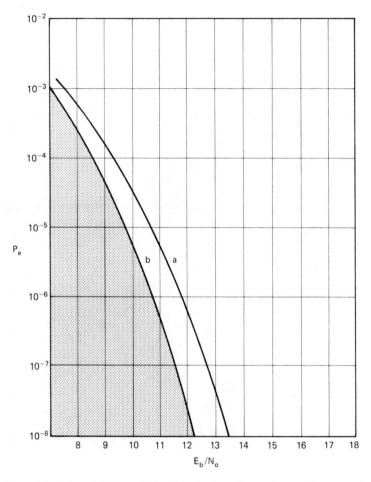

Figure 9.8 Effect of different filter positioning in a regenerative satellite system for QPSK modulated signals. System block diagrams are given in Fig. 9.7

   However, the lack of a coherent carrier for demodulation implies that **DPSK** requires a higher carrier-to-noise ratio for the same bit error rate as that of the coherent PSK. One of the most difficult design tasks in the *implementation* of a high-speed DPSK demodulator on board the satellite is in achieving a highly accurate and stable delay element of one symbol. Temperature-dependent delay variations result in phase errors of the reference signal and so deteriorate the bit-error-rate performance. Although this delay-element stability problem can be reduced by implementing it at IF, this approach somewhat reduces the weight-saving advantage of DPSK that arises due to the omission of the carrier recovery circuitry. Recent developments in microwave integrated-circuit technology have resulted in a delay filter that can be used as a one-symbol delay unit in a DQPSK (differential quartenary PSK) demodulator. This has a C/N degradation less than 0.2 dB, at BER $= 10^{-4}$, in comparison with that of a DQPSK modem using an ideal delay line.

Although it has been proposed that DQPSK be used for the uplink and QPSK for the downlink, in a number of regenerative satellites, only the DQPSK for both the up- and downlinks are considered in this chapter. The major signal-processing operations required in the digital computer simulation are illustrated in Fig. 9.9. The simulation may be performed in the equivalent baseband system of the satellite channel. The conversions from the time domain to the frequency domain are performed

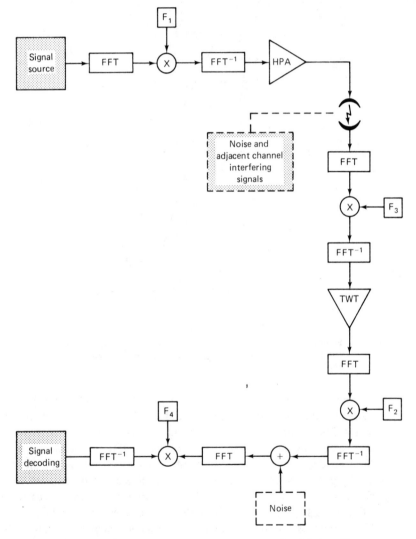

Figure 9.9 Signal processing in the computer simulation of a digital satellite system. (After [Feher et al., 1981], with permission of the CRC.)

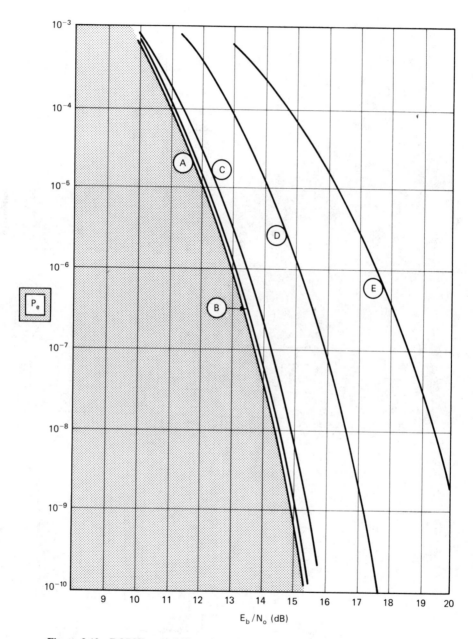

Figure 9.10 DQPSK probability of error through an INTELSAT-V system. A, calculated; B, simulated (A and B are ideal); C, linear (with filter imperfection); D, regenerative; E, cascaded HPA and TWT. (After [Feher et al., 1981], with permission of the CRC.)

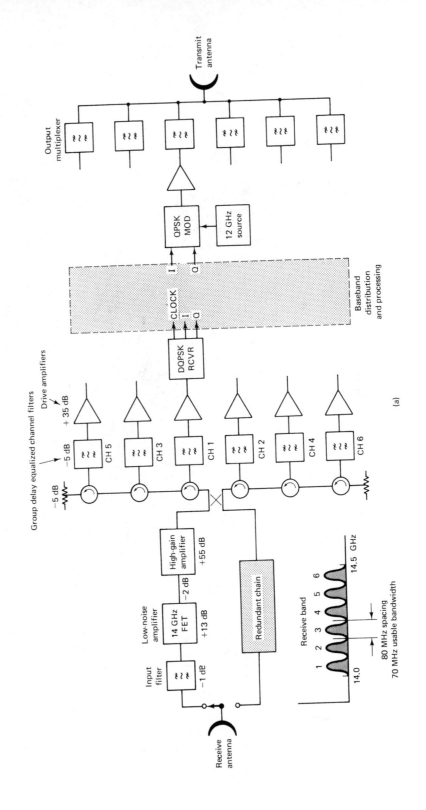

(a)

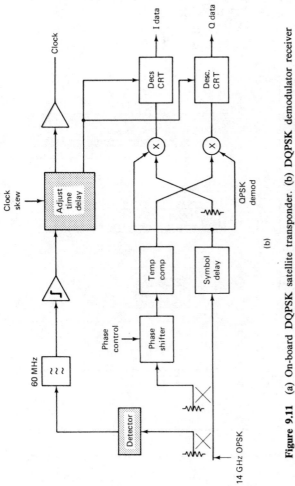

Figure 9.11 (a) On-board DQPSK satellite transponder. (b) DQPSK demodulator receiver block diagram. After [Childs et al., 1981], with permission from COMSAT Laboratories.

(b)

by means of the fast Fourier transform (FFT), while the frequency domain-to-time domain conversions are performed by the inverse fast Fourier transform (FFT^{-1}). Typical results are shown in Fig. 9.10. From the results it is evident that the cascaded nonlinear satellite channel has the worst performance, while the regenerative system has a significantly improved $P_e = f(E_b/N_o)$ performance.

The block diagram of a typical DQPSK on board regenerative satellite transponder and of an on-board radio-frequency DQPSK demodulator is shown in Fig. 9.11. This configuration is for 120-Mb/s QPSK-14 GHz uplinks having DQPSK on-board demodulators [Childs et al., 1981]. *Solve Problems 9.4 and 9.5.*

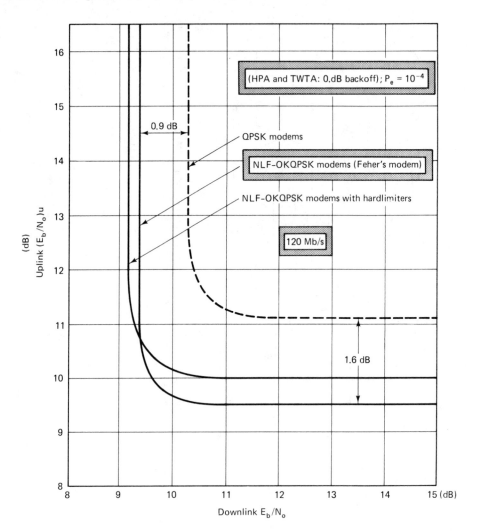

Figure 9.12 Comparative performance of regenerative satellite systems using NLF-OKQPSK and conventional QPSK modems.

9.4 PERFORMANCE OF REGENERATIVE SATELLITE SYSTEMS USING NLF-OKQPSK (FEHER'S QPSK) AND CONVENTIONAL QPSK MODEMS

The uplink and downlink E_b/N_o requirements for regenerative satellites using conventional QPSK and nonlinearly filtered (NLF) offset-keyed NLF-OKQPSK (Feher's QPSK) systems are illustrated in Fig. 9.12. The associated modems are described in Chapter 4. For the illustrated 120-Mb/s system the earth station high-power amplifier HPA and the satellite traveling-wave-tube amplifier TWTA are assumed to operate in **saturation** (i.e., 0-dB output power backoff). The E_b/N_o requirements for a $P_e = 10^{-4}$ are shown for the NLF-OKQPSK and conventional QPSK modems when transmitted through a cascaded HPA and TWTA having the characteristics of an INTELSAT-V system and for ideal hard-limited amplifiers which do not contain AM/PM. From these results it is evident that the **nonlinearly filtered** OKQPSK modem has a better performance (lower E_b/N_o requirement) than that of a conventional QPSK.

PROBLEMS

9.1. Assume that the available energy per bit-to-noise density ratio, E_b/N_o, of an uplink equals the E_b/N_o of the downlink; that is, $(E_b/N_o)_u = (E_b/N_o)_d = 12$ dB. A 60-Mb/s QPSK satellite system is required. Assuming that the output amplifiers operate in a linear mode (e.g., 6 dB output-power backoff) and that the performance degradation due to hardware imperfections is negligible, determine:
(a) The system P_e performance in a conventional nonregenerative satellite link.
(b) The system P_e performance in a regenerative link.
(*Hint:* The effective improvement of the regenerative link over the conventional systems is less than 3 dB. Why is it so?)

9.2. Draw the attenuation characteristics of a square-root raised-cosine ($\sqrt{\alpha = 0.4}$)-shaped transmit filter having an $x/\sin x$ amplitude equalizer. Assume that 120-Mb/s data are transmitted using a QPSK modulator. Compare this frequency response with the INTELSAT-V satellite input multiplex filter shown in Fig. 9.3(a). Which filter will degrade the QPSK system performance to a greater degree? Why?

9.3. Assume that the conventional satellite P_e performance curves given in Fig. 9.6 are measured in one of your laboratory-built prototype systems. Your E_b/N_o specifications are the same as in the INTELSAT-V satellite system, indicated by the shaded area in the figure. Present your ideas for performance improvement of the cascaded-nonlinear system.

9.4. Present a summary of the advantages and disadvantages of differential demodulators when compared to coherent demodulators. Pay particular attention to DQPSK burst-operated (TDMA) systems.

9.5. In the uplink of a planned regenerative system, DQPSK demodulation is used, whereas in the downlink coherent QPSK demodulation is deployed. Is this a logical choice? Why?

10

SINGLE-CHANNEL-PER-CARRIER (SCPC) PREASSIGNED AND DEMAND-ASSIGNED, SPADE, DIGITAL SATELLITE EARTH STATIONS

10.1 INTRODUCTION TO FREQUENCY-DIVISION MULTIPLE-ACCESS, (SCPC-FDMA) DIGITAL SATELLITE SYSTEMS

In Chapter 8, **Dr. S. J. Campanella** and **Dr. D. Schaefer** described time-division multiple-access (TDMA) digital satellite communications systems. The bulk of *high-capacity* INTELSAT satellite system traffic will be handled by TDMA systems. The INTELSAT-V time-division multiple-access system provides efficient power and spectrum utilization for a relatively small (up to 32) number of traffic stations. If the number of earth stations is larger, single-channel-per-carrier (SCPC) frequency-division multiple-access (FDMA) techniques might provide a more efficient system utilization. These systems are operational in many countries and provide reliable, cost-effective commercial and military, domestic and international voice and data services. Both the principles and applications of these systems are described in this chapter.

Alternative systems, such as spread-spectrum multiple-access (SSMA) and pulse-address multiple-access (PAMA) systems are suitable for certain military and nonconventional communication systems applications. However, these somewhat specialized systems are not as frequently used as FDMA-SCPC and TDMA systems. A good text on spread-spectrum systems is [Dixon, 1976]. To limit the size of this volume, we describe only TDMA and SCPC-FDMA systems.

The signal path in a multiple-access satellite network is shown in Fig. 10.1.

422

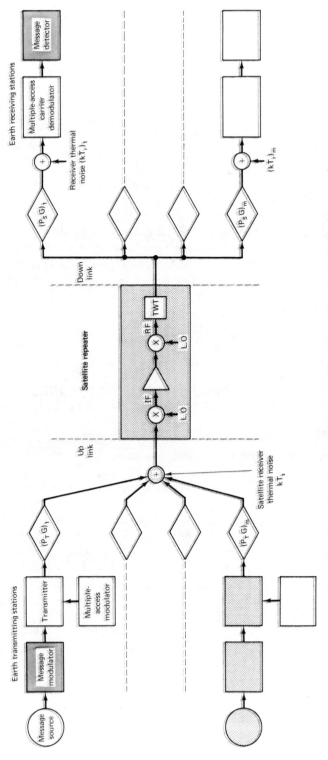

Figure 10.1 Signal flow in a multiple-access satellite system. The diamond enclosing a generic term $(P_T G)_m$ represents the effective receive power at the satellite repeater input. (After [Schwartz et al., 1966a], with permission from the IEEE © 1966.)

In this somewhat generalized diagram, paired communications links are shown; however, in some circumstances one transmitter may be linked to communicate with more than one receiver. The message waveforms are modulated onto each multiple-access carrier waveform. The multiple-access modulator *"addresses"* the modulated message to the receiver. In time-division multiple-access (TDMA) systems the multiple-access modulator provides the time-gating function, which locates each transmission burst within its preassigned time slot. In frequency-division multiple-access (FDMA) systems, the multiple-access modulator determines the carrier frequency of the modulated signal. In single-channel-per-carrier (SCPC) and FDMA systems, each modulated signal has a separate carrier frequency. In spread-spectrum multiple-access (SSMA) systems a pseudo-random key stream generator provides the carrier addressing function. In SSMA systems the message-modulated bandwidth is typically very small relative to the RF bandwidth. Finally, in pulse-address multiple-access (PAMA) systems, the multiple-access modulator generates a series of pulses within specific frequency bands. Each waveform can be specified by dividing the time-frequency plane into a matrix.

To summarize, a categorization of multiple-access techniques is illustrated in Fig. 10.2 [Schwartz et al., 1966].

10.2 SINGLE-CHANNEL-PER-CARRIER FREQUENCY-DIVISION MULTIPLE-ACCESS (SCPC-FDMA) DIGITAL SATELLITE SYSTEMS

10.2.1 Preassigned and Demand-Assigned Operation

Many commercial satellite communications systems in operation in the 1970s and early 1980s made use of full-time dedicated frequency-division-multiplexed/frequency-modulated (FDM/FM) carriers with either single or multiple destinations. In both cases, however, extensive use is made of voice circuits preassigned between two given points in the systems. This preassignment of circuits provides efficient system operation for large earth stations. From a traffic point of view, as the number of circuits per group is decreased, the utilization of satellite circuits for a given grade of service becomes increasingly inefficient, eventually becoming impractical when the link has a small traffic requirement.

In FDM/FM satellite systems, all earth stations transmitting to the satellite have their output power controlled so that the satellite high-power amplifier continues to operate in its linear region. As the received power at the satellite input decreases, the high-power amplifier is further backed off from saturation, and its radiated output power reduces. As the number of FDM/FM networks increases, the satellite channel capacity decreases.

Lightly loaded links also known as *thin-route* systems have a problem because of their loading; a pool of satellite circuits is shared among all earth stations concerned.

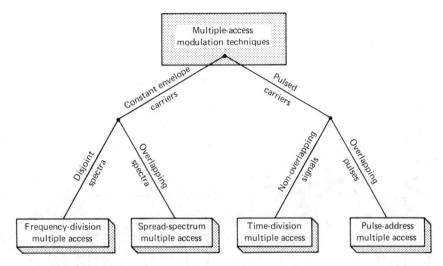

Figure 10.2 Categorization of multiple-access modulation techniques. (After [Schwartz et al., 1966a], with permission from the IEEE © 1966.)

The circuits are then assigned on *demand,* forming a temporary connection on a per circuit basis between any two pool-member earth stations within the region covered by the satellite. At the end of the communication the circuits are returned to the demand-assigned satellite pool.

An example of a preassigned multidestination FDM/FM system is illustrated in Fig. 10.3. The signal flow in a demand-assignment FDMA (DA-FDMA) system is shown in Fig. 10.4. Long-distance telephone calls originating in country A enter a telephone exchange or transit center, CT, and are multiplexed, as shown in Fig. 10.3, in a 60-channel (5×12 channel) baseband. Country A transmits on a single FM carrier f_A with a 60-channel capacity. The 60 channels have been preassigned in groups of 12 channels to be received by five other countries. Countries B, C, D, E, and F require an FM receive chain operating on frequency f_A in order to receive telephone traffic from A. Conversely, country A must have a receive chain for return paths from the countries concerned. In this system if all 12 channels to country F in country A's baseband are being used, a new call entering the telephone exchange, CT, in country A would receive a busy signal from the satellite system even though the remaining 48 channels in the baseband might be unoccupied. Nevertheless, for large-capacity links between two earth stations, the statistics of call placements are such that few calls are lost due to overload [Puente and Werth, 1971].

A map and a summary chart illustrating the geographic locations and channel assignment method for INTELSAT stations in the early 1970s is given in Figs. 10.5 and 10.6 In this system 820 links are possible. Considering that a large proportion of the connections carry light traffic, the satellite operating in an FDM/FM/FDMA mode, as described in Fig. 10.3, could not accommodate the increasing number of connections in an efficient manner. In Fig. 10.6, the stars at the intersections of

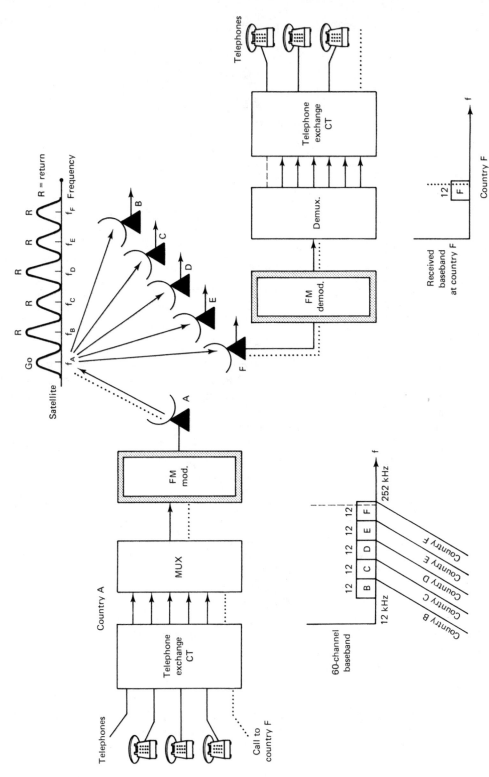

Figure 10.3 Preassigned-multidestination FDM/FM/FDMA carriers. CT, transit center. (After [Puente and Werth, 1971], with permission from the IEEE © 1971.)

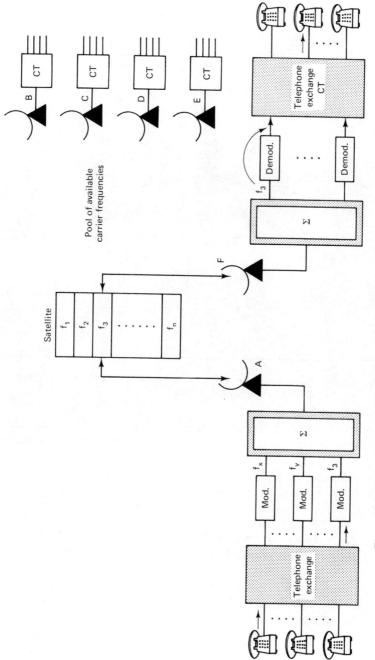

Figure 10.4 Voice-signal flow in demand-assignment FDMA systems. (After [Puente and Werth, 1971], with permission from the IEEE © 1971.)

427

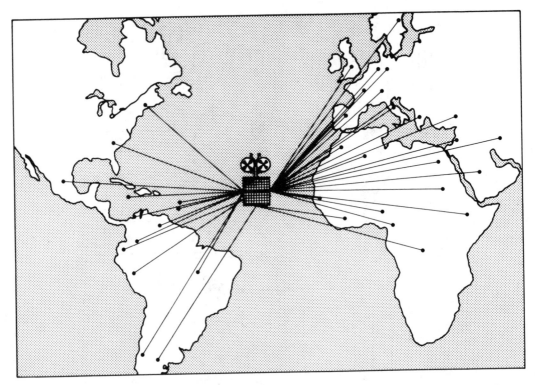

Figure 10.5 Atlantic Basin interconnection of earth stations in 1970s. (After [Puente and Werth, 1971], with permission from the IEEE © 1971.)

two countries represent traffic in excess of 12 circuits (preassigned channels), the dots indicate traffic less than 12 telephony circuits (demand-assigned channels). A figure of 12 circuits has been taken arbitrarily for illustrative purposes since it represents a standard group.

The difference between preassigned and demand-assigned systems is that for the demand-assigned systems, the channels are assigned on a temporary basis.

10.2.2 Description of the INTELSAT Spade Demand-Assignment Multiple-Access (DAMA) System

The best known operational demand-assigment multiple-access (DAMA) system is called SPADE (*Single-channel per carrier PCM multiple-Access Demand-assigned Equipment*). This equipment was designed at COMSAT laboratories for INTELSAT-IV and later INTELSAT satellites. The goals of the SPADE project were stated as follows [Edelson and Werth, 1972; Cacciami, 1971].

1. To provide efficient service to light traffic links
2. To handle overflow traffic from medium-capacity preassigned links

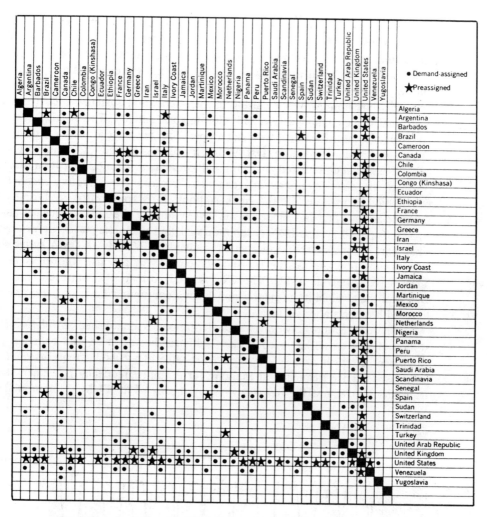

Figure 10.6   Summary of traffic density among INTELSAT stations in 1970s. (After [Puente and Werth, 1971], with permission from the IEEE © 1971.)

3. To allow establishment of a communication link from any earth station to any other earth station within the same zone on demand

4. To utilize satellite capacity efficiently by assigning circuits individually

5. To make optimum use of existing earth station equipment

SPADE increased the traffic handling capability of INTELSAT satellites and made the establishment of new small capacity links more economical. Even the least dense traffic links can now be accommodated with no penalty to the system.

In the SPADE system a specific satellite radio-frequency band is divided on the basis of assigning a single data or PCM converted voice channel per RF carrier. The SPADE-DAMA system, using QPSK modulation, is fully variable, allowing all circuits to be selected by any station on demand. Neither end of a channel is permanently connected with any given terminal, and the channels are paired to form two-way circuits within the demand assignment pool. Also, the satellite *power* utilization efficiency is improved by using a voice-activation system which turns on the individual single-channel carrier only during voice activity. The SPADE system does not require a central control for system operation, but instead uses a demand-assignment signaling and switching (DASS) unit for self-assignment of channels. This channel self-assignment is based on continually updated channel allocation status data provided via a common signaling channel. This *common signaling channel* (CSC) information is used by all the earth stations. It is transmitted in a time-division multiple-access (TDMA) mode, in the lowest edge of the radio-frequency band (see Figs. 1.10 and 10.7). The TDMA common signaling channel signal occupies a relatively small part of the available transponder bandwidth (160 kHz). TDMA systems which occupy only one part of the transponder bandwidth are also known as *thin-route* TDMA [Nuspl et al., 1981]. The CSC is used both to evaluate each earth station DASS of the availability of pool channels on a continuous basis and also to establish links with other stations [Intelsat, 1969].

The connection between two earth stations is established as follows: If earth station A has to set up a connection with earth station B, the common control equipment of earth station A examines its table of available frequencies. It selects an unused carrier frequency at random, for example frequency f_{55}. Earth station A transmits the request for frequency f_{55} by means of its TDMA common signaling

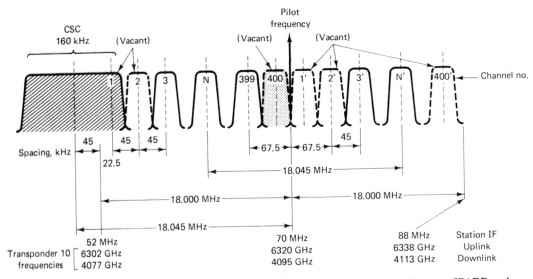

Figure 10.7 Radio-frequency allocation in the INTELSAT demand-assignment, SPADE, and preassigned SCPC systems.

channel. This request is transmitted in the time slot assigned to earth station A. If the requested channel is available, station B sends a message to station A, to transmit the QPSK modulated message at a center frequency f_{55}. Stations A and B assign to their frequency synthesizers the corresponding f_{55} frequencies until the termination of the call, which is detected by the voice-activation circuitry. Thereafter, this pair of carrier frequencies is pooled for the next call between any other two stations. Occasionally, earth station A requests a certain frequency and before its request reaches its destination total delay (propagation plus equipment is about 280 ms), that frequency may be assigned to another earth station. Station A detects this and makes a new request for a different channel.

The frequency selected is provided to the channel unit by means of a frequency synthesizer which is capable of generating any of the 800 discrete frequencies required. This is used for both the outgoing carrier and the received signal local oscillator. Channel pairings are based on the common use of the synthesizer for receive and transmit signals.

As shown in Fig. 10.7, the QPSK modulated ($f_b = 64$ kb/s) channels are divided into higher and lower channel groups, separated by the pilot frequency. One *telephone channel* requires a matched pair of higher and lower RF channels for the transmit and receive directions. The total number of channels is 399, of which the pairs 1–1' and 2–2' are not used commercially.

The block diagram of the SPADE demand-assigned terminal is shown in Fig. 10.8. Analog or digital voice detectors are used to gate the modulated channel carrier on or off. This conserves satellite power as a function of actual speech activity. The digital bit stream in and out of the PCM voice codec is synchronized by the transmit–receive synchronizer. This unit provides the timing, buffering, and framing functions. The common IF subsystem interfaces with the earth station up- and down-converters [INTELSAT, 1969].

10.2.3 The INTELSAT SCPC-QPSK-FDMA Preassigned System

In 1974 INTELSAT issued the performance specifications for a medium-speed data system utilizing single channel per carrier (SCPC) which uses a frequency assignment scheme that is *completely compatible* with the SPADE demand assignment system (INTELSAT, BG/T-5-21E, W/1/74]. This INTELSAT SCPC system is operational in many countries, is an outgrowth of the SPADE system, and has the same features as SPADE except for the demand assignment capability.

In the SCPC system, individual RF carriers are allocated either to a 64-kb/s PCM converted voice channel or to a forward error-correction-coded data channel (48 or 56 kb/s) on a preassigned basis. The control and logic functions in an SCPC terminal are considerably simpler than those of a SPADE terminal, providing significant earth station cost savings.

In the SCPC system the signal processing is QPSK/FDMA, with PCM modulation for voice. The three functional configurations of the INTELSAT standard SCPC earth station terminal are:

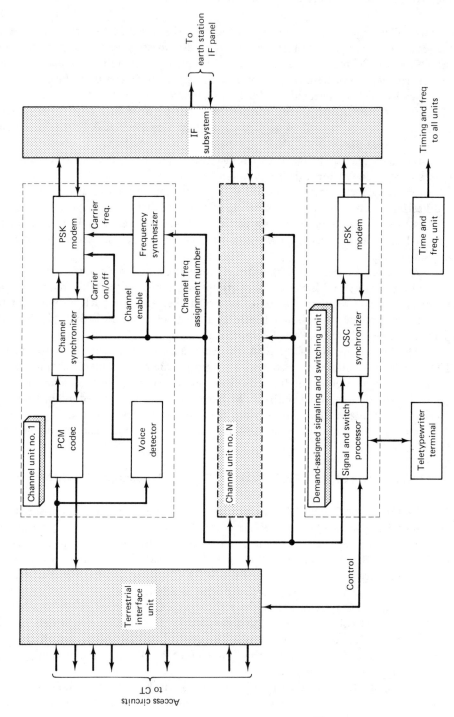

Figure 10.8 INTELSAT standard SPADE demand-assigned terminal functional block diagram.

1. Digitized voice using the conventional SPADE-type voice-channel unit (PCM, 64 kb/s)

2. Digital data at 48 or 50 kb/s using rate $\frac{3}{4}$ convolutional coding (with corresponding encoded data rated 64 and 66 kb/s)

3. Digital data at 56 kb/s using rate $\frac{7}{8}$ convolutional coding

The functional block diagram of a single-channel-per-carrier earth station is shown in Fig. 10.9. The earth station GCE (ground communications equipment) contains the up- and down-converters, filters, and amplifiers. An SCPC earth station may operate in the SPADE satellite transponder. In Fig. 10.10 a fully equipped INTELSAT standard SCPC terminal is shown. The specifications for this system, in use in many countries, are summarized in Table 10.1 and Fig. 10.11.

The principles and performance of individual building blocks of this system (such as the PCM codec, QPSK modem, and forward error-correcting subsystems) have been described in previous chapters. The PCM encoder/decoder, described in Chapter 2, is a 7-bit sample encoder having a sampling rate of 8 ksamples/s. The eighth bit is reserved for synchronization purposes. INTELSAT specified the use of $A = 87.6$ law companding.

The function of the data codecs is:

1. To generate appropriate parity bits for forward error control and to interface with the modulator

2. To accept the QPSK demodulated signal and provide correct synchronization and decommutation (resolve the demodulator carrier ambiguity problem)

3. To use the parity bit information for error correction

The convolutional codecs in the SCPC system may operate as $\frac{3}{4}$ or $\frac{7}{8}$ rate error correcting codecs which utilize *threshold decoding*. The principles of modem design are presented in Chapter 4; those of error-correcting codecs can be found in Chapter 6.

The $\frac{3}{4}$ convolutional encoder/threshold decoder subsystem is a self-synchronizing unit designed specifically for use with a coherent QPSK modem. The principle of operation of this subsystem is of interest because of its application in a large number of combined modem/codec networks system and therefore a description follows.

3/4 Convolutional encoder/threshold decoder and modem ambiguity resolution subsystem. The encoder block diagram is shown in Fig. 10.12. The input serial data stream is fed to a serial-to-parallel converter, the outputs of which enter three shift registers. These registers, in turn, are tapped and their outputs summed modulo-2 to form the parity bit. The four data streams (three from the serial-to-parallel converter, one is the parity bit) are commutated to form two data streams, which serve as inputs to the in-phase and quadrature channels of the PSK modulator.

The first functional element of the decoder (also shown in Fig. 10.12) is an

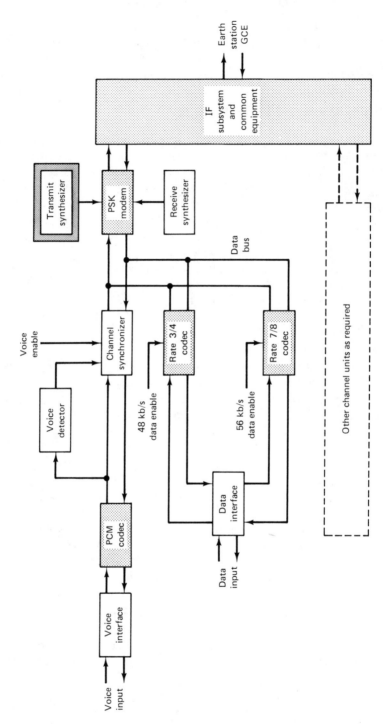

Figure 10.9 INTELSAT single-channel-per-carrier (SCPC) preassigned system.

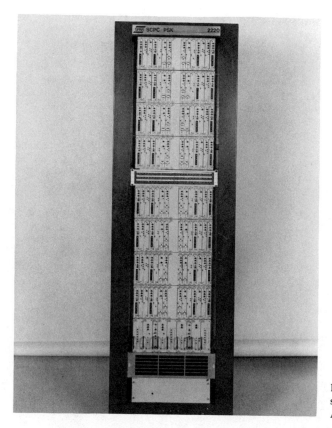

Figure 10.10 Fully equipped INTELSAT standard SCPC terminal. (Courtesy of Spar Aerospace Limited, Canada.)

ambiguity resolver and decommutator. The four-phase PSK signals have eight ambiguous states relative to the in-phase and quadrature channel units, and there are also ways in which to decommutate the signals for a total of 16 possibilities. The ambiguity resolver cycles through the possible bit combinations until it receives an indication from the *syndrome register* (containing a measure of the error rate) that the correct combination has been identified. The output of the ambiguity resolver/decommutator is four parallel channels, three data and one parity.

The three data bit channels are processed in what amounts to a reproduction of the encoder circuitry to form a second parity bit stream which is compared modulo-2 to the parity stream from the decommutator. Bit patterns in the syndrome register which indicate errors in one or more of the three data bit streams are interpreted by threshold detectors which, when activated (a threshold of three), correct the data bit streams and reset the syndrome register. The threshold detectors are also used to detect errors when the correct combination of the aforementioned 16 possible bit combinations has been decommutated. The last element of the decoder is a 48 or 56 kb/s parallel-to-serial bit stream converter. The effect of the decoder on channel

TABLE 10.1 SUMMARY OF THE MOST IMPORTANT VOICE AND DATA CHANNEL UNIT SPECIFICATIONS OF THE INTELSAT STANDARD PCM/QPSK/FDMA TERMINAL

Voice channel unit		Data channel unit	
Frequency response	300 to 3400 Hz (meets INTELSAT BG-9-21E Rev. 2 spec)	Bit rate	48/50 kb/s with rate $\frac{3}{4}$ codec; 56 kb/s with rate $\frac{7}{8}$ codec
Input/output impedance	600 Ω balanced	Interface standard	CCITT V-35, WE303
Return loss	$\geqslant$26 dB	Data scrambler	CCITT V-35 using 20 bit register with taps at stages 3 and 20
Transmit level	0 to −16 dBm, presettable		
Receive level	−4 to +7 dBm, presettable	Input/output signals	Data and clock
Encoding	7 bits PCM; companding law; $A = 87.6$; sampling rate: 8 kHz	Transmission rate	64 kb/s for 48 and 56 kb/s data 66.6 kb/s for 50 kb/s data
		Modulation/ demodulation	4-phase coherent PSK
Carrier control	Voice activated (VOX)		
VOX threshold	−20 to −32 dBm0, presettable	Nominal BER at $E_b/N_o = 13.2$ dB without coding and scrambling	$\leqslant 1 \times 10^{-6}$
Signal to quantization noise ratio (using white noise)	$\geqslant$30 dB for input levels above −30 dBm	Nominal BER at $E_b/N_o = 13.2$ dB with coding	$\leqslant 1 \times 10^{-9}$ (without scrambling)
Idle noise	$\leqslant$ −61 dBm0p		$\leqslant 3 \times 10^{-9}$ (with scrambling)
Overload point	+2 dBm0		
		Frequency synthesizer channel spacing	45-kHz steps
Transmission rate	64 kb/s		
Modulation/ demodulation	4-phase coherent PSK	Spurious output	$\geqslant$50 dB below carrier
E_b/N_o for modem BER of 1×10^{-4} with adjacent channel and worst frequency offset	$\leqslant$11.2 dB		
Frequency synthesizer channel spacing	45-kHz steps		
Spurious output	$\geqslant$50 dB below carrier		

Courtesy of Spar Aerospace Limited, Canada.

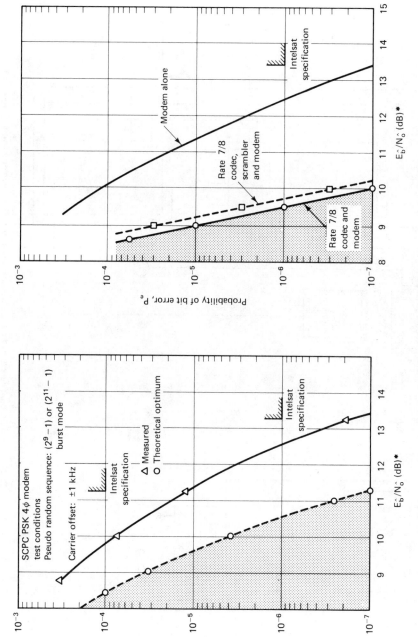

Figure 10.11 Typical performance curves for uncoded and convolutional encoded INTELSAT standard SCPC systems. (Courtesy of Spar Aerospace Limited, Canada.)
*E_b represents the encoded bit energy)

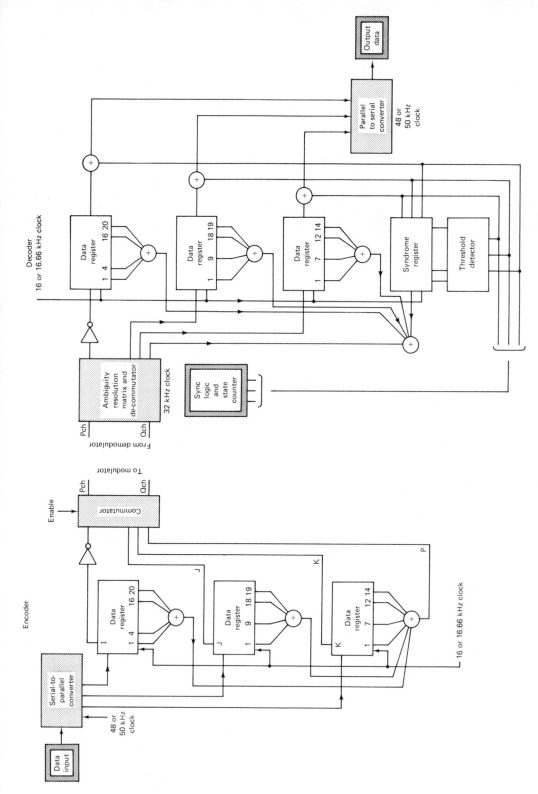

Figure 10.12 Convolutional encoder ($\frac{3}{4}$ rate), threshold decoder and modem ambiguity subsystem for INTELSAT standard and other SCPC systems. (Courtesy of Spar Aerospace Limited, Canada.)

error rate can be seen from Fig. 10.11. Both the decoder and encoder utilize a variety
of bit rates which are provided by the bit timing recovery system.

The 3/4 and 7/8 rate convolutional encoder/threshold decoder subsystems, as
specified by INTELSAT, have a coding gain of 10^3 at a threshold P_e of 10^{-7} (see
Fig. 10.11). This coding gain is achieved in an additive white Gaussian noise environment. Unfortunately, the source data rate has to be reduced from 64 kb/s to 64 $\times$
3/4 or 48 kb/s and to 64 $\times$ 7/8 or 56 kb/s. For a description of the coding gain
of convolutional encoders/threshold decoders in a complex Gaussian noise and sinusoidal interference environment, see [Brind'Amour and Feher, 1980].

Example 10.1

The available bit energy to thermal noise density ratio, E_b/N_o, at the receiver input of
a preassigned single-channel-per-carrier (SCPC) system is 10 dB. Coherent QPSK modulation is used for the transmission of the $f_b = 64$ kb/s rate data signal. Assuming that
due to adjacent-channel interference, intermodulation products, and hardware imperfections the model operates about 2dB of the theoretical curve, determine:

(a) The probability of error performance of the uncoded $f_b = 64$ kb/s rate PCM voice
channel, P_{eu}

(b) The probability of error of the 7/8 rate convolutional encoded threshold decoded
56-kb/s data, P_{ec}

Solution   (a) The measured performance of an uncoded QPSK modem having an approximate 2-dB degradation is illustrated in Fig. 10.11. From this performance curve
we conclude that the probability of error of the uncoded voice channel is $P_{eu} = 10^{-4}$
(approximately).

(b) If instead of 64-kb/s PCM converted voice, a 56-kb/s data source encoded in a 7/8
rate convolutional encoder is transmitted, the channel transmission rate remains 64 kb/s
(see Fig. 10.9). For an $E_b/N_o = 10$ dB the probability of error performance of the
7/8 convolutional encoded threshold decoded system, having a modem that operates
within 2 dB of the theoretical performance, is obtained from Fig. 10.11. It is $P_{ec} =$
10^{-7}. Thus a performance improvement of 10^3 is obtained. Note that the descrambler
increases the error rate to approximately 2×10^{-7}; that is, a very small performance
degradation is introduced by the descrambler due to error multiplication.

    ■

10.3 SYSTEM CAPACITY AND TRADE-OFFS IN SPADE AND SCPC SYSTEMS

10.3.1 Operation with INTELSAT Standard (30-m) Antenna Earth Stations

An attractive feature of the demand assignment SPADE and the preassigned SCPC
systems is the high achievable channel capacity per transponder relative to FDM/
FM/FDMA operation. When voice activation is used on each carrier, there is sufficient
satellite power to support more than 800 SPADE-type channels in one INTELSAT-

IV transponder using standard earth stations [McClure, 1970]. Since each 64-kb/s rate QPSK modulated channel occupies 45 kHz of RF bandwidth, it is evident that the 36-MHz transponder is *bandwidth limited* to 800 channels (actually 794 channels after housekeeping channels are excluded).

The voice-activation feature provides a 4-dB satellite power saving. Voice-activated carriers can be used only when a single channel is assigned to each carrier, such as in the SPADE and in the SCPC systems. In frequency-division-multiplexed frequency-modulated multiple-access (FDM/FM/FDMA) systems, such as that shown in Fig. 10.3, this power saving, due to voice activation, could not be achieved. A power-saving advantage could also be achieved by using *analog* single-channel FM carriers, but QPSK is more spectrally efficient in terms of single-channel operation, thus the *digital SCPC approach provides more channels per transponder.* Calculations show that single-channel-per-carrier FM/FDMA with voice activation provided a maximum of 450 channels per carrier [Edelson and Werth, 1972]. However, the complexity and cost of analog earth station equipment has been, during the 1970s and early 1980s, lower than that of digital systems. For this reason a number of countries use analog FM single-channel-per-carrier systems for their domestic traffic.

In Table 10.2 a summary of the channel capacity of the INTELSAT-IV global-beam transponder using *standard* earth stations is given. INTELSAT-IV standard

TABLE 10.2 SUMMARY OF THE CHANNEL CAPACITY OF AN INTELSAT-IV GLOBAL-BEAM TRANSPONDER USING INTELSAT STANDARD DIGITAL EARTH STATIONS

Modulation technique	Access mode	Transponder capacity (channels)
FDM/FM	Single access/single carrier	900
PCM/PSK	Single access/single carrier (64 Mb/s)	1000
FDM/FM/FDMA	Multiple access (average, mix of various size carriers)	450
FDM/FM/FDMA	Multiple access, 14 24-channel carriers	336
PCM/PSK/FDMA (voice)	Single channel/carrier (64 kbps)	With voice-activated carrier: 800 Without voice-activated carrier: 350 Without voice-activated carrier, with 3/4 coding: ~550
PCM/PSK/FDMA (1×10^{-6} data)	Single channel/carrier (64 kb/s)	Without coding: 250 With 3/4 Coding: 450
PCM/PSK/FDMA (1×10^{-6} data)	24 multiplexed data channels/carrier (total rate equal to 1.544 Mb/s)	With coding: 450

Source: B. I. Edelson and A. M. Werth, *COMSAT Tech. Rev.*, Spring 1972, with permission.

earth stations are those that comply with the INTELSAT requirements in ICSC-45-13 (nominally 30-m diameter antenna and $G/T \geqslant 40.7$ dB/K).

10.3.2 Operation and Trade-offs with Small Earth Station Systems

The SCPC operation makes it possible to adjust the transmitted power of each modulated carrier to an appropriate level for the size of the earth station for which it is destined. To provide the same signal quality, the power level destined to smaller earth stations has to be increased. Since only the carriers assigned to smaller earth stations are increased, the number of channels per satellite transponder is higher than if power for all the links had to be increased.

If all stations in a SPADE network using an INTELSAT-IV global-beam transponder are *uncoded* QPSK, had a $G/T = 35$ dB/K, the capacity would be reduced to 400 channels and the system would be *power limited*. If, however, 3/4 convolutional coding threshold coding is introduced, the capacity is increased to 600 channels. With 10-m antenna earth stations using uncooled low-noise receivers, a $G/T = 29$ dB/K is typical. In this case the transponder capacity of the uncoded SPADE system would be only 150 channels; with rate 3/4 coding, the capacity would be 250 channels; and with rate one-half coding, up to 350 channels. If a small-station regional SPADE or SCPC network were to operate through an INTELSAT-IV transponder connected to a spot-beam antenna, it could trade the 12 dB additional satellite antenna gain for a *corresponding reduction* in the size of earth station antennas. If a regional network with earth station antennas of the order of 5 m and $G/T = 21$ dB/K is required, the INTELSAT-IV spot beam transponder could provide approximately 250 channels without coding. The advantages of spot-beam operation and of convolutional-encoded threshold decoded QPSK transmission for small earth stations are illustrated in Figs. 10.13 and 10.14.   •

10.4 NEW MODULATION TECHNIQUES FOR LOW-COST POWER-EFFICIENT EARTH STATIONS

The need for power- and bandwidth-efficient modulation schemes has led to the extensive use of phase-shift keying (PSK) in digital transmission systems. Since the power spectra of PSK signals exhibit sidelobes that may interfere with adjacent channels, a certain amount of filtering is necessary at the transmitter. However, this filtering results in considerable envelope fluctuation, which leads to considerable spectrum spreading due to the AM/AM and AM/PM nonlinear effects of the transmit high-power amplifiers (HPAs). These nonlinearities tend to restore the spectral sidelobes that have been previously removed.

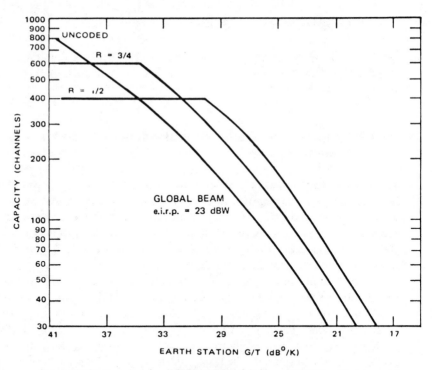

Figure 10.13 Capacity of an INTELSAT-V (*global beam*) 36-MHz-wide transponder in a single channel per carrier (QPSK mode of operation, SPADE, and SCPC). For earth station *G/T* larger than 37 dB/K, the system operates in a *bandwidth-limited* region; thus coding is not desirable. For lower *G/T,* the system operates in a *power-limited* region; thus coding increases the system capacity. (From [Edelson and Werth, 1972], with permission from COMSAT.)

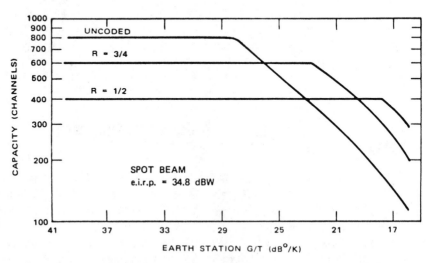

Figure 10.14 INTELSAT-IV—single-channel-per carrier transponder capacity (*spot beam*). See also description of Fig. 10.13. (From [Edelson and Werth, 1972], with permission from COMSAT.)

As a precaution against this spectral spreading, the transmit HPA may have to operate *below* saturation in an approximately linear zone with the associated power penalty. Post-HPA filtering to control spectrum spreading may become impractical for direct RF modulation at a high radio frequency-to-bit rate ratio due to the high selectivity requirements (e.g., 6 GHz to 64 kb/s). It is also discouraged because of **frequency agility** requirements, which are of particular interest in a number of applications. One solution to this problem is to use modulation techniques which *do not* restore the sidelobes. This is required to avoid the harmful effect of adjacent-channel interference in an SCPC mode of operation. The *tamed frequency modulation* (TFM) technique, introduced by researchers of Philips laboratories [Jager and Dekker, 1978], leads to a low spectral spreading. A class of generalized MSK signals known as *continuous-phase frequency-shift-keyed* (CPFSK) signals lead also to low spectral restoration [Galko and Pasupathy, 1981].

In Sections 4.8 and 4.9 the principles of the nonlinearly switched offset-keyed QPSK (NLF-OKQPSK) modulation techniques as initially proposed by Feher are presented. For power-efficient single-channel-per-carrier earth station designs, this new modulation technique is particularly suitable, since the spectral spreading intro-

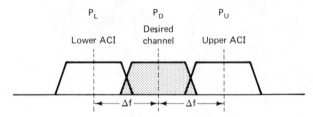

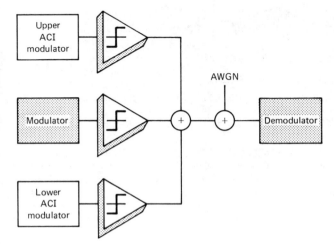

Figure 10.15 Simplified block diagram of a hard-limited multichannel SCPC system.

duced by a saturated output amplifier (in first approximation a hard limiter) is much lower than in conventional MSK and QPSK systems (see Fig. 4.42). A simplified single-channel-per-carrier system block diagram illustrating one required channel, together with the upper and lower adjacent channel transmitters, is shown in Fig. 10.15. The most significant adjacent-channel interference, which enters into the demodulator of the desired channel, is created by the adjacent channels. Thus Fig. 10.15 is a good representation of the SCPC system environment.

The results of performance comparisons of conventional QPSK, OQPSK, MSK, and Feher's QPSK [also known as intersymbol-jitter free (IJF-OQPSK)] are plotted in Fig. 10.16. For a spectral efficiency of about 1.3 b/s/Hz, the IJF-OQPSK (same as the NLF-OKQPSK) hard-limited single-channel-per-carrier system has a degradation of only 1.3 dB, whereas the conventional QPSK is degraded by more than 4 dB. Thus it is evident that Feher's QPSK could lead to the use of saturated HPA systems and an overall reduction in system cost [Le-Ngoc et al., 1982].

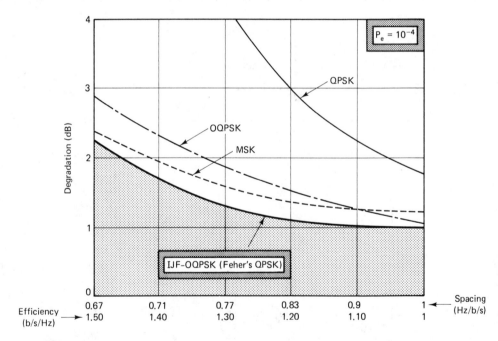

Figure 10.16 Performance comparison of QPSK, OQPSK, MSK, and IJF-OQPSK (Feher's QPSK) in hard-limited multichannel SCPC systems. Degradation is compared to the theoretical optimum performance (i.e., $E_b/N_o = 8.4$ dB, $P_e = 10^{-4}$). It is assumed that the power of the adjacent channels equals the desired signal power, that is, $P_U = P_D = P_L$. (After [Le-Ngoc et al., 1982].) Intersymbol-interference and jitter free (IJF) signals are generated by nonlinear switching filters (NLSF or NLF) techniques patented by Feher. (After [Feher "Filter" patent].)

PROBLEMS

10.1. Describe the difference between FDM/FM/FDMA, single-channel-per-carrier preassigned (SCPC) and demand-assigned (SPADE) systems.

10.2. In certain SCPC systems, error correction coding is used, whereas in others it is not in use. Explain why.

10.3. What are the advantages of *saturated* high-power-amplifier operation? What are the disadvantages?

10.4. Assume that the transmit HPA of your earth station operates saturated and that it can be approximated by an ideal hard limiter. The bit rate of your single-channel-per-carrier system is $f_b = 64$ kb/s and the upper and lower adjacent signal power equals the power of the desired channel (Figs. 10.15 and 10.16). The available E_b/N_o at the demodulator input is 11 dB. Assume that due to miscallenous hardware imperfections (for all modem designs) a 1-dB degradation from the theoretical curve is measured. A system $P_e = 10^{-4}$ is required. How many 64-kb/s channels could you transmit in the INTELSAT-V transponder (72 MHz bandwidth) if you used (a) QPSK; (b) OKQPSK; (c) MSK; (d) **Feher's QPSK (IJF-OKQPSK)**?

10.5. For the system parameters outlined in Problem 10.4, could you increase the system capacity by using forward error-correction coding? Describe the type of error-correcting codecs you would specify. Under which conditions (or modified conditions) would you use these codecs?

REFERENCES

AHMED, N. U., and M. ABOUD: "Optimum Design of a Feedback Controller for a Phase-Locked Loop," *IEEE Trans. Commun.*, Vol. COM-23, No. 12, December 1975.

AHMED, N. U., and S. H. WONG: "On Optimum Phase-Locked Loops," *Int. J. Electron.*, Vol. 36, No. 2, 1974.

AHUJA, B. K., M. A. COPELAND, and C. H. CHAN: "A Sampled Analog MOS LSI Adaptive Filter," *IEEE Trans. Commun.*, Vol. COM-27, No. 2, February 1979.

American Telephone and Telegraph Company: "Telecommunications Transmission Engineering," Bell System Center for Technical Education, Winston-Salem, N.C., 1977.

ANGELLO, P. S., M. C. AUSTIN, M. FASHANO, and D. F. HORWOOD: "MSK and Offset Keyed QPSK through Band Limited Satellite Channels," Proc. 4th Int. Conf. Digital Satellite Commun., Montreal, October 1978.

ANTONIOU, A.: *Digital Filters,* McGraw-Hill, New York, 1980.

AOYAMA, T., I. MANO, K. WAKABAYASHI, R. MARUTA, and A. TOMOZAWA: "120-Channel Transmultiplexer Design and Performance," *IEEE Trans. Commun.*, Vol. COM-28, No. 9, September 1980.

ARUNACHALAM, V., and K. FEHER: "Cascaded Non-linear Satellite Channel-Computer Simulation," Digital Communications Research Group, University of Ottawa, Rep. No. 101, 1980a.

ARUNACHALUM, V., and K. FEHER: Computer Simulation Program, Univ. of Ottawa, No. 102, 1980b.

ASHGHAR, M. M., Y. SENUMA, and R. PINEZ: "Work of the World Plan Committee for the

Development of Telecommuncations," Paris 1980, and "The Evolution of Telephone Traffic," *Telecommun. J.,* ITU, Geneva, September 1980.

ASTRAIN, S.: "Intelsat and International Digital Satellite Communications: Issues and Opportunities," Proc. 4th Int. Conf. Digital Satellite Commun., Montreal, October 1978.

ASTRAIN, S.: "Keynote Address," Proc. 5th Int. Digital Satellite Commun. Conf., Genoa, Italy, March 23–26, 1981.

ATOBE, N., Y. K. MATSUMOTO, and Y. TAGASHIRA: "One Solution for Constant Envelope Modulation," Proc. 4th Int. Digital Satellite Commun. Conf., Montreal, October 1978.

BANTIN, C. C. and R. G. LYONS: "The Evaluation of Satellite Link Availability," *IEEE Trans. Commun.,* Vol. COM-26, No. 6, June 1978.

BARGELLINI, P. L.: "Commercial U.S. Satellites," *IEEE Spectrum,* Vol. 16, No. 10, October 1979.

Bell Laboratories—Members of Technical Staff: "Transmission Systems for Communications," Bell Telephone, 1971.

BENEDETTO, S., E. BIGLIERI, and R. DAFFARA: "Performance Prediction for Digital Satellite Links—A Volterra Series Approach," 4th Int. Conf. Digital Satellite Commun., Montreal, October 23–25, 1978.

BENNETT, W. R.: "Statistics of Regenerative Digital Transmission," *Bell Syst. Tech. J.,* Vol. 37, November 1958, pp. 1501–1542.

BENNETT, W. R.: *Introduction to Signal Transmission,* McGraw-Hill, New York, 1970.

BENNETT, W. R., and J. R. DAVEY: *Data Transmission,* McGraw Hill, New York, 1965.

BERLEKAMP, E. R.: *Algebraic Coding Theory,* McGraw-Hill, New York, 1968.

BLANSCHILD, R. A., P. A. TUCCI, H. T. RUSSELL, D. M. PURINTON, and E. MURTHI: "A Single Chip PCM Codec Using I²L," IEEE, NTC-78, Vol. 3, Birmingham, Ala., December 3–6, 1978.

BLEVIS, B. C.: "The History, Development and Future of Satellite Communications in Canada," *Can. J. Inf. Sci.,* Vol. 4, May 1979.

BRADY, P. T.: "A Statistical Analysis of On-Off Patterns in Sixteen Conversations," *Bell Syst. Tech. J.,* Vol. 47, 1968, pp. 73–92.

BRIND'AMOUR, A., and K. FEHER: "Design and Evaluation of a Convolutional Codec in Additive White Gaussian Noise, Sinusoidal Interference, and Intersymbol Interference Environment," *IEEE Trans. Commun.,* Vol. COM-28, No. 3, March 1980.

BYLANSKI, P., and D. INGRAM: *Digital Transmission Systems,* Peter Peregrinus Ltd. (IEE), Huddershield, England, 1976.

CACCIAMANI, JR., E. R.: "The SPADE System as Applied to Data Communications and Small Earth Station Operation," *COMSAT Tech. Rev.,* Vol. 1, No. 1, Fall 1971.

CACCIAMANI, JR., E.R., and C. J. WOLEJSZA, JR.: "Phase-Ambiguity Resolution in a Four-Phase PSK Communications System," *IEEE Trans. Commun.,* Vol. COM-19, No. 12, December 1971.

CACCIAMANI, JR., E.R., "The Spade System as Applied to Data Communications and Small Earth Station Operations," Comsat Tech. Rev., Vol. 1, No. 1, 1971.

CAMPANELLA, S. J.: "Digital Speech Interpolation," COMSAT Tech. Rev., Vol. 6, No. 1, Spring 1976.

CAMPANELLA, S. J.: "Synchronization of SS/TDMA On-Board Clocks," Int. Telemetering Conf., October 1980a (abstract only).

CAMPANELLA, S. J.: "TDMA System Terrestrial Interfaces," INTELSAT TDMA Seminar Notes, ESS-TDMA-1-8 H/11/80, Intelsat, Washington, D.C., November 1980b.

CAMPANELLA, S. J., F. ASSAL and A. BERMAN, "On-Board Communication Processing Technology," Symposium on Transportation and Communications, Genoa, Italy, October 1980.

CAMPANELLA, S. J., and K. HODSON: "Open Loop TDMA Frame Acquisition and Synchronization," Proc. 4th Int. Conf. Digital Satellite Commun., Montreal, October 1978.

CAMPANELLA, S. J. and ROGER J. COLLEY, "Network Control for Multibeam TDMA and SS/TDMA," IEEE Communication Society Transactions, Special Issue on Digital Satellite Communications, 1983.

CAMPANELLA, S. J. and T. INUKAI: "SS-TDMA Frame Synchronization," ICC'81, Denver, Colo.

CAMPANELLA, S. J., H. C. SUYDERHOUD, and M. ONFRY: "Analysis of an Adaptive Impulse Response Echo Canceller," *COMSAT Tech. Rev.*, Vol. 2, No. 1, Spring 1972.

CAMPANELLA, S. J., F. ASSAL, and A. BERMAN: "On-board Regenerative Repeater," IEEE ICC-77, Chicago, June 1977, pp. 6.2–121 to 6.2–125.

CAMPANELLA, S. J., R. J. COLBY, B. A. PONTANO, H. SUYDERHOUD, and M. ONFRY: "The INTELSAT TDMA Field Trial," *COMSAT Tech. Rev.*, Vol. 9, No. 2A, Fall 1979.

CARLSON, J.: *Communication Systems*, 2nd ed., McGraw-Hill, New York, 1975.

CARTER, C. R., "Survey of Synchronization Techniques for a TDMA Satellite-Switched System," *IEEE Trans. Commun.*, Vol. COM-28, No. 8, August 1980.

CCIR: "Digital Interface Characteristics between Satellite and Terrestrial Networks," Rep. No. 707, Vol. 4, International Telecommunications Union, Geneva, 1978a.

CCIR: "Energy Dispersal Techniques for Use with Digital Signals," Annex III to Rep. No. 384–3, CCIR Vol. 4, Geneva, 1978b.

CCIR: "Methods for Determining Interference in Terrestrial Radio-Relay Systems and Systems in the Fixed Satellite Service," Rep. No. 388–3, Vol. 4, CCIR International Radio Consultative Committee, International Telecommunications Union, Geneva, 1978c.

CCIR, Recommendations and Reports of the CCIR, Kyoto 1978: "Fixed Service Using Communication Satellites," Vol. 4, International Telecommunications Union, Geneva, 1978d.

CCIR, Recommendations and Reports of the CCIR, Kyoto, 1978: "Fixed Service Using Radio-Relay Systems. Frequency Sharing and Coordination between Systems in the Fixed Satellite Service and Radio-Relay Systems," Vol. 9, International Telecommunications Union, Geneva, 1978e.

CHANG, P. Y. and O. SHIMBO: "Input Power Assignment of Multicarrier Systems from Given Output Power Levels," *IEEE Trans. Commun.*, Part II, October 1979.

CHETNIK, F.: "Bandwidth Efficient Modulation and Detection," IEEE Can. Commun. Power Conf., Montreal, October 1976.

CHETNIK, F., and R. S. DAVIES: "Application of Regenerative Repeaters for Communication Satellites," Intelcom-77, Vol. I, Atlanta, October 1977.

CHIAO, J. T., and F. CHETNIK: "Satellite Regenerative Repeater Study," IEEE Can. Commun. Power Conf., Montreal, October 1976, pp. 222–225.

CHILDS, W. H., P. A. CARLTON, R. EGRI, C. E. MAHLE, and A. E. WILLIAMS: "A 14

GHz Regenerative Receiver for Spacecraft Application," Proc. 5th Int. Digital Satellite Commun. Conf., Genoa, Italy, March 1981.

CHRISTIAN, E., and EISENMANN, E.: *Filter Design Tables and Graphs,* Wiley, New York, 1966.

CLARK, A. P.: *Advanced Data-Transmission Systems,* Wiley–Halsted Press, New York, 1977.

CLARK, G. C., JR., and J. B. CAIN: *Error-Correction Coding for Digital Communications,* Plenum Press, New York, 1981.

CLEWER, R.: "Report on the Status of Development of the High Speed Digital Satellite Modem," RML-009-79-24, Spar Aerospace Limited, St. Anne de Bellevue, P.Q., Canada, November 1979.

COATES, R. F. W.: *Modern Communication Systems,* Macmillan, London, 1975.

COHEN, P. and D. HACCOUN: "Structure des conversations sur les voies radio-téléphoniques mobiles," *Can. Electr. Eng. J.,* Vol. 5, No. 4, October 1980.

COHN-SFETCU, S., and J. DOYLE: "A Low-Cost Real-Time Service Digital Signal Processor," *IEEE Trans. Commun.,* Vol. COM-26, No. 5, May 1978.

COLAVITO, C., and G. PALADIN: "Transmission Media for a 34 Mb/s Hieararchical Level," Proc. World Telecommun. Forum, International Telecommuncations Union, Geneva, October 1975.

COOPER, G. R. and C. D. MCGILLEM: *Probabilistic Methods of Signal and System Analysis,* Holt, Rinehart and Winston, 1971.

COSTA, B. M., and VENETSANOPOULOS, A. N.: "Recursive Implementation of Factorable Two-Dimensional Digital Filters," *Can. Electri. Eng. J.,* Vol. 4, No. 3, July 1979.

CROWELL, R.: "Telecommuncations Technology Working Paper," Office of Technology Assessment, Congress of the United States, April 1980; also published in excerpted form in *Telecommun. J.,* International Telecommunications Union, Geneva, June 1980, pp. 388–391.

CUCCIA, C. L.: *The Handbook of Digital Communications,* EW Communications, Inc., Palo Alto, Calif., 1979.

CUCCIA, L., and C. HELLMAN: "Status Report: The Low-Cost Low-Capacity Earth Terminal," *Microwave Syst. News,* June–July 1975.

CUCCIA, R., R. DAVIES, and E. MATHEWS: "Baseline Considerations of Beam Switched SS-TDMA Satellites Using Baseband Matrix Switching," IEEE ICC-77, pp. 6.3.126–6.3.131.

CULBERTSON, A. F.: "The Role of Digital Microwave Systems in World Telecommunications," Proc. World Telecommun. Forum, International Telecommunications Union, Geneva, October 1975.

DAMMANN, C. L., L. D. MCDANIEL, and C. L. MADDOX: "D2 Channel Bank-Multiplexing and Coding," *Bell Syst. Tech. J.,* October 1972.

DE BUDA, R.: "Coherent Demodulation of Frequency-Shift Keying with Low Deviation Ratio," *IEEE Trans. Commun.,* Vol. COM-20, No. 6, June 1972.

DE BUDA, R.: "Fast FSK Signals and Their Demodulation," *Can. Electr. Eng. J.,* Vol. 1, No. 1, 1976.

DEVIEUX, C.: "Spectral Spreading," *COMSAT J. Tech. Rev.,* September 1974.

DICKS, J. L., and M. P. BROWN, JR.: "Intelsat V Satellite Transmission Design," Proc. Int. Conf. Commun., ICC-78, Vol. 1, Toronto, June 1978.

DIXON, R. C.: *Spread Spectrum Systems,* Wiley, New York, 1976.

DODDS, D. E., A. M. SENDYK, and D. B. WOHLBERG: "Error Tolerant Adaptive Algorithms for Delta-Modulation Coding," *IEEE Trans. Commun.,* Vol. COM-28, No. 3, March 1980.

DRISCOLL, M. M.: "Two-Stage Self-Limiting Series Mode Type Quartz-Crystal Oscillator Exhibiting Improved Short-Term Frequency Stability," *IEEE Trans. Instrum. Meas.,* Vol. IM-22, No. 6, June 1973.

DUBOIS, E.: "Effects of Digital Demodulation on Component Coding of NTSC Color Signals," *IEEE Trans. Commun.,* Vol. COM-27, No. 9, September 1979.

DUPONTEIL, D.: "Critères de Nyquist," NT/TCR/EFT/578, Centre National d'Etudes des Télécommunications (CNET), Issy-les-Moulineaux, France, August 1978.

DUTTWEILER, D. L.: "Bell's Echo Killer Chip," *IEEE Spectrum,* October 1980.

DYSART, H. G., and N. D. GEORGANAS: "NEWCLUST: An Algorithm for the Topological Design of Two-Level, Multidrop Teleprocessing Networks," *IEEE Trans. Commun.,* Vol. COM-26, No. 1, January 1978.

EDELSON, B. I., and A. M. WERTH: "SPADE System Progress and Application," *COMSAT Tech. Rev.,* Vol. 2, No. 1, Spring 1972.

Electronics-Staff of Satellite Systems Engineering, Inc.: "More Communications in Orbit," *Electronics,* September 11, 1980, pp. 148–151.

ERIC, M. J.: "Intermodulation Analysis of Nonlinear Devices for Multiple Carrier Inputs," CRC Report, Department of Communications, No. 1234, Ottawa, Canada, November 1972.

European Space Research and Technology Center: "Performance Analysis of Uplinks for Regenerative Satellite Repeaters," Call For Competitive Offers No AO/1–1079/79/NL/DG, Noordwijk, Netherlands, March 1979.

FANG, R., and O. SHIMBO: "Unified Analysis of a Class of Digital Systems in Additive Noise and Interference," *IEEE Trans. Commun.,* Vol. COM-21, No. 10, October 1973.

FEHER, K.: "Timing Technique for NRZ Data Signals," U.S. Patent No. 3,944,926 issued March 16, 1976. Canada Patent No. 210–237. RCA Docket No. 67.504.

FEHER, K.: *Digital Modulation Techniques in an Interference Environment,* Vol. 9 of *Encyclopedia on EMC,* Don White Consultants, Inc., Gainesville, Va., 1977.

FEHER, K.: "On Bit Timing Transmission and Spectral Shaping of Digital Signals," *Can. Electr. Eng. J.,* July 1978, pp. 32–34.

FEHER, K.: "In Service Jitter Measurement Technique," Canadian Patent Development Limited Disclosure, File No. 265–6977–1, Ottawa, July 30, 1979.

FEHER, K.: "Filter," Canadian Patent Disclosure 327,365, filed May 10, 1979; U.S. Patent No. 4,339,724, July 13, 1982.

FEHER, K.: *Digital Communications: Microwave Applications,* Prentice-Hall, Englewood Cliffs, N.J., 1981.

FEHER, K. and D. CHAN: "PSK Combiners for Fading Microwave Channels," *IEEE Trans. Commun.,* Vol. COM-72, No. 6, June 1974.

FEHER, K., and R. DECRISTOFARO: "Transversal Filter Design and Application in Satellite Communications," *IEEE Trans. Commun.,* Vol. COM-24, No. 11, November 1976, pp. 1262–67.

FEHER, K., and J. HUANG: "PAM Microwave Transmission in Coloured Gaussian Noise Environment," *Radio Electron. Eng.* (IRE), April 1977, pp. 167–171.

FEHER, K., and M. MORRIS: "Simultaneous Transmission of Digital PSK and of Analog Television Signals," *IEEE Trans. Commun.*, Vol. COM-23, No. 12, December 1975.

FEHER, K., and G. S. TAKHAR: "A New Symbol Timing Recovery Technique for Burst Modem Applications," *IEEE Trans. Commun.*, Vol. COM-26, No. 1, January 1978, pp. 100–108.

FEHER, K., R. GOULET, and S. MORISSETTE: "Order Wire Transmission in Digital Microwave Systems," *IEEE Trans. Commun.*, Vol. COM-22, No. 5, May 1974.

FEHER, K., R. GOULET, and S. MORISSETTE: "1.544 Mbit/s Data above FDM Voice (DUV) Microwave Transmission," *IEEE Trans. Commun.*, Vol. COM-23, No. 11, November 1975.

FEHER, K., M. ROUX, R. GOULET, and S. MORISSETTE: "Comparison of PSK and Multilevel AM Data Modulators," *Int. J. Electron.*, Vol. 41, No. 2, 1976, pp. 153–158.

FEHER, K., M. EL-TORKY, R. DECRISTOFARO, and M. SWAMY: "Optimum Pulse Shaping Application of Binary Transversal Filters Used in Satellite Communications," *Radio Electron. Eng.* (IRE), June 1977.

FEHER, K., T. LE-NGOC, R. CLEWER, and A. GUIBORD: "Fast Acquisition Methods for High Speed OQPSK Satellite Systems," Proc. IEEE Int. Electr. and Electron. Expo., Toronto, September 1979a.

FEHER, K., R. TETARENKO, P. R. HARTMAN, and V. K. PRABHU: "Digital Communications by Radio," *IEEE Trans. Commun.*, Special Issue on Digital Radio, December 1979b.

FEHER, K., V. ARUNACHALAM, H. GIRARD, T. LE-NGOC, A. GUIBORD, M. WACHIRA, and J. Y. C. HUANG: "Regenerative Transponders for More Efficient Digital Satellite Systems—Phase I-II-Report," prepared for Department of Communications, Communications Research Centre, 03SU-36100-9-9521, SU79–00146, Ottawa, Canada, April 1980 and April 1981 (Part II).

FEHER, K., V. ARUNACHALAM, M. WACHIRA, D. PRENDERGAST, T. LE-NGOC, J. HUANG, P. HILL, P. AMLEKAR, and H. GIRARD: "Regenerative Transporters for More Efficient Digital Satellite Systems," prepared for the Department of Communications, Communications Research Centre CRC Report for Contract No. 09SU.36101–0–3212, Serial No. OSU80–000197, Parts I and Part II, Ottawa, Ontario, March 1981.

FEHER, K. and S. KATO: "X-PSK: Crosscorrelated-bandlimited Constant Envelope PSK Modulation"—Patent Disclosure, Canada and USA March, 1982, and IEEE-ICC-1982.

FORNEY, G. D.: "Coding and Its Application in Space Communications," *IEEE Spectrum,* June 1970.

FORSEY, R. J., GOODING, V. E., MCLANE, P. J., and CAMPBELL, L. L.: "*M*-ary PSK Transmission via a Coherent Two-Link Channel Exhibiting AM-AM and AM-PM Nonlinearities," *IEEE Trans. Commun.*, Vol. COM-26, No. 1, January 1978.

FRANKS, L. E.: "Further Results on Nyquist's Problem in Pulse Transmission," *IEEE Trans., Commun.*, Vol. COM-16, No. 4, April 1968, pp. 337–340.

FRANKS, L. E.: *Signal Theory,* Prentice-Hall, Englewood Cliffs, N.J., 1969.

FRANKS, L. E., ed.: *Data Communication: Fundamentals of Baseband Transmission,* Benchmark Papers, Dowden, Hutchinson & Ross, Stroudsburg, Pa., 1974.

FRANKS, L. E.: "Carrier and Bit Synchronization in Data Communication—A Tutorial Review," *IEEE Trans. Commun.*, Vol. COM-28, No. 8, August 1980, pp. 1107–1121.

FRANKS, L. E., and J. P. BUBROUSKI: "Statistical Properties of Timing Jitter in a PAM Timing Recovery Scheme," *IEEE Trans. Commun.*, Vol. COM-22, No. 7, July 1974, pp. 913–920.

FREENY, S. L.: "TDM/FDM Translation as an Application of Digital Signal Processing," *IEEE Commun. Mag.*, January 1980.

GAGLIARDI, R.: *Introduction to Communications Engineering*, Wiley, New York, 1978.

GALKO, P., and S. PASUPATHY: "On a Class of Generalized MSK," Proc. 1981 IEEE Int. Commun. Conf., Denver, Colo., June 1981.

GALLAGER, R. G.: *Information Theory and Reliable Communication,*" Wiley, New York, 1968.

GARDNER, F. M.: *Phaselock Techniques*, Wiley, New York, 1979.

GARDNER, F. M.: "Self-Noise in Synchronizers," *IEEE Trans. Commun.*, Vol. COM-28, No. 8, August 1980, pp. 1159–1163.

GENDRON, M., and K. FEHER: "Une nouvelle famille de filtre non-linéaire," *Can. Electr. Eng. J.*, Vol. 4, No. 1, January 1979.

GEORGANAS, N. D., S. G. S. SHIVA, P. K. VERMA, and J. S. JAWANDA: "Evaluation of the Mean Error-Free Interval of a Noisy Data Channel," *IEEE Trans. Commun.*, Vol. COM-26, No. 1, January 1978.

GERBER, E. A., and R. A. SYKES: "Quartz Crystal Units and Oscillators," *Time and Frequency: Theory and Fundamentals*, NBS Monograph 140, U.S. Department of Commerce, May 1974, Chap. 2.

GIBSON, R.: "Satellite Communications I: ESA at the Crossroads," *IEEE Spectrum*, March, 1980.

GIBSON, I. B., and I. F. BLAKE: "Decoding the Binary Golay Code with Miracle Octad Generators," *IEEE Trans. Inf. Theory*, Vol. IT-24, No. 2, March 1978.

GIBSON, R., and W. LUKSCHI: "The European Space Agency and Its Programmes, in Particular Its Telecommunications Programme," ITU 3rd World Telecommun. Forum, Geneva, September 1979.

GITLIN, R. D., and J. SALZ: "Timing Recovery in PAM Systems," *Bell Syst. Tech. J.*, Vol. 50, pp. May–June 1971, 1645–1669.

GLASGAL, R.: *Advanced Techniques in Data Communications*, Artech House, Inc., Dedham, Mass., 1976.

GLAVE, F. E., and A. S. ROSENBAUM: "An Upper Bound Analysis for Coherent Phase-Shift Keying with Cochannel, Adjacent-Channel, and Intersymbol Interference," *IEEE Trans. Commun.*, Vol. COM-23, No. 6, June 1975.

GODIER, I.: "DRS-8A Digital Radio for Long Haul Transmission," Proc. IEEE Int. Conf. Commun. IEE-1977, Chicago, June 1977.

GOLOMB, S. W.: *Digital Communications with Space Applications*, Prentice-Hall, Englewood Cliffs, N.J., 1964.

GOODE, B.: "Demand Assignment of the SBS TDMA Satellite Communications System," EASCON '78, Washington, D.C.

GOODING, V. E., and P. J. McLANE: "Jointly Optimal Filters for Data Transmission over Multiroute Systems," *IEEE Trans. Inf. Theory*, Vol. IT-26, No. 4, July 1980.

GOULD, R. G., and G. K. HELDER: "Transmission Delay and Echo Suppression," *IEEE Spectrum*, April 1970.

GRAY, P. R., and D. G. MESSERSCHMITT: "Intergrated Circuits for Telephony," *Proc. IEEE*, August 1980.

GREBENE, A. B.: "The Monolithic Phase-Locked Loop—A Versatile Building Block," *IEEE Spectrum,* March 1971.

GREGG, W. D.: *Analog and Digital Communication,* Wiley, New York, 1977.

GRINICH, V. H., and H. G. JACKSON: *Introduction to Integrated Circuits,* McGraw-Hill, New York, 1975.

GRONEMEYER, S., and A. MCBRIDE: "MSK and Offset QPSK Modulation," *IEEE Trans. Commun.,* Vol. COM-24, No. 8, August 1976.

GTE Lenkurt: "Satellite Communications Update, Part I," *GTE Lenkurt Demodulator,* Vol. 28, No. 1, January–February, 1979a.

GTE Lenkurt: "VLSI Codecs," *GTE Lenkurt Demodulator* Vol. 28, No. 1, January–February, 1979b.

HACCOUN, D.: "A Markov Chain Analysis of the Sequential Decoding Metric," *IEEE Trans. Inf. Theory,* Vol. IT-26, No. 1, January 1980.

HAGIWARA, S., N. SATA, and A. TOKIMASA: "1.544 Mb/s PCM-FDM Converters over Coaxial and Microwave Systems," *Fujitsu Sci. Tech. J.,* September 1976.

HAMMING, R. W.: *Coding and Information Theory,* Prentice-Hall, Englewood Cliffs, N.J., 1980.

HANSON, B. A., and R. W. DONALDSON: "Subjective Evaluation of an Adaptive Differential Voice Encoder with Oversampling and Entropy Coding," *IEEE Trans. Commun.,* Vol. COM-26, No. 2, February 1978.

HARRINGTON, E. A.: "Issues in Terrestrial/Satellite Network Synchronization," *IEEE Trans. Commun.,* Vol. Com-27, No. 11, November 1979.

HARRIS, R. A.: "Transmission Analysis and Design for the ECS Systems," Proc. 4th Int. Conf. Digital Satellite Commun., INTELSAT, Montreal, October 23–25, 1978.

HARSHMAN, J. V., and R. W. MATHEWS: "An Integrated DAMA System," *Telecommunications,* September 1979.

HATCH, R. W., and A. E. RUPPEL: "New Rules for Echo Suppressors in the DDD Network," *Bel Lab. Rec.,* Vol. 52, 1974.

HATZIGEORGIOU, S.: "Integrate Sample and Dump Receiver and Relationships between E_b/N_o and S/N," Eng. Rep. No.-701, Spar Aerospace Limited, St. Anne de Bellevue, P.Q. Canada, July 1980.

HAYKIN, S. S.: *Communication Systems,* Wiley, New York, 1978.

HELLER, J. A., and I. M. JACOBS: "Viterbi Decoding for Satellite and Space Communication," *IEEE Trans. Commun. Technol.,* October 1971, pp. 835–848.

HEWLETT-PACKARD: "Biasing and Driving Considerations for PIN Diode RF Switches and Modulators," Application Note 914, Hewlett-Packard, Palo Alto, Calif., January 1967.

HOEBER, C. F.: "INTELSAT V System Design," WESCON, Western Electronic Show and Convention, IEEE, September 1977.

HUANG, J. C. Y.: "On Bandwidth Efficient Spectral Shaping Methods and Digital Modulation Techniques in Linear and Nonlinear Channels," Ph. D. thesis, Concordia University, Montreal, April 1979.

HUANG, J., and K. FEHER: "Performance of QPSK, OKQPSK and MSK through Cascaded Nonlinearity and Bandlimiting," Proc. IEEE Int. Conf. Commun., ICC-79, Boston, June, 1979a.

HUANG, J. C. Y., and K. FEHER: "On Partial Response Digital Radio Systems," *IEEE Trans. Commun.,* Vol. COM-27, No. 11, November 1979b.

HUANG, J. C. Y., K. FEHER, and M. GENDRON: "Techniques to Generate ISI and Jitter Free Bandlimited Nyquist Signals and a Method to Analyze Jitter Effects," *IEEE Trans. Commun.,* Vol. COM-27, No. 11, November 1979.

HUANG, T. C., J. K. OMURA, and L. BIEDERMAN: "Comparison of Conventional and Regenerative Satellite Communication System Performance," ICCC'79, Boston, June 1979, pp. 58.3.1–58.3.5.

IEEE Transactions on Communications: Special Issue on Satellite Communication, October 1979.

Infospar-Editor: "Preparation for Primacy," *Infospar Quarterly Review,* Spar Aerospace Limited, Toronto, Vol. 11, No. 2, July 1979.

INTELSAT: "Construction Details for an INTELSAT Demand Assigned Multiple Access Terminal (SPADE)," ICCSC/T-31, 20E w/6/69, Washington, D.C., 1969.

INTELSAT: "SCPC System Specification," BG/T-5–21E, W/1/74, January 7, 1974, Washington, D.C., 1974.

INTELSAT: "SCPC/PSK (4ϕ) and SCPC/PCM/PSK (4ϕ) System Specification," BG-9–21E H/S/74 (Rev. 2), INTELSAT, Washington, D.C., October 20, 1976.

INTELSAT: "INTELSAT TDMA/DSI System Specification (TDMA/DSI Traffic Terminals)," BG-42–65E B/6/80, Intelsat, Washington, D.C., June 26, 1980.

International Telecommunications Union: "Transmission Systems," *Economic and Technical Aspects of the Choice of Transmission Systems Gas 3 Manual,* ISBN 92–61–00211–0, Vol. 1, Geneva, 1976.

International Telecommunications Union: "Table of Artificial Satellites Launched in 1978," *Telecommun. J.,* Vol. 46–V, Geneva, 1979.

IRVINE, R. G.: *Operational Amplifier: Charateristics and Applications,* Prentice-Hall, Englewood Cliffs, N.J., 1981.

ISHIO, H., M. WASHIO, M. INOKUCHI, S. SEKI, et al.: "A New Multilevel Modulation and Demodulation System for Carrier Digital Transmission," Proc. IEEE Int. Conf. Commun., ICC-76, Philadelphia, June 1976.

ISMAIL, M. G., and R. J. CLARKE: "Facsimile Compression Using a Classified Adaptive Block/Run Length Coding Scheme," Proc. IEEE-NTC, Houston, December 1980.

JAGER, F., and C. B. DEKKER: "Tamed Frequency Modulation: A Novel Method to Achieve Spectrum Economy in Digital Transmission," *IEEE Trans. Commun.,* Vol. COM-26, No. 5, May 1978.

JAKES, W. C.: *Microwave Mobile Communications,* Wiley, 1974.

JANKOWSKI, J.: "General Description—TDMA Operation," INTELSAT Seminar-ESS-TDMA-1-9, H/11/80, Washington, D.C., November 1980.

JAYANT, S. N.: "Digital Coding of Speech Waveforms: PCM, DPCM and DM Quantizers," *Proc. IEEE,* May 1975.

JONES, M. E., and M. R. WACHS: "Optimum Filtering for QPSK in Bandwidth-Limited Nonlinear Satellite Channels," *COMSAT Tech. Rev.,* Vol. 9, No. 2A, Fall 1979.

KAHWA, T. J., and N. D. GEORGANAS: "A Hybrid Channel Assignment Scheme in Large-Scale, Cellular-Structured Mobile Communication Systems," *IEEE Trans. Commun.,* Vol. COM-26, No. 4, April 1978.

KAMAL, S. S., and S. A. MAHMOUD: "A Study of Users' Buffer Variations in Random Access Satellite Channels," *IEEE Trans. Commun.,* Vol. COM-27, No. 6, June 1979.

KAMANGAR, F. A., and K. R. RAO: "Adaptive Coding of NTSC Component Video Signals," Proc. IEEE-NTC, Houston, December 1980.

KANEKO, H., and T. ISHIGURO: "Digital Television Transmission Using Bandwidth Compression Techniques," *IEEE Commun. Mag.,* July 1980.

KEELTY, M., and K. FEHER: "On-Line Pseudo-error Monitors for Digital Transmission Systems," *IEEE Trans. Commun.,* Vol. COM-26, No. 8, August 1978.

KOBAYASHI, H.: "Simultaneous Adaptive Estimation and Decision Algorithm for Carrier Modulated Data Transmission Systems," *IEEE Trans. Commun.,* Vol. COM-19, No. 6, June 1971, pp. 268–280.

KOGA, K., T. MURATANI, and A. OGAWA: "Onboard Regenerative Repeaters Applied to Digital Satellite Communications," *Proc. IEEE,* Vol. 65, No. 3, March 1977, pp. 401–410.

KOLL, V. G., and S. B. WEINSTEIN: "Simultaneous Two-Way Data Transmission over a Two-Wire Line," *IEEE Trans. Commun.,* Vol. COM-19, No. 2, February 1971.

KUH, E. S., and D. O. PEDERSON: *Principles of Circuit Synthesis,* McGraw-Hill, New York, 1959.

KURIHARA, H., R. KATOH, H. KOMIZO, and H. NAKAMURA: "Carrier Recovery Circuit with Low Cycle Skipping Rate for CPSK/TDMA Systems," Proc. 5th Int. Conf. Digital Satellite Commun., Genoa, March 1981.

KWAN, R. K.: "Modulation and Multiple Access Selection for Satellite Communications," IEEE, NTC-78, Vol. 3, Birmingham, Ala., December 3–6, 1978.

LATHI, B. P.: *An Introduction to Random Signals and Communication Theory,* International Textbook Co., Scranton, Pa., 1968.

LAWTON, J. G.: "Comparison of Binary Data Transmission Systems," Proc. 2nd Natl. Conf. Military Electron., 1958, pp. 54–61.

LEE, Y. W.: *Statistical Theory of Communication,* Wiley, New York, 1960.

LEE, Y. S.: Simulation Analysis for Differentially Coherent Quarternary PSK Regenerative Repeater," *COMSAT Tech. Rev.,* Vol. 7, No. 2, Fall 1977, pp. 447–474.

LENDER, A.: "The Duobinary Technique for High Speed Data Transmission," *IEEE Trans. Commun. Electron.,* Vol. 82, May 1963.

LENDER, A.: "Correlative Digital Communications Techniques," *IEEE Trans. Commun. Technol.,* December 1964.

LENDER, A.: "Correlative (Partial Response) Techniques and Applications to Digital Radio Systems," in K. Feher, *Digital Communications: Microwave Applications,* Prentice-Hall, Englewood-Cliffs, N.J., 1981, Chap. 7.

LE-NGOC, T., and K. FEHER: "A Digital Approach to Symbol Timing Recovery Systems," *IEEE Trans. Commun.,* Vol. COM-28, No. 12, December 1980.

LE-NGOC, T., K. FEHER, and H. PHAM-VAN: "New Modulation Techniques for Low-Cost Power and Bandwidth Efficient Satellite Earth Stations," *IEEE Trans. Commun.,* Vol. COM-30, No. 1, January 1982.

LESTER, R. M.: "The Role of the Satellite System in the Canadian Digital Network," Proc. 3rd World Telecommun. Forum, Part 2, International Telecommunication Union, Geneva, September 1979.

LIAO, S. Y.: *Microwave Devices and Circuits,* Prentice-Hall, Englewood Cliffs, New Jersey, 1980.

LIN, S.: *An Introduction to Error-Correcting Codes,* Prentice-Hall, Englewood Cliffs, N.J., 1970.

LINDSEY, W. C.: *Synchronous Systems in Communications and Control,* Prentice-Hall, Englewood Cliffs, N.J., 1972.

LINDSEY, W. C., and M. K. SIMON: "Data-Aided Carrier Tracking Loop," *IEEE Trans. Commun.,* Vol. COM-19, No. 4, April 1971, pp. 157–168.

LINDSEY, W., and M. SIMON: *Telecommuncations System Engineering,* Prentice-Hall, Englewood Cliffs, N.J., 1973.

LINUMA, K., Y. LIJIMA, T. ISHIGURO, H. KANEKO, and S. SHIGAKI: "Interframe Coding for 4 MHz Color Television Signals," *IEEE Trans. Commun.,* Vol. COM-23, No. 12, December 1975.

LOMBARD, D.: "Utilisation du CELTIC sur des liaisons par câble sous-marin et par satellite," ITU 3rd World Telecommun. Forum, Geneva, September 1979.

LOMBARD, D., M. BIC, and L. IMBEAUX: "An FDMA Digital Data Transmission System Using DSI Equipment and a VITERBI Decoder," Proc. 4th Int. Conf. Digital Satellite Commun., Montreal, October 23–25, 1978.

LUCKY, R. W., J. SALZ, and J. WELDON: *Principles of Data Communication,* McGraw-Hill, New York, 1968.

LUM, Y. F.: "Satellite Communications I and II," ELG 5174 course notes, Dept. of Electrical Engineering, University of Ottawa, Ottawa, September 1978.

LUNDQUIST, L.: "Modulation Techniques for Band and Power Limited Satellite Channels," Proc. 4th Int. Conf. Digital Satellite Commun., Montreal, October 23–25, 1978.

LUNDQUIST, L., M. LOPRIORI, and F. M. GARDNER: "Transmission of 4ϕ-Phase-Shift-Keyed Time-Division Multiple Access over Satellite Channels," *IEEE Trans. Commun.,* Vol. COM-22, No. 9, September 1974, pp. 1354–1360.

LUNSFORD, J.: "Satellite Position Determination and Acquisition Window Accuracy in the INTELSAT TDMA System," COMSAT Lab. Tech. Memorandum CL-28–81.

LYONS, R. G.: "Signal and Interference Output of a Bandpass Nonlinearity," *IEEE Trans. Commun.,* Vol. COM-27, No. 6, June 1979.

MANCIANTI, M., U. MENGALI, and R. REGGIANNINI: "A Fast Start-Up Algorithm for Channel Parameter Acquisition in SSB-AM Data Transmission," ICC'79, Boston, 1979.

MARK, J. W., and S. F. W. NG: "A Coding Scheme for Conflict-Free Multiaccess Using Global Scheduling," *IEEE Trans. Commun.,* Vol. COM-27, No. 9, September 1979.

MARSTEN, R. B.: "Service Needs and Systems Architecture in Satellite Communications," *IEEE Commun. Soc. Mag.,* May 1977.

MARTIN, J.: *Future Developments in Telecommunications,* Prentice-Hall, Englewood Cliffs, N.J., 1977a.

MARTIN, J.: *Telecommuncations and the Computer,* Prentice-Hall, Englewood Cliffs, N.J., 1977b.

MARTIN, J.: *Communications Satellite Systems,* Prentice-Hall, Englewood Cliffs, N.J., 1978.

MASSEY, J. L.: *Threshold Decoding,* MIT Press, Cambridge, Mass., 1963.

MATYAS, R.: "Effect of Noisy Phase References on Coherent Detection of FFSK Signals," *IEEE Trans. Commun.,* Vol. COM-26, No. 6, June 1978.

MATYAS, R., and P. J. McLANE: "Decision-Aided Tracking Loops for Channels with Phase Jitter and Intersymbol Interference," *IEEE Trans. Commun.*, Vol. COM-22, No. 8, August 1974, pp. 1014–1023.

McCLURE, R. B.: "The Effect of Earth Station and Satellite Parameters on the SPADE System," IEE Conf. Earth Station Technol., IEEE Conf. Publ. 72, London, October 1970.

McGLYNN, D. R.: *Distributed Processing and Data Communications*, Wiley, New York, 1978.

McLANE, P. J.: "A Residual Intersymbol Interference Error Bound for Truncated-State Viterbi Detectors," *IEEE Trans. Inf. Theory*, Vol. IT-26, No. 5, September 1980.

MENGALI, U.: "Synchronization of QAM Signals in the Presence of ISI," *IEEE Trans. Aerosp. Electron. Syst.*, Vol. AES-12, September 1976, pp. 556–560.

MENGALI, U.: "Joint Phase and Timing Acquisition in Data Transmission," *IEEE Trans. Commun.*, Vol. COM-25, No. 10, October 1977, pp. 1174–1185.

METZGER, L. S.: "On-Board Satellite Signal Processing," IEEE NTC'78, December 1978, pp. 8.1.1–8.1.5.

MEYERS, M. H., and L. E. FRANKS: "Joint Carrier Phase and Symbol Timing Recovery for PAM Systems," *IEEE Trans. Commun.*, Vol. COM-28, No. 8, August 1980, pp. 1121–1129.

MINAMI, T., T. MURAKONI, and T. ICHIKAWA: "An Overview of the Digital Transmission Network in Japan," Proc. IEEE Int. Conf. Commun., Vol. 1, ICC-78, Toronto, June 1978.

MIYA, K.: *Satellite Communications Engineering*, Lattice Co., Japan, 1975.

MORAIS, D. H.: "Digital Modulation Techniques for Terrestrial Point-to-Point Microwave Systems," Ph.D. thesis, Dept. of Electrical Engineering, University of Ottawa, Ottawa, 1981.

MORAIS, D., and K. FEHER: "Bandwidth Efficiency and Probability of Error Performance of MSK and OKQPSK Systems," *IEEE Trans. Commun.*, Vol. COM-27, No. 12, December 1979.

MORAIS, D. H., and K. FEHER: "The Effects of Filtering and Limiting on the Performance of QPSK Offset QPSK and MSK Systems," *IEEE Trans. Commun.*, Vol. COM-28, No. 12, December 1980.

MORAIS, D., and K. FEHER: "NLA QAM—Nonlinearly Amplified QAM Modulation Technique. . . ," *IEEE Trans. Commun.*, Vol. COM-30, No. 3, March 1982.

MORAIS, D., A. SEWERINSON, and K. FEHER: "The Effects of the Amplitude and Delay Slope Components of Frequency Selective Fading on QPSK, Offset QPSK and 8-PSK Systems," *IEEE Trans. Commun.*, Vol. COM-27, No. 12, December 1979.

MORGAN, L. W.: "Communications Satellites—1973 to 1983," IEEE, ICC-78, Vol. 1, Toronto, June 4–7, 1978.

MORKHOFF N.: "Communications and Microwave," *IEEE Spectrum*, January 1980.

MORRIS, M. J., and K. FEHER: "Development in Canadian and International Data above Voice/Video (DAV) Telecommunications Networks," *Can. Electr. Eng. J.*, January 1980.

MUMFORD, M. W., and E. H. SCHEIBE: *Noise Performance Factors in Communication Systems*, Horizon House, Dedham, Mass., 1968.

MURANO, K., Y. MOCHIDA, S. AMANO, and T. KINOSHITA: "Multiprocessor Architecture for Voiceband Data Processing (Application to 9600 BPS Modem)," Proc. IEEE Int. Commun. Conf., ICC-79, Boston, June 1979.

MURATANI, T., H. SAITOH, K. KOGA, Y. MIZUNO, and Y. J. S. SNYDER: "Application of

FEC Coding to the Intelsat TDMA Systems," Proc. 4th Int. Conf. Digital Satellite Commun., Intelsat, Montreal, October 1978.

NOORDANUS, J.: "New Digital Phase Modulation Methods to Establish Digital Voice Transmission in Mobile Radio Networks, with Optimum Spectrum Efficiency," ITU 3rd World Telecommun. Forum, Geneva, September 1979.

NUSPL, P. P., R. G. LYONS, and R. BEDFORD: "SLIM TDMA Project-Development of Versatile 3 Mb/s TDMA Systems," Proc. 5th Int. Conf. Digital Satellite Commun., Genoa, Italy, March 1981.

NYQUIST, H.: "Certain Topics in Telegraph Transmission Theory," *Trans. AIEE*, Vol. 47, February, 1928, copyright IEEE, 1974.

O'NEAL, J. B.: "Waveform Encoding of Voiceband Data Signals," *Proc. IEEE*, February 1980.

OPPENHEIM, A. V., and R. W. SCHAFER: *Digital Signal Processing*, Prentice-Hall, Englewood Cliffs, N.J., 1975.

OUZAS, N., and S. HAYKIN: "Performance of Digital Data Transmission Using Single-Sideband Modulation over Nonlinear Channels," *Can. Electr. Eng. J.*, Vol. 5, No. 4, October 1980.

PANTER, P. F.: *Modulation, Noise and Spectral Analysis*, McGraw-Hill, New York, 1965.

PAPOULIS, A.: *Probability, Random Variables and Stochastic Processes*, McGraw-Hill, New York, 1965.

PARES, J., and V. TOSCER: "Systèmes des communications par satellites," École Nationale de Paris, 1977.

PASUPATHY, S., M. A. YONGACOGLU, and J. B. TALLER: "Carrier Phase Error in Single Sideband-Partial Response Systems," *IEEE Trans. Commun.*, Vol. COM-28, No. 12, December 1980, p. 2010.

PEEBLES, Z. P.: *Communication System Principles*, Addison-Wesley, Reading, Mass., 1976.

PELTON, J.: "An Overview of Satellite Communications," Satellite Commun. Users Conf., Denver, Colo., August 1980, and *Satellite Communi.*, October 1980.

PERERA, A., K. FEHER, and M. SWAMY: "Design of Cascaded Digital and Analog Filters to Meet with Requirements on ISI and Band-Limitation in Low and Medium Capacity Digital Transmission Systems," Proc. IEEE Int. Electr. and Electron. Expo., Toronto, September 1979.

PERILLAN, L., and T. R. ROWBOTHAM: "INTELSAT VI SS-TDMA System Definition and Technology Assessment," Proc. 5th Int. Conf. Digital Satellite Commun., Genoa, Italy, 1981.

PETERSON, W. W.: *Error Correcting Codes*, MIT Press, Cambridge, Mass., 1961.

PONTANO, B., G. FORCINA, J. DICKS, and J. PHIEL: "Description of the INTELSAT TDMA/ DSI System," Proc. 5th Int. Conf. Digital Satellite Commun., Genoa, Italy, 1981.

POTTS, J. B.: "WARC-1979: Its Meaning and Impact on the Fixed Satellite Service," *IEEE Commun. Mag.*, September 1979.

PRABHU, V. K.: "Error Rate Considerations for Coherent Phase-Shift Keyed Systems with Co-channel Interference," *Bell Syst. Tech. J.*, March 1969.

PRABHU, V. K.: "The Detection Efficiency of 16-ary QAM," *Bell Syst. Tech. J.*, April 1980.

PRENDERGAST, D.: "Design and Evaluation of a Sixteen State APK Quasi Universal Data Modem," M.Sc.A. thesis, University of Ottawa, 1981.

PRITCHARD, W. L.: "Satellite Communication—An Overview of the Problems and Programs," *Proc. IEEE*, Special Issue on Satellite Communications, March 1977.

PROMHOUSE, G., and S. E. TAVARES: "The Minimum Distance of All Binary Cyclic Codes of Odd Lengths from 69 to 99," *IEEE Trans. Inf. Theory*, Vol. IT-24, No. 4, July 1978, p. 438.

PUENTE, J. G., and A. M. WERTH: "Demand-Assigned Service for the INTELSAT Global Network," *IEEE Spectrum*, January 1971.

RAPUANO, R. A., and N. SHIMOSAKI: "Synchronization of Earth Stations to Satellite-Switched Sequence," AIAA 4th Commun. Satellite Syst. Conf., Paper No. 72–545, 1972.

RODEN, M. S.: *Analog and Digital Communication Systems*, Prentice-Hall, Englewood Cliffs, N.J., 1979.

ROSENBAUM, A. S., and F. E. GLAVE: "An Error Probability Upper Bound for Coherent Phase-Shift Keying with Peak-Limited Interference," *IEEE Trans. Commun.*, Vol. 22, No. 1, January 1974.

ROSSITER, P., R. CHANG, and T. KANION: "Echo Control Considerations in an Integrated Satellite Terrestrial Network," 4th Int. Conf. Digital Satellite Commun., Montreal, October 23–25, 1978.

SABLATASH, M., and J. R. STOREY: "Determination of Throughputs, Efficiencies and Optimal Block Lengths for an Error-Correction Scheme for the Canadian Broadcast Telidon System," *Can. Electr. Eng. J.*, Vol. 5, No. 4, October 1980.

SCHMIDT, W. G.: "The Application of TDMA to the INTELSAT IV Satellite Series," *COMSAT Tech. Rev.*, Vol. 3, No. 2, Fall 1973, pp. 257–276.

SCHWARTZ, M.: *Computer Communications Network Design and Analysis*, Prentice-Hall, Englewood Cliffs, N.J., 1977.

SCHWARTZ, J. W., J. M. AEIN, and J. KAISER: "Modulation Techniques for Multiple Access to a Hard-Limiting Satellite Repeater," *Proc. IEEE*, Vol. 54, May, 1966a.

SCHWARTZ, M., W. R. BENNET, and S. STEIN: *Communication Systems and Techniques*, McGraw-Hill, New York, 1966b.

SCIULLI, J. A.: "Transmission-Delay Effects on Satellite Communications," Telecommunications Techniques Corp. Rep., Rockville, Md., 1978.

SEGUIN, G.: "Linear Ensembles of Codes," *IEEE Trans. Inf. Theory*, Vol. IT-25, No. 4, July 1979.

SETZER, R.: "Echo Control for RCA Americom Satellite Channels," RCA Engineer-25-1, June–July 1979.

SHANMUGAN, K. S.: *Digital and Analog Communication Systems*, Wiley, New York, 1979.

SHANNON, C. E.: "A Mathematical Theory of Communications," *Bell Syst. Tech. J.*, 1948, Part 1, pp. 379–423, Part 2, pp. 623–656.

SIMON, M. K.: "Tracking Performance of Costas Loops with Hard-Limited In-Phase Channel," *IEEE Trans., Commun.*, Vol. COM-26, No. 4, April 1978a, pp. 420–432.

SIMON, M. K.: "Optimum Receiver Structures for Phase Multiplexed Modulations," *IEEE Trans. Commun.*, Vol. COM-26, No. 6, June 1978b, pp. 865–872.

SIMON, M. K.: "On the Optimallity of the MAP Estimation Loop for Carrier Phase Tracking BPSK and QPSK Signals," *IEEE Trans. Commun.*, Vol. COM-27, No. 1, January 1979, pp. 158–164.

SIMON, M. K., and J. K. SMITH: "Offset Quadrature Communications with Decision-Feedback Carrier Synchronization," *IEEE Trans. Commun.,* Vol. COM-22, No. 10, October 1974.

SIVO, J. N.: "Satellites Using the 30/20 GHz Band," NASA Technical memorandum, Lewis Research Center, Cleveland, Ohio, 1980 (prepared for IEEE-NTC, Houston, November 30–December 4, 1980).

SKOLNIK, M. I.: *Introduction to Radar Systems,* McGraw-Hill, New York, 1962.

SONDHI, M. M.: "An Adaptive Echo Canceller," *Bell Syst. Tech. J.,* Vol. 46, 1967, pp. 487–511.

SONDHI, M. M., D. A. BERKELEY: "Silencing Echoes on the Telephone Network," *Proc. IEEE,* August 1980.

SPILKER, J. J.: *Digital Communications by Satellite,* Prentice-Hall, Englewood Cliffs, N.J., 1977.

STARK, H., and F. B. TUTEUR: *Modern Electrical Communications,* Prentice-Hall, Englewood Cliffs, N.J., 1979.

STIFFLER, V. I.: *Theory of Synchronous Communications,* Prentice-Hall, Englewood Cliffs, N.J., 1971.

STREMLER, G. F.: *Introduction to Communication Systems,* Addison-Wesley, Reading, Mass., 1977 (second printing, 1979).

SUNDE, E. D.: "Pulse Transmission by AM, FM, and PM in the Presence of Phase Distortion," *Bell Syst. Tech. J.,* March 1961.

SUNDE, E. D.: *Communications Systems Engineering Theory,* Wiley, New York, 1969.

SUYDERHOUD, H. G., M. ONUTRY, and S. J. CAMPANELLA: "Echo Control in Telephone Communications," IEEE Natl Telecommun. Conf., November 29–December 1, 1976.

SZENTIRMAI, G.: "FILSYN—A General Purpose Filter Synthesis Program," *Proc. IEEE,* Vol. 65, No. 10, 1977.

TACK, T., and M. MOONS: "Optimised DAMA System for Thin Route Networks Using SCPC Satellite Communication," ITU 3rd World Telecommun. Forum, Geneva, September 1979.

TAUB, H., and D. L. SCHILLING: *Principles of Communication Systems,* McGraw-Hill, New York, 1971.

TAYLOR, D. P., and D. C. C. CHEUNG: "A Decision-Directed Carrier Recovery Loop for Duobinary Encoded Offset QPSK Signals," *IEEE Trans. Commun.,* Vol. COM-27, No. 2, February 1979.

TAYLOR, D. P., and P. HETRAKUL: "Receiver Structure for Saturating Channels," *IEEE Trans. Commun.,* Vol. COM-26, No. 2, February 1978.

TESCHER, A. G.: "Adaptive Coding of NTSC Component Video Signals," *Proc. IEEE,* NTC, Houston, December 1980.

THOMAS, C. M., M. Y. WEIDNER, and S. H. DURRANI: "Digital Amplitude Phase Keying with *M*-ary Alphabets," *IEEE Trans. Commun.,* Vol. COM-22, No. 2, February 1974.

TOWNES, X., KON, X., AGRAWAL, X., O'NEAL, and G. COOPER: "Performance of an ADPCM/TASI System," Proc. Int. Commun. Conf., IEEE-ICC-1980, Seattle, June 1980.

TURIN, G. L.: "An Introduction to Matched Filters," *IRE Trans. Inf. Theory,* June 1960.

UNGERBOECK, G.: "Adaptive Maximum Likelihood Receiver for Carrier-Modulated Data Transmission Systems," *IEEE Trans. Commun.,* Vol. COM-22, No. 5, May 1974, pp. 624–636.

VAN TREES, H. L.: *Detection, Estimation and Modulation Theory,* Part 1, Wiley, New York, 1968.

VAN TREES, H. L.: *Detection, Estimation and Modulation Theory,* Parts 2 and 3, Wiley, New York, 1971.

VAN TREES, H. L., ed.: *Satellite Communications,* IEEE Press Selected Reprint Series, IEEE Press, New York, 1979a.

VAN TREES, H. L.: "The Communications System," Introduction to Part III of *Satellite Communications,* H. L. Van Trees, ed., IEEE Press Selected Reprint Series, IEEE Press, New York, 1979b.

VERMA, S. N., and M. RAMASASTRY: "Digital Speech Interpolation Applications for Domestic Satellite Communications," Proc. IEEE Natl. Telecommun. Conf., NTC, 1978.

VITERBI, A. J.: "Error Bounds for Convolutional Codes and an Asymptically Optimum Decoding Algorithm," *IEEE Trans. Inf. Theory,* April 1967, pp. 260–269.

VITERBI, A.: *Principles of Coherent Communications,* McGraw-Hill, New York, 1966.

VITERBI, A. J., and J. K. OMURA: *Principles of Digital Communications and Coding,* McGraw-Hill, New York, 1979.

WAIT, J. V., L. P. HUELSMAN, and G. A. KORN: *Introduction to Operational Amplifier Theory and Applications,* McGraw-Hill, New York, 1975.

WEBER, W. J., III: "Differential Encoding for Multiple Amplitude and Phase Shift Keying Systems," *IEEE Trans. Commun.,* Vol. COM-26, No. 3, March 1978.

WEBER, C. L., and W. K. ALEM: "Demod-Remod Coherent Tracking Receiver for QPSK and SQPSK," *IEEE Trans. Commun.,* Vol. COM-28, No. 12, December 1980a, pp. 1945–1954.

WEBER, C. L., and W. K. ALEM: "Performance Analysis of Demod–Remod Coherent Receiver for QPSK and SQPSK Input," *IEEE Trans. Commun.,* Vol. COM-28, No. 12, December 1980b, pp. 1954–1968.

WEINBERG, L.: *Network Analysis and Synthesis,* McGraw-Hill, New York, 1957.

WEINSTEIN, S. B.: "Echo Cancellation in the Telephone Network," *IEE Commun. Soc. Mag.,* January 1977.

WHYTE, J. S.: "The United Kingdom Telecommunications Strategy," Proc. First Int. Telecommun. Union, Intelcom-77, Atlanta, October 1977.

WILLIAMS, A. B.: *Active Filter Design,* Artech House, Inc., Dedham, Mass., 1975.

WOLEJSZA, C. J.: "Effects of Oscillator Phase Noise on PSK Demodulation," *COMSAT Tech. Rev.,* Vol. 6, Spring 1976.

WOLEJSZA, C. J., and D. CHAKRABORTY: "TDMA Modem Design Criteria," *COMSAT Tech. Rev.,* Vol. 9, No. 2A, Fall 1979.

WOOD, W. A.: "Modulation and Filtering Techniques for 3 Bits/Hertz Operation in the 6 GHz Frequency Band," Proc. IEEE Int. Conf. Commun., Chicago, June 1977.

WOZENCRAFT, J. M., and I. M. JACOBS: *Principles of Communication Engineering,* Wiley, New York, 1965.

YAMAMOTO, H.: "Design PSK Modulators for Multigigabit Rates," *Microwaves,* August 1980.

YAZDANI, H., K. FEHER, and W. STEENAART: "Constant Envelope Bandlimited BPSK Signal," *IEEE Trans. Commun.* Vol. COM-28, No. 6, June 1980.

YEH, L. P.: "Geostationary Satellite Orbital Geometry," *IEEE Trans. Commun.,* Vol. COM-20, No. 4, April 1972.

ZIEMER, R. E., and W. H. TRANTER: *Principles of Communications,* Houghton Mifflin, Boston, 1976.

INDEX